Derivatives

Derivatives: Mathematical Foundations for Finance Students is written for students of finance and actuarial science who want to understand derivatives not just as formulas but as ideas that make sense. The goal is to build the mathematical foundations up step by step – starting from the familiar territory of first-year undergraduate probability and statistics and moving all the way to stochastic calculus and the principles of risk-neutral pricing, which form the heart of modern financial mathematics.

This text will enable the reader to:

- understand models of asset price behavior and apply stochastic calculus
- use stochastic simulations in R to investigate processes and price options
- price options with the binomial model
- derive solutions to the Black-Scholes equation using risk-neutral pricing
- price interest-rate derivatives using Black's model, short-rate models and the Heath–Jarrow–Morton (HJM) framework
- tackle quantitative finance exam questions.

It incorporates a range of learning features to aid student understanding, including boxed examples, end-of-chapter summaries, selected questions from Society of Actuaries (US) Quantitative Finance examinations and further reading suggestions. The book is also supported by a suite of digital learning resources, including PowerPoint slides, multiple choice questions, instructor manual/advice document for lecturers and a test bank.

This book will appeal to both undergraduate and postgraduate students who wish to understand the principles of stochastic calculus and option pricing.

Declan French is Professor of Finance at Queen's Business School, Queen's University Belfast, United Kingdom.

"Declan French has produced an exceptional text, combining rigour and practical relevance. Using examples and simulations, the book moves with remarkable accessibility from core calculus and probability to stochastic calculus, Itô's formula, and risk-neutral pricing. French's ability to connect theory with real-world applications makes this an invaluable resource for finance and actuarial students."

John O.S. Wilson, *Professor, University of St Andrews, UK*

"This excellent textbook makes quantitative finance simple and intuitive. Its mathematical content is succinct, engaging, and focused on key concepts, making it ideal for new undergraduates studying option pricing or derivatives modules who want a solid grounding in mathematics, statistics, and probability to support their understanding."

Danny McGowan, *Professor of Business, Durham University, UK*

Derivatives

Mathematical Foundations for Finance Students

Declan French

Routledge
Taylor & Francis Group
LONDON AND NEW YORK

Designed cover image: William_Potter / Getty Images ®

First published 2026
by Routledge
4 Park Square, Milton Park, Abingdon, Oxon OX14 4RN

and by Routledge
605 Third Avenue, New York, NY 10158

Routledge is an imprint of the Taylor & Francis Group, an informa business

British Library Cataloguing-in-Publication Data
A catalogue record for this book is available from the British Library

ISBN: 978-1-032-76219-7 (hbk)
ISBN: 978-1-032-76218-0 (pbk)
ISBN: 978-1-003-47762-4 (ebk)

DOI: 10.4324/9781003477624

Typeset in TimesNewRoman
by SPi Technologies India Pvt Ltd (Straive)

Access the Support Material: www.routledge.com/9781032762180

To Clare, Sean and Niamh for giving meaning to the meanings.

Contents

About the author

Declan French is Professor of Finance at Queen's Business School, United Kingdom. His research has generated over 40 publications in economics and finance. Seven research students have successfully obtained PhDs under his supervision. He has taught Stochastic Processes in Finance for over 10 years. He is Associate Editor for the journal, *International Review of Financial Analysis.*

Notation

$:=$	is defined as
$[X]^+$	value X if positive and zero otherwise
$\frac{dy}{dx}$ or $f'(x)$	derivative of function $y = f(x)$
$\frac{d^n y}{dx}$ or $f^n(x)$	nth order derivative of function $y = f(x)$
$\frac{\partial z}{\partial x}$	partial derivative of function $z = f(x, y)$ with respect to x
$\log x$	natural logarithm of x
$\mathbb{E}X$	expectation of random variable X
$\phi(z)$	density function of the standard normal distribution Z
$\Phi(z)$	cumulative distribution function of the standard normal distribution Z
$N(\mu, \sigma^2)$	normal distribution with mean μ and variance σ^2
$M_X(r)$	moment-generating function of random variable X, where r a real number
$\{X_t\}$	stochastic process
X_t	value of stochastic process at t
W or $\{W_t : t \geq 0\}$ or $\{W_t\}_{t\geq 0}$	standard Brownian motion

$\mathbb{E}(X_T \mid \mathcal{F}_t)$ or $\mathbb{E}_t X_T$	conditional expectation of the value of process X at time T given filtration $\mathcal{F}$
$\mathbb{Q}$	risk-neutral measure
$\tilde{W}$	standard Brownian motion under the risk-neutral measure
Θ	Radon-Nikodým derivative
$P_{t,T}$	price at time t of a zero-coupon bond with principal \$1 at maturity T
$R_{t,T}$	spot rate from time t to maturity T
$F_{t,S,T}$	forward rate at time t from time S to later time T
r_t	short rate at time t
$f_{t,T}$	instantaneous forward rate at time t for maturity T
$\mathbb{Q}^T$	T-forward measure

Preface

This book was written for students of finance and actuarial science who want to understand derivatives not just as formulas but as ideas that make sense. The mathematics of derivatives is often considered impenetrable. The goal here is to build up the mathematical foundations step by step – starting from the familiar territory of first-year undergraduate probability and statistics and moving all the way to stochastic calculus and the principles of risk-neutral pricing which form the heart of modern financial mathematics.

Many excellent books on derivatives exist but they often assume a background in pure mathematics. This text takes a different approach. It aims to connect the dots in a clear and logical way showing how the mathematics naturally develops as we move from basic calculus to Brownian motion, Itô's formula and risk-neutral pricing. By proceeding in gradual and connected steps, the book seeks to provide sufficient technical rigour for practical financial applications. Key results are supported with proofs or at least with a guide to their logic.

I owe particular thanks to the Society of Actuaries for permitting the inclusion of selected questions from their Quantitative Finance examinations. These problems have provided a valuable source of realistic and well-crafted exercises that test both conceptual understanding and technical skill.

I also wish to express sincere gratitude to the many students of my Stochastic Processes in Finance course at Queen's University Belfast. Their enthusiasm, insightful questions, and constructive feedback have greatly influenced both the structure and tone of this text. Many of the explanations and examples in this book grew directly out of our classroom discussions.

My hope is that this book helps readers see that the mathematics of derivatives does not have to be mysterious – there are clear logical steps leading to the important formulae. This book should help readers develop not only the mathematical proficiency required to study derivatives but also the confidence to connect the theory to real-world financial problems.

Declan French
Queen's University Belfast
October 2025

Part I

Background

1 Calculus background

Overview

Much of this text is concerned with stochastic processes and stochastic calculus. The random behavior of asset prices over time will be modeled by stochastic processes. Stochastic calculus is then required to manipulate these processes in useful ways such as for portfolio management or option pricing. Stochastic calculus is an extension of ordinary calculus to deal with continuous processes involving uncertainty over time.

Before tackling stochastic calculus, a review of key ideas in classical calculus is provided in this chapter. A solid understanding of this material is essential for studying stochastic calculus because stochastic calculus builds on – and extends – the concepts of classical calculus in a probabilistic setting.

The content is aimed at students who need a refresher having forgotten elements from earlier in their studies or alternatively students who may have a limited background in mathematics.

This chapter is not intended to be a comprehensive overview of the topics presented. Instead, it provides the basic ideas and applications needed at various points in the course. A number of worked examples are provided but ultimately details are kept fairly brief. A selection of references is provided at the end to develop a broader understanding if the reader feels necessary.

1.1 Limits

The concept of limits is the basis for understanding derivatives and integrals.

Definition 1.1. *Given a function $f(x)$, the limit of $f(x)$ as x approaches a is the value of $f(x)$ as x gets very close to a. It is written as*

$$\lim_{x \to a} f(x)$$

The function f does not necessarily have to be defined at a.

A limit exists when the function approaches a single finite value.

DOI: 10.4324/9781003477624-2

Therefore $\lim_{x \to 0} \frac{1}{x}$ does not exist as the function does not approach a finite value.

Consider the situation where there is a jump in the function at a. We will get different answers for the limit whether we approach a from values of x less than a or from values of x greater than a. As a result, the function does not approach a single finite value at a. The limit then does not exist.

For example, letting $f(x) = \begin{cases} 2 & x \leq 1 \\ 3 & x > 1 \end{cases}$ then $\lim_{x \to 1} f(x)$ does not exist.

The limit of $f(x)$ approaching a from values of x less than a is written as $\lim_{x \to a^-} f(x)$, and the limit approaching a from values of x greater than a is written as $\lim_{x \to a^+} f(x)$.

Example 1.1.

(i) $\lim_{x \to 1} (x + 1) = 2$

(ii) $\lim_{x \to 0} x^2 - 9 = -9$

□

The following results can be proven for functions $f(x)$ and $g(x)$:

(a) The limit of a sum or difference is the sum or difference of the limits

$$\lim_{x \to a} \left(f(x) \pm g(x) \right) = \lim_{x \to a} f(x) \pm \lim_{x \to a} g(x)$$

(b) The limit of a product is the product of the limits

$$\lim_{x \to a} f(x)g(x) = \lim_{x \to a} f(x) \lim_{x \to a} g(x)$$

Example 1.2.

(i) $\lim_{x \to 2} (x^2 + 3x) = \lim_{x \to 2} x^2 + \lim_{x \to 2} 3x = 4 + 6 = 10$

(ii) $\lim_{x \to 1} (x + 2)(x + 1) = \lim_{x \to 1} (x + 2) \lim_{x \to 1} (x + 1) = 3 \times 2 = 6$

□

Instead of evaluating the limit of $f(x)$ as x approaches a number a, we can allow x to become larger than any number. This is written as

$$\lim_{x \to \infty} f(x)$$

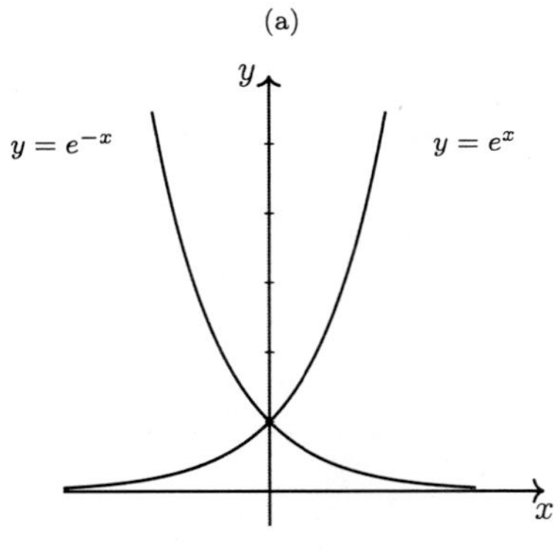

Figure 1.1 Graphs of the exponential function and the natural logarithm function. (a) $y = e^x$ and $y = e^{-x}$ (b) $y = \log x$

Example 1.3.

(i) $\lim_{x\to\infty} \frac{1}{x} = 0$

(ii) $\lim_{x\to\infty} \frac{2x-3}{4x+9} = \lim_{x\to\infty} \frac{2-3/x}{4+9/x} = \frac{1}{2}$

□

We can also let x become less than any number. This is written as

$$\lim_{x\to-\infty} f(x)$$

The exponential function $y = e^x$ is graphed on the left below along with $y = e^{-x}$. We can infer

$$\lim_{x\to-\infty} e^x = 0$$

and

$$\lim_{x\to\infty} e^{-x} = 0$$

The natural logarithm function $y = \log_e x$ is graphed on the right. The shorthand *log* will refer to the logarithm to the base e everywhere in this text.

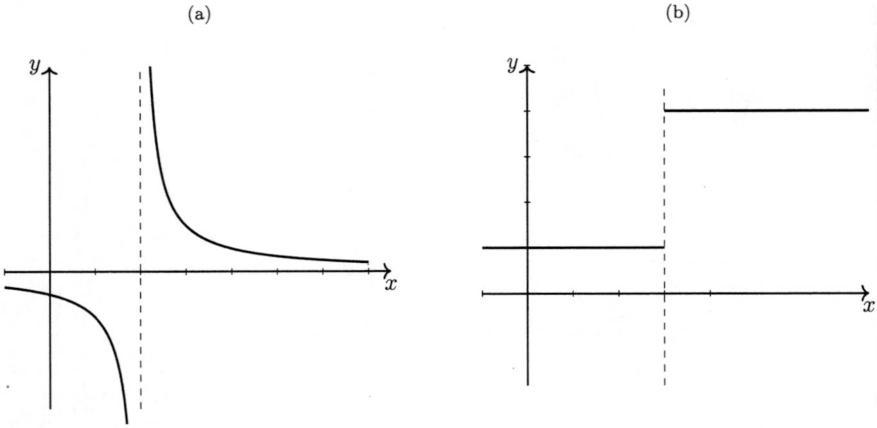

Figure 1.2 Functions that are not continuous. (a) $f(x) = \frac{1}{x-2}$ (b) $f(x) = \begin{cases} 1 & x\leq 3 \\ 4 & x>3 \end{cases}$

1.2 Continuity

A continuous function is a function with no breaks, jumps or holes in its graph.

More formally, function f is continuous at a if these three conditions hold:

(a) $f(a)$ is defined
(b) $\lim_{x\to a} f(x)$ exists
(c) $\lim_{x\to a} f(x) = f(a)$

f is a continuous function if it is continuous at every real number.

Example 1.4.

(i) $f(x) = x^2 + 2x - 5$ *is a continuous function.*

(ii) $f(x) = |x|$ *is a continuous function.*

(iii) $f(x) = \frac{1}{x-2}$ *is not continuous at* $x = 2$ *as* $f(2)$ *is not defined.*

(iv) $f(x) = \begin{cases} 1 & x \leq 3 \\ 4 & x > 3 \end{cases}$ *is not continuous as the limit does not exist since* $\lim_{x\to 3^-} f(x) = 1$ *and* $\lim_{x\to 3^+} f(x) = 4$.

□

1.3 Differentiation

Let f be a function. Draw two points on the graph of the function $y = f(x)$ at x_0 and $x_0 + h$ labeled A and B (Figure 1.3). The slope of the line segment

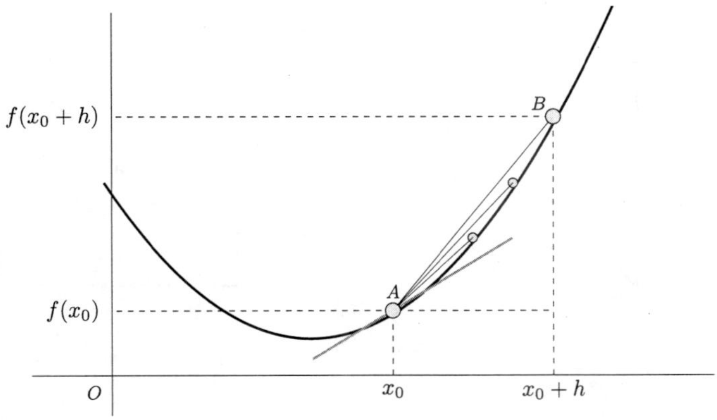

Figure 1.3 Derivative as slope of tangent to curve

connecting these points is

$$\frac{f(x_0 + h) - f(x_0)}{h}$$

Let h approach 0 so B travels down the graph toward A. The slope of the line segment becomes the slope of the tangent to the curve at A. This is the derivative of $y = f(x)$ at $x = x_0$.

Definition 1.2. *The derivative of a function $f(x)$ at $x = x_0$ is the slope of the tangent to the curve at that point*

$$f'(x_0) :- \lim_{h \to 0} \frac{f(x_0 + h) - f(x_0)}{h}$$

A function $y = f(x)$ is *differentiable* at x_0 if the derivative at x_0 exists. The derivative $f'(x_0)$ is also written as

$$\frac{df(x_0)}{dx}$$

In order to find the limit in Definition 1.2 at any point the function f must necessarily be continuous at that point. However, continuity is not a sufficient condition for differentiability.

For example, the function $y = |x|$ is continuous but not at $x = 0$ since the limit does not exist. From Figure 1.4, it can be seen that approaching 0 from the left-hand side of the slope of the tangent to the curve is always –1, while approaching 0 from the right-hand side of the slope of the tangent is always 1. The slope does not approach a single finite value; therefore, the limit in the definition does not exist.

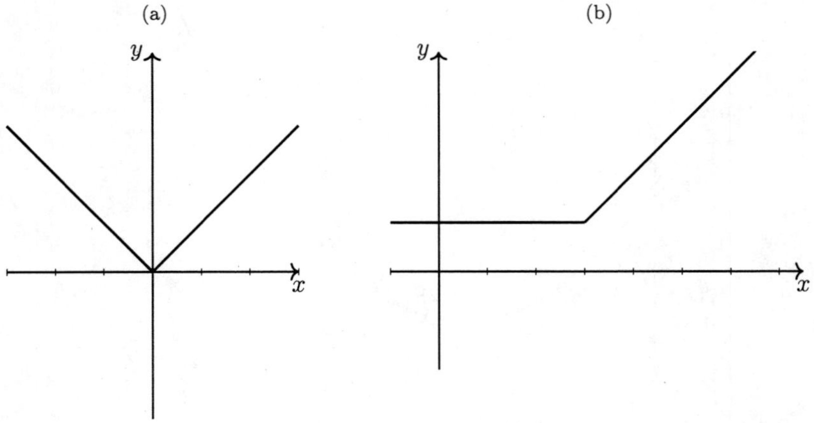

Figure 1.4 Functions that are continuous but not everywhere differentiable. (a) $f(x) = |x|$ (b) $f(x) = \left\{ \begin{matrix} \frac{1}{x \le 3} & \frac{x-2}{x>3} \end{matrix} \right.$

Informally, also the curve has to be smooth at x_0 for f to be differentiable at x_0 that is no abrupt changes in the direction. Figure 1.4 shows another function that is continuous but not everywhere differentiable.

1.4 Rules for differentiation

Some basic rules for differentiating functions follow from the definition of the derivative given in the previous section.

Given f and g functions of x; c and n constants:

(a) derivative of a constant $\frac{dc}{dx} = 0$
(b) derivative of a power $\frac{dx^n}{dx} = nx^{n-1}$
(c) derivative of a sum or difference $\frac{d(f(x) \pm g(x))}{dx} = \frac{df(x)}{dx} \pm \frac{dg(x)}{dx}$
(d) derivative of the exponential function $\frac{de^x}{dx} = e^x$
(e) derivative of the natural logarithm $\frac{d \log x}{dx} = \frac{1}{x}$

Example 1.5. *Differentiating the following functions, we get*

(i) $f(x) = x^3$, *then* $f'(x) = 3x^2$

(ii) $f(x) = x^4 + 3x$ *then,* $f'(x) = 4x^3 + 3$

(iii) $f(x) = x^2 - 2x - 5$ *then,* $f'(x) = 2x - 2$

(iv) $f(x) = 3e^x + 6 \log x$ *then,* $f'(x) = 3e^x + \frac{6}{x}$

□

For products and quotients of functions of x, we need the following rules. Given u and v functions of x:

(a) product rule

$$\frac{d(uv)}{dx} = u\frac{dv}{dx} + v\frac{du}{dx}$$

(b) quotient rule

$$\frac{d}{dx}\left(\frac{u}{v}\right) = \frac{v\frac{du}{dx} - u\frac{dv}{dx}}{v^2}$$

Example 1.6. *Differentiating the following functions, we get*

(i) $f(x) = x^3 \log x$, *then*

$$f'(x) = x^3\,\frac{d(\log x)}{dx} + \log x\,\frac{d(x^3)}{dx} = x^3\left(\frac{1}{x}\right) + \log x\,(3x^2) = x^2 + 3x^2 \log x$$

(ii) $f(x) = \left(\frac{e^x}{x^2}\right)$, *then*

$$f'(x) = \frac{x^2\frac{d(e^x)}{dx} - e^x\frac{d(x^2)}{dx}}{(x^2)^2} = \frac{x^2e^x - e^x(2x)}{x^4} = \frac{e^x(x-2)}{x^3}$$

□

The chain rule allows us to differentiate composite functions. That is, functions inside other functions.

If $y = g\Big(f(x)\Big)$, then using the substitution $u = f(x)$

$$\frac{dy}{dx} = \frac{dy}{du} \times \frac{du}{dx} \tag{1.1}$$

Example 1.7. *Differentiating the following functions, we get*

(i) $f(x) = (3x+1)^5$, *then letting* $u = 3x+1$

$$f'(x) = 5(3x+1)^4 \times (3) = 15(3x+1)^4$$

(ii) $f(x) = e^{x^2+3x}$, *then letting* $u = x^2 + 3x$

$$f'(x) = (2x + 3)e^{x^2+3x}$$

(iii) $f(x) = \log(5x^4 - x^2)$, *then letting* $u = 5x^4 - x^2$

$$f'(x) = \frac{20x^3 - 2x}{5x^4 - x^2}$$

(iv) $f(x) = \sqrt{4x^2 + 1}$, *then letting* $u = 4x^2 + 1$

$$f'(x) = \frac{1}{2}(4x^2 + 1)^{-\frac{1}{2}} \times (8x) = \frac{4x}{\sqrt{4x^2 + 1}}$$

□

1.5 Higher-order derivatives

Higher derivatives are derivatives of derivatives. Given a function $f(x)$ the first derivative is

$$\frac{df(x)}{dx} \text{ or } f'(x)$$

Differentiating the first derivative, we get the second derivative

$$\frac{d^2f(x)}{dx^2} = \frac{d}{dx}\left(\frac{df(x)}{dx}\right) \text{ or } f''(x)$$

Differentiating the second derivative, we get the third derivative and so on

$$\frac{d^3f(x)}{dx^3} = \frac{d}{dx}\left(\frac{d^2f(x)}{dx^2}\right) \text{ or } f^3(x)$$

Example 1.8.

(i) The second derivative of $f(x) = x^5$ *is* $20x^3$ *since*

$$f''(x) = \frac{d(5x^4)}{dx} = 20x^3$$

(ii) The third derivative of $f(x) = \log x$ *is* $2x^{-3}$ *since*

$$f^3(x) = \frac{d^2}{dx^2}\left(\frac{1}{x}\right) = \frac{d(-x^{-2})}{dx} = 2x^{-3}$$

(iii) The third derivative of $f(x) = \sqrt{x}$ *is* $\frac{3}{8}x^{-\frac{5}{2}}$ *since*

$$f^3(x) = \frac{d^2}{dx^2}\left(\frac{1}{2}x^{-\frac{1}{2}}\right) = \frac{d}{dx}\left(-\frac{1}{4}x^{-\frac{3}{2}}\right) = \frac{3}{8}x^{-\frac{5}{2}}$$

(iv) The fourth derivative of $f(x) = e^x$ *is* e^x *since*

$$f^4(x) = \frac{d^3(e^x)}{dx^3} = \frac{d^2(e^x)}{dx^2} = \frac{d(e^x)}{dx} = e^x$$

□

1.6 Partial derivatives

So far, we have only considered functions of one variable. Consider a function of two variables $z = f(x, y)$, then the partial derivative measures how f changes with respect to one variable while keeping the other variable constant.

Definition 1.3. *The partial derivative of a function* $f(x, y)$ *with respect to* x *at* (x_0, y_0) *is*

$$\frac{\partial f(x_0, y_0)}{\partial x} :- \lim_{h \to 0} \frac{f(x_0 + h, y_0) - f(x_0, y_0)}{h}$$

and the partial derivative of $f(x, y)$ *with respect to* y *at* (x_0, y_0) *is*

$$\frac{\partial f(x_0, y_0)}{\partial y} :- \lim_{k \to 0} \frac{f(x_0, y_0 + k) - f(x_0, y_0)}{k}$$

The rules to partially differentiate are the same as for ordinary differentiation except we must remember to treat the other variable as a constant.

Example 1.9. *Let* $f(x, y) = x^2y + 4y^3$

(i) Treating y *as a constant, the partial derivative of* f *with respect to* x *is*

$$\frac{\partial f}{\partial x} = 2xy$$

(ii) Treating x as a constant, the partial derivative of f with respect to y is

$$\frac{\partial f}{\partial y} = x^2 + 12y^2$$

□

Example 1.10. *Let* $f(x, y) = 3x^2y + 2xy^2 + 5y$

(i) The partial derivative of f with respect to x is

$$\frac{\partial f}{\partial x} = 6xy + 2y^2$$

(ii) The partial derivative of f with respect to y is

$$\frac{\partial f}{\partial y} = 3x^2 + 4xy + 5$$

□

Partial derivatives are defined similarly for functions of three or more variables.

1.7 Higher-order partial derivatives

Higher-order partial derivatives are partial derivatives of partial derivatives. Given a function $f(x, y)$, the second-order partial derivatives are

$$\frac{\partial^2 f}{\partial x^2} = \frac{\partial}{\partial x}\left(\frac{\partial f}{\partial x}\right)$$
$$\frac{\partial^2 f}{\partial x \partial y} = \frac{\partial}{\partial x}\left(\frac{\partial f}{\partial y}\right)$$
$$\frac{\partial^2 f}{\partial y \partial x} = \frac{\partial}{\partial y}\left(\frac{\partial f}{\partial x}\right)$$
$$\frac{\partial^2 f}{\partial y^2} = \frac{\partial}{\partial y}\left(\frac{\partial f}{\partial y}\right)$$

Exercise 1.1.
If $f(x, y) = 2x^2 + 4xy$*, find*

$$\frac{\partial f}{\partial x}, \frac{\partial f}{\partial y}, \frac{\partial^2 f}{\partial x \partial y} \text{ and } \frac{\partial^2 f}{\partial y \partial x}$$

Proof.

$$\frac{\partial f}{\partial x} = 4x + 4y$$
$$\frac{\partial f}{\partial y} = 4x$$
$$\frac{\partial^2 f}{\partial x \partial y} = \frac{\partial}{\partial x}(4x) = 4$$
$$\frac{\partial^2 f}{\partial y \partial x} = \frac{\partial}{\partial y}\left(4x + 4y\right) = 4$$

□

Exercise 1.2.
If $f(x, y) = x^2y + 3xy^2$, find all four second-order partial derivatives.

Proof.

$$\frac{\partial^2 f}{\partial x^2} = \frac{\partial}{\partial x}(2xy + 3y^2) = 2y$$
$$\frac{\partial^2 f}{\partial x \partial y} = \frac{\partial}{\partial x}(x^2 + 6xy) = 2x + 6y$$
$$\frac{\partial^2 f}{\partial y \partial x} = \frac{\partial}{\partial y}(2xy + 3y^2) = 2x + 6y$$
$$\frac{\partial^2 f}{\partial y^2} = \frac{\partial}{\partial y}(x^2 + 6xy) = 6x$$

□

In these exercises, the order of differentiation for the mixed partials does not matter. This is always the case. For any two differentiable function $f(x, y)$

$$\frac{\partial^2 f}{\partial x \partial y} = \frac{\partial^2 f}{\partial y \partial x}$$

1.8 Total differential

The total differential of a function represents the total change in the function's value due to changes in all of its independent variables.

The definition of the derivative of a function $y = f(x)$ at $x = x_0$ can be rewritten as

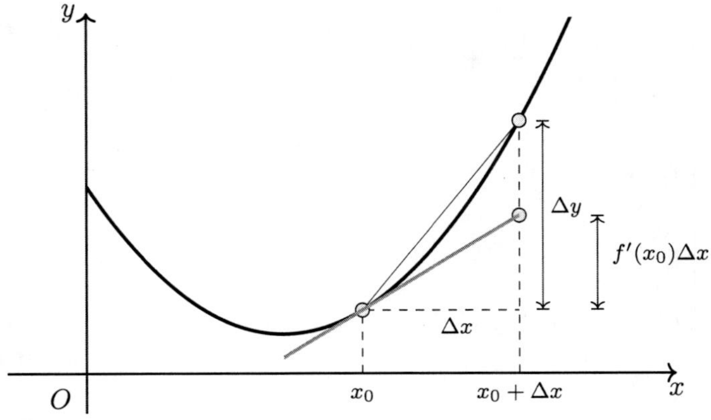

Figure 1.5 Total differential in one variable

$$f'(x_0) = \lim_{\Delta x \to 0} \frac{\Delta y}{\Delta x}$$

or

$$\frac{\Delta y}{\Delta x} \approx f'(x_0)$$

where the approximation improves as $\Delta x \to 0$ (Figure 1.5). Assuming that the derivative exists the increment Δy is then

$$\Delta y \approx f'(x_0)\Delta x$$

In the limit, we can write

$$dy = f'(x_0)dx \tag{1.2}$$

This is the total differential in one variable.

For a function of two variables $z = f(x, y)$, the change in z moving from (x_0, y_0) to $(x_0 + \Delta x, y_0 + \Delta y)$ is equal to the sum of the changes in z keeping y constant and then x constant. Using the definition of partial derivatives from above then[1]

$$\Delta z \approx \frac{\partial f}{\partial x}(x_0, y_0)\,\Delta x + \frac{\partial f}{\partial y}(x_0, y_0)\,\Delta y \tag{1.3}$$

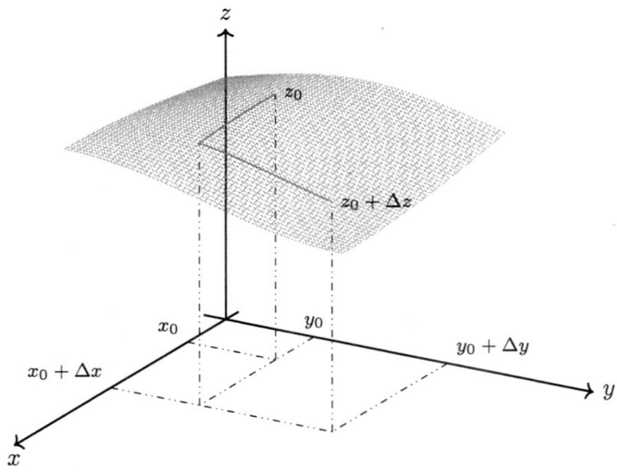

Figure 1.6 Total differential in two variables

where the approximation improves as $\Delta x \to 0$ and $\Delta y \to 0$ (Figure 1.6). In the limit

$$dz = \frac{\partial f}{\partial x}(x_0, y_0)dx + \frac{\partial f}{\partial y}(x_0, y_0)dy$$

We will sometimes use the notation

$$dz = \frac{\partial f}{\partial x}dx + \frac{\partial f}{\partial y}dy \quad \text{or} \quad dz = \frac{\partial z}{\partial x}dx + \frac{\partial z}{\partial y}dy$$

We can also derive the chain rule in two variables.

If x and y are functions of time t. Dividing Equation (1.3) by Δt and taking the limit give

$$\frac{dz}{dt} = \frac{\partial z}{\partial x}\frac{dx}{dt} + \frac{\partial z}{\partial y}\frac{dy}{dt}$$

These definitions can be extended to three or more variables.

1.9 Geometric series

A geometric series has the form;

$$a + ar + ar^2 + ar^3 + ...$$

where a is the *first term* and r is the *common ratio.*

The nth partial sum S_n of the geometric series is the sum of the first n terms

$$S_n = a + ar + ar^2 + ar^3 + ... + ar^{n-1}$$

Then

$$\begin{aligned} S_n &= a + ar + ar^2 + ar^3 + ... + ar^{n-1} \\ \Longrightarrow rS_n &= ar + ar^2 + ar^3 + ar^4 + ... + ar^n \\ \Longrightarrow S_n - rS_n &= a - ar^n \\ \Longrightarrow S_n &= \frac{a(1 - r^n)}{1 - r} \end{aligned}$$

If $|r| < 1$, then the limit of S_n as the number of terms tends to infinity is

$$S_\infty = \frac{a}{1 - r}$$

The series is said to *converge* to $\frac{a}{1-r}$.

If $|r| > 1$, the limit does not exist. In this case, the series *diverges*.

Example 1.11. *(i) Given* $1 + \frac{1}{2} + \frac{1}{4}....$ *the sum to infinity is*

$$S_\infty = \frac{1}{1 - 1/2} = 2$$

(ii) Given $0.3 + 0.03 + 0.003 + 0.0003 + ...$ *the sum to infinity is*

$$S_\infty = \frac{0.3}{1 - 1/10} = \frac{1}{3}$$

□

1.10 Taylor expansion

Assume that $f(x)$ can be represented as

$$f(x) = \sum_{n=0}^{\infty} a_n x^n = a_0 + a_1 x + a_2 x^2 + ...$$

in an interval around $x = 0$.

Then repeated differentiation gives

$$\begin{aligned} f'(0) &= a_1 \\ f''(0) &= 2a_2 \\ f^3(0) &= 3 \times 2a_3 \\ f^4(0) &= 4 \times 3 \times 2a_3 \text{ etc.} \end{aligned}$$

Substituting for the coefficients back into the original series

$$f(x) = f(0) + f'(0)x + \frac{f''(0)}{2!}x^2 + \frac{f^3(0)}{3!}x^3 + \frac{f^4(0)}{4!}x^4 + ...$$

The series on the right is known as the *Taylor expansion* (or *Taylor series*) of $f(x)$.

The following result will be used in a number of places in this text.

Example 1.12. *The Taylor series for the exponential function around $x = 0$ is*

$$e^0 + xe^0 + \frac{x^2}{2!}e^0 + \frac{x^3}{3!}e^0 + \dots$$

$$1 + x + \frac{x^2}{2!} + \frac{x^3}{3!} + \dots \tag{1.4}$$

since the first and all higher-order derivatives of e^x are e^x.

It can be shown that this Taylor series converges to e^x for every x. Hence, for every x

$$e^x = 1 + x + \frac{x^2}{2!} + \frac{x^3}{3!} + \dots$$

□

When f is a function of x and y, the Taylor expansion in two variables uses partial derivatives

$$f(0,0) + x\frac{\partial}{\partial x}f(0,0) + y\frac{\partial}{\partial y}f(0,0) + \frac{1}{2!}\left(x^2\frac{\partial^2}{\partial x^2}f(0,0) + 2xy\frac{\partial^2}{\partial x \partial y}f(0,0) + y^2\frac{\partial^2}{\partial y^2}f(0,0)\right) + \dots \tag{1.5}$$

1.11 Variation

The *total variation* V of a function $y = f(x)$ on an interval $[a, b]$ is a measure of the total amount of rising and falling of the curve over the interval.

More precisely, the total variation is found by partitioning the interval into n segments (Figure 1.7) and summing up the absolute value of the dashed segments Δy_k.

$$V_n = |\Delta y_1| + |\Delta y_2| + \cdots + |\Delta y_k| + \cdots + |\Delta y_n|$$

Then find the least upper bound of V_n over all possible partitions. The total variation V is this bound if it exists.

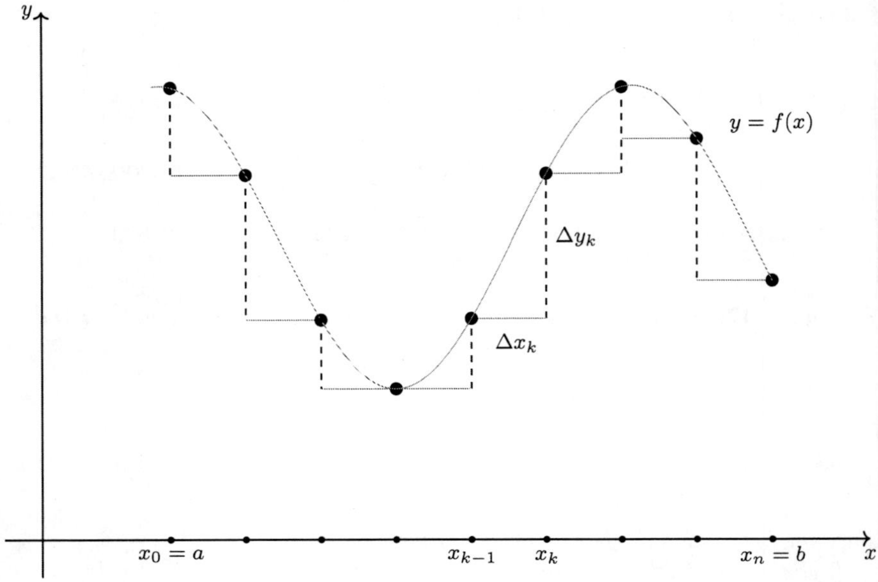

Figure 1.7 Total variation

If a bound does not exist, the function is said to have *unbounded variation* on the interval.

Most continuous functions that you will encounter in ordinary calculus are bounded variation. The function in Figure 1.8 has unbounded variation on any interval including $x = 0$ since the number of up and down movements of the curve becomes infinite approaching zero.

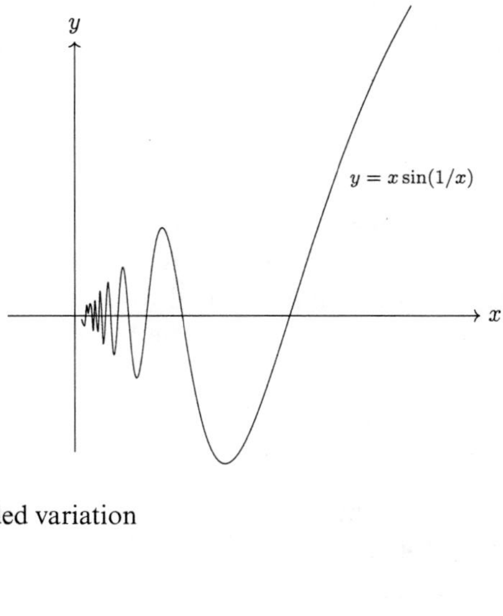

Figure 1.8 Unbounded variation

1.12 Integration

In this section, integration is first defined as the area under a curve, and then in the next section, it is demonstrated to be the reverse of differentiation.

Given the curve $y = f(x)$ in Figure 1.9, we want to find the area under the curve between the vertical lines $x = a$ and $x = b$ and the x-axis.

The area can be approximated by dividing it into strips (Figure 1.10). Partition the interval $[a, b]$ into n subintervals

$$[x_0 = a, x_1], [x_1, x_2], ..., [x_{n-1}, x_n = b]$$

not necessarily of equal width. Then approximate the shaded area by drawing two rectangles for each subinterval $[x_{k-1}, x_k]$ with

base $\Delta x_k = x_k - x_{k-1}$ and height m_k for the first rectangle

where m_k is the minimum of $f(x)$ over the subinterval and

base $\Delta x_k = x_k - x_{k-1}$ and height M_k for the second rectangle

where M_k is the maximum of $f(x)$ over the subinterval.

The area under the graph should be between the sum of the area of the first set of rectangles (*lower sum*) and the sum of the area of the second set of rectangles (*upper sum*).

$$\sum_{k=1}^{n} m_k \Delta x_k \leq \text{Area} \leq \sum_{k=1}^{n} M_k \Delta x_k$$

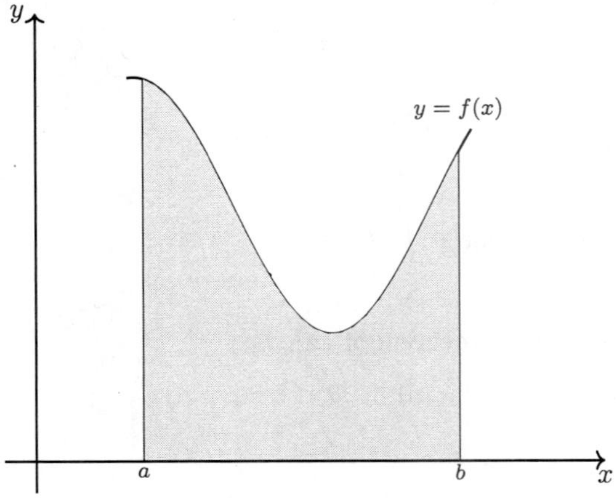

Figure 1.9 Area under a curve

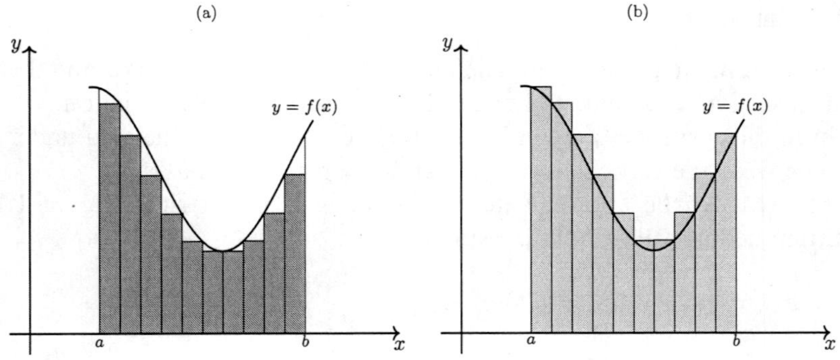

Figure 1.10 Approximations to area under curve. (a) Lower sum (b) Upper sum

As extra points are added to the partition of the interval $[a, b]$, the number of rectangles increases. As long as $f(x)$ is continuous, the lower and upper sums will get closer and closer to the area under the curve. As $n \to \infty$, these sums will close in on a common limit. The common limit will also exist for some discontinuous functions.[2] We refer to functions where the common limit exists as being *integrable*.

The common limit is denoted as

$$\int_a^b f(x)dx$$

where a and b are called the *lower* and *upper limits*, and $f(x)$ is called the *integrand*.

The following properties then follow:

$$\int_a^a f(x)dx = 0$$

and for $a \le b \le c$

$$\int_a^c f(x)dx = \int_a^b f(x)dx + \int_b^c f(x)dx$$

1.13 The fundamental theorem of calculus

In order to evaluate the integral of $f(x)$ from a to b

$$\int_a^b f(x)dx$$

as defined in the previous section, we need the *fundamental theorem of calculus*.

This theorem states that integration and differentiation are reverse processes. It comes in two different parts which are given below without proof. In simple terms, the first part says that the derivative of the integral is the original function while the second part says that the integral of the derivative is the original function.

This is the first fundamental theorem of calculus.

Theorem 1.1. *Let f be integrable on $[a, b]$ and define F as*

$$F(x) :- \int_a^x f(t)\, dt$$

If f is continuous, then F is differentiable and

$$F'(x) = f(x)$$

This is the second fundamental theorem of calculus.

Theorem 1.2. *Let f be integrable on $[a, b]$ and define $F'(x) :- f(x)$, then*

$$\int_a^b f(x)\, dx = F(b) - F(a)$$

We will use the shorthand of $F(x)\Big|_a^b$ for $F(b) - F(a)$.

1.14 Rules of integration

Now that we know integration and differentiation are reverse processes, we can develop rules for integration. The second fundamental theorem of calculus tells us that if we know a function F whose derivative is f, then the integral of f is $F(b) - F(a)$. So we can evaluate integrals of f as long as we know what F is.

Example 1.13.

(*i*) *Since*

$$\frac{d}{dx}\left(\frac{x^3}{3}\right) = x^2$$

then

$$\int_1^3 x^2\, dx = \frac{x^3}{3}\Big|_1^3 = \frac{3^3}{3} - \frac{1^3}{3} = \frac{26}{3}$$

(ii) *Since*

$$\frac{d}{dx}(x^2 + x) = 2x + 1$$

then

$$\int_{-1}^{1} (2x+1)\,dx = (x^2+x)\Big|_{-1}^{1} = (1^2+1) - \Big((-1)^2+(-1)\Big) = 2-0 = 2$$

(iii) *Since*

$$\frac{d}{dx}(-e^{-x}) = e^{-x}$$

then

$$\int_{0}^{\infty} e^{-x}\,dx = -e^{-x}\Big|_{0}^{\infty} = -e^{-\infty} - (-e^0) = 0 - (-1) = 1$$

□

In Example 1.13 (i), our choice of F could have been

$$\frac{x^3}{3} + C$$

for any constant C as the derivative is still equal to x^2. However, it would not have made any difference to the integral result as

$$\left(\frac{x^3}{3} + C\right)\Big|_{1}^{3} = \left(\frac{3^3}{3} + C\right) - \left(\frac{1^3}{3} + C\right) = \frac{26}{3}$$

When the lower and upper limits of integration are not specified, it is conventional to specify the constant C in the integral. An integral where the limits are not specified is known as an *indefinite* integral. Integrals where the limits are specified as in the given example are *definite* integrals.

Using the fundamental theorem of calculus, a number of basic rules of integration are established.

Given f and g functions of x; a and n constants

(a) $\int f(x) \pm g(x)\,dx = \int f(x)\,dx \pm \int g(x)\,dx$

(b) $\int ax^n \, dx = \frac{a}{n+1} x^{n+1} + C$

(c) $\int e^x \, dx = e^x + C$

Example 1.14.

(i) $\int x^3 \, dx = \frac{1}{4} x^4 + C$

(ii) $\int 8x^2 + 7x \, dx = \frac{8}{3} x^3 + \frac{7}{2} x^2 + C$

(iii) $\int 3x^2 + 2x + 5 \, dx = x^3 + x^2 + 5x + C$

(iv) $\int 3e^x - 6x^4 \, dx = 3e^x - \frac{6}{5} x^5 + C$

(v) $\int 4 + 3\sqrt{x} \, dx = 4x + 2x^{\frac{3}{2}} + C$

□

There are a number of techniques for dealing with more difficult integrals.

Integration by substitution is a method used to simplify integrals by making a change of variable – it is equivalent to the chain rule in differentiation. It relies on the following theorem.

Theorem 1.3. *If f and g' are continuous, then*

$$\int_a^b f(g(x))g'(x) \, dx = \int_{g(a)}^{g(b)} f(u) \, du$$

For example, we are given

$$\int 2x(x^2 + 1)^3 \, dx$$

then letting $u = g(x) = x^2 + 1$ and replacing $g'(x) \, dx = 2x \, dx$ by du, the integral becomes

$$\int u^3 \, du = \frac{u^4}{4} + C = \frac{(x^2 + 1)^4}{4} + C$$

Example 1.15.

(i) *Using the substitution* $u = x^3 - 7 \implies du = 3x^2\,dx$

$$\int 3x^2(x^3-7)^4\,dx = \int u^4\,du = \frac{u^5}{5} + C = \frac{(x^3-7)^5}{5} + C$$

(ii) *Using the substitution* $u = x^2 + 1 \implies \frac{1}{2}du = x\,dx$

$$\int x\sqrt{x^2+1}\,dx = \int \frac{1}{2}\sqrt{u}\,du = \frac{1}{3}u^{\frac{3}{2}} + C = \frac{1}{3}(x^2+1)^{\frac{3}{2}} + C$$

(iii) *Using the substitution* $u = 5x + 2 \implies \frac{1}{5}du = dx$

$$\int (5x+2)^3\,dx = \int \frac{1}{5}u^3\,du = \frac{1}{20}u^4 + C = \frac{1}{20}(5x+2)^4 + C$$

(iv) *Using the substitution* $u = 1 - x \implies du = -dx$

$$\int \frac{1}{(1-x)^2}\,dx = -\int \frac{1}{u^2}\,du = \frac{1}{u} + C = \frac{1}{1-x} + C$$

(v) *Using the substitution* $u = ax + b \implies \frac{1}{a}du = dx$*, where a and b are constants*

$$\int e^{ax+b}\,dx = \int \frac{1}{a}e^u\,du = \frac{1}{a}e^u + C = \frac{1}{a}e^{ax+b} + C$$

(vi) *Using the substitution* $u = e^x + 3 \implies du = e^x dx$

$$\int \frac{e^x}{(e^x+3)^3}\,dx = \int \frac{du}{u^3} = -\frac{1}{2u^2} + C = -\frac{1}{2(e^x+3)^2} + C$$

(vii) *Using the substitution* $u = \frac{-x^2}{2} \implies -du = xdx$

$$\int xe^{-x^2/2}\,dx = -\int e^u du = -e^u + C = -e^{-x^2/2} + C$$

(viii) *Using the substitution* $u = 1 + 2x \implies \frac{1}{2}du = dx$ *and if* $x = 0$ *or* $x = 3$*, then* $u = 1$ *or* $u = 7$

$$\int_0^3 \sqrt{1+2x}\,dx = \frac{1}{2}\int_1^7 \sqrt{u}\,du = \frac{1}{3}u^{\frac{3}{2}}\Big|_1^7 = \frac{1}{3}(7^{\frac{3}{2}} - 1)$$

(ix) *Using the substitution* $u = x^2 \implies \frac{1}{2}du = xdx$ *and if* $x = 0$ *or* $x = 2$, *then* $u = 0$ *or* $u = 4$

$$\int_0^2 xe^{x^2}\, dx = \frac{1}{2}\int_0^4 e^u\, du = \frac{1}{2}e^u\Big|_0^4 = \frac{1}{2}(e^4 - 1)$$

□

From Section 1.4, $\frac{d}{dx}\log x = \frac{1}{x}$ which suggests that

$$\int \frac{1}{x}\, dx = \log x + C$$

As $\log x$ is only defined for $x > 0$, then this result holds for all x positive.
If x is negative, then let $-u = x \implies -du = dx$

$$\text{For } x < 0, \int \frac{1}{x}\, dx = -\int \frac{1}{-u}\, du = \int \frac{1}{u}\, du = \log u + C = \log(-x) + C$$

Combining results for $x < 0$ and $x > 0$ together, we have

$$\int \frac{1}{x} dx = \log|x| + C \text{ for } x \neq 0$$

Example 1.16.

(i) *Using the substitution* $u = x^2 + 1 \implies \frac{1}{2}du = x\, dx$

$$\int \frac{x}{x^2+1}\, dx = \frac{1}{2}\int \frac{1}{u}\, du = \frac{1}{2}\log|u| + C = \frac{1}{2}\log(x^2+1) + C$$

There is no need for the absolute value sign in the final answer as $x^2 + 1$ *is always positive*

(ii) *Using the substitution* $u = ax + b \implies \frac{1}{a}du = dx$, *where a and b are constants*

$$\int \frac{1}{ax+b}\, dx = \frac{1}{a}\int \frac{1}{u}\, du = \frac{1}{a}\log|u| + C = \frac{1}{a}\log|ax+b| + C$$

(iii) *Using the substitution* $u = ae^x + b \implies \frac{1}{a}du = e^x\, dx$, *where a and b are constants*

$$\int \frac{e^x}{ae^x+b}\, dx = \frac{1}{a}\int \frac{1}{u}\, du = \frac{1}{a}\log|u| + C = \frac{1}{a}\log|ae^x+b| + C$$

□

1.15 Double integrals

The value of the integral

$$\int_a^b f(x)\,dx$$

was the area under the curve $y = f(x)$ for x in $[a, b]$.

The double integral

$$\iint_R f(x, y)\,dA$$

can be interpreted as the volume under the surface $z = f(x, y)$ for (x, y) in the region R.

This can be seen from Figure 1.11. The volume of interest is the volume standing on the region R in the xy-plane. The volume of the thin column drawn is approximately the base ΔA times the height $f(x, y)$. The sum of all these volumes approximates the volume under the curve and in the limit as the number of these columns becomes infinite the sum gives the exact volume. The limiting value is the double integral.

The area of the element ΔA is given by $\Delta x\,\Delta y$ or $\Delta y\,\Delta x$ and so the double integral can be written as

$$\iint_R f(x, y)\,dA = \iint_R f(x, y)\,dy\,dx = \iint_R f(x, y)\,dx\,dy$$

If the region R is enclosed by $x = a, x = b, y = c, y = d$, then

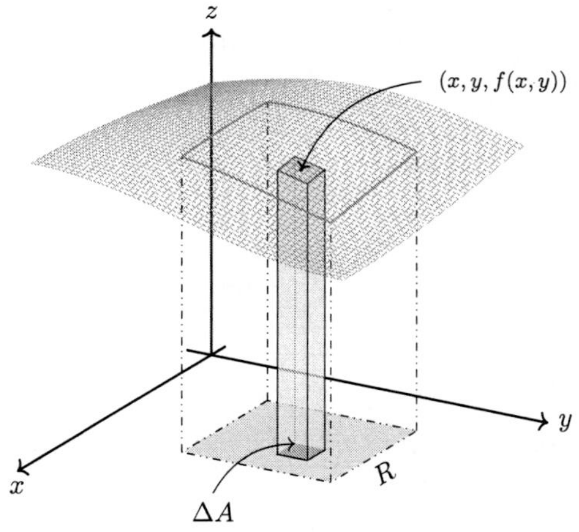

Figure 1.11 Double integrals

$$\iint_R f(x,y)\,dA = \int_a^b \int_c^d f(x,y)\,dy\,dx = \int_c^d \int_a^b f(x,y)\,dx\,dy$$

That is we can find the volume by first allowing dA to move along a thin strip parallel to the y-axis. This gives the volume of a very thin three-dimensional slice. Then we sum up these slices by moving parallel to the x-axis.

Or using the second set of limits we can first allow dA to move along a thin strip parallel to the x-axis then sum up the volume of these slices by then moving parallel to the y-axis.

Example 1.17.

$$\int_0^1 \int_0^2 (x+y)\,dy\,dx = \int_0^1 (xy+\frac{1}{2}y^2)\Big|_0^2 dx = \int_0^1 (2x+2)\,dx = (x^2+2x)\Big|_0^1 = 3$$

The region R in the xy-plane does not have to be a rectangle.

The region in Figure 1.12 (a) is enclosed by $x = a, x = b, y = g_1(x), y = g_2(x)$ and the double integral becomes

$$\iint_R f(x,y)\,dA = \int_a^b \int_{g_1(x)}^{g_2(x)} f(x,y)\,dy\,dx \tag{1.6}$$

Such a region is said to be *vertically simple*. So here we are allowing dA to move along a thin strip parallel to the y-axis until it hits the bounding curves $y = g_1(x)$ and $y = g_2(x)$. Then sum up the volume of the three-dimensional slices standing on this region by moving parallel to the x-axis.

The region in Figure 1.12 (b) is enclosed by $y = c, y = d, x = h_1(y), x = h_2(y)$ and is said to be *horizontally simple*. The double integral over this region is

$$\iint_R f(x,y)\,dA = \int_c^d \int_{h_1(x)}^{h_2(x)} f(x,y)\,dx\,dy \tag{1.7}$$

In this case, we are allowing dA to move along a thin strip parallel to the x-axis until it hits the bounding curves $x = h_1(y)$ and $x = h_2(y)$ and then summing up volumes by moving parallel to the y-axis.

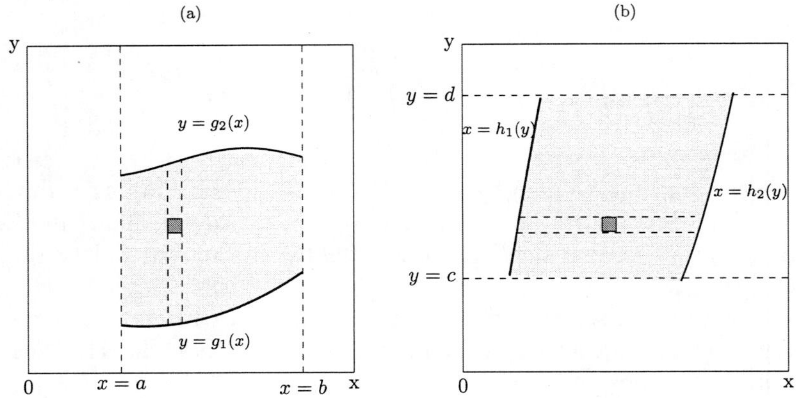

Figure 1.12 Iterated integrals. (a) Vertically simple region R (b) Horizontally simple region R

Exercise 1.3. *Demonstrate that*

$$\int_a^b \int_a^x f(x,y)\,dy\,dx = \int_a^b \int_y^b f(x,y)\,dx\,dy \tag{1.8}$$

and hence evaluate

$$\int_2^5 \int_2^x (xy - x)\,dy\,dx$$

in two ways.

Proof. The region in Figure 1.13 (a) is vertically simple and enclosed by $x = a, x = b, y = x, y = a$. The double integral over this region is

$$\int_a^b \int_a^x f(x,y)\,dy\,dx$$

This region is also horizontally simple.

In Figure 1.12 (b), the same region is enclosed by $y = a, y = b, x = y, x = b$ and the double integral becomes

$$\int_a^b \int_y^b f(x,y)\,dx\,dy$$

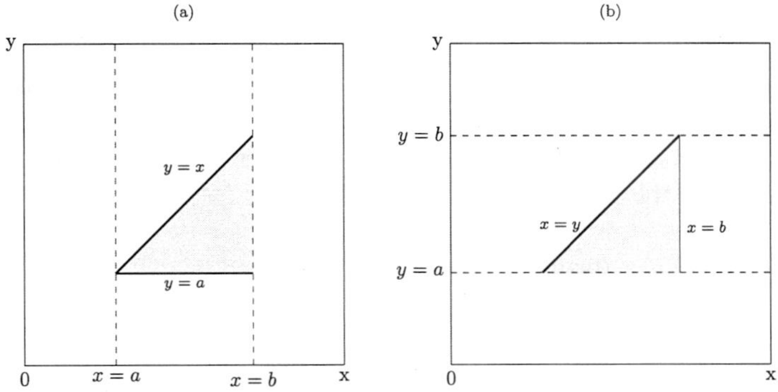

Figure 1.13 Changing the order of integrals. (a) Vertically (b) Horizontally

Therefore,

$$\int_a^b \int_a^x f(x,y)\,dy\,dx = \int_a^b \int_y^b f(x,y)\,dx\,dy$$

Evaluating the integral given

$$\int_2^5 \int_2^x (xy - x)\,dy\,dx = \int_2^5 \left(\frac{1}{2}xy^2 - xy\right)\Big|_2^x dx = \int_2^5 \frac{1}{2}x^3 - x^2 - 2x + 2x$$

$$dx = \frac{1}{8}x^4 - \frac{1}{3}x^3\Big|_2^5 = \frac{297}{8}$$

Changing the order of the integrals

$$\int_2^5 \int_2^x (xy - x)\,dy\,dx = \int_2^5 \int_y^5 (xy - x)\,dx\,dy$$

Evaluating this new integral

$$\int_2^5 \int_y^5 (xy - x)\,dx\,dy = \frac{1}{2}\int_2^5 (x^2y - x^2)\Big|_y^5 dy = \frac{1}{2}\int_2^5 25y - 25 - y^3 + y^2$$

$$dy = \frac{25}{4}y^2 - \frac{25}{2}y - \frac{y^4}{8} + \frac{y^3}{6}\Big|_2^5 = \frac{297}{8}$$

□

1.16 Solving first-order differential equations

A differential equation is an equation that involves an unknown function and one or more of its derivatives.

The *order* of a differential equation is the highest derivative of the unknown function that appears in the equation.

Given an unknown function $y = f(x)$,
a differential equation where the highest derivative is

$$\frac{dy}{dx}$$

is first order, while a differential equation where the highest derivative is

$$\frac{d^2y}{dx^2}$$

is second order, etc.

A very simple example is given by

$$\frac{dy}{dx} = x^2$$

This is a first-order differential equation since the highest derivative is one. The solution is very easy to find. It is simply the function $y = f(x)$ which when differentiated gives x^2 or, in other words, the integral

$$y = \frac{x^3}{3} + C$$

For other differential equations, we can refer back to integration by substitution. For example, given

$$\frac{dy}{dx} = xy$$

then rearranging and integrating both sides with respect to x

$$\frac{1}{y}\frac{dy}{dx} = x \implies \int \left(\frac{1}{y}\frac{dy}{dx}\right) dx = \int x\,dx$$

By Theorem 1.3, this is equivalent to

$$\int \frac{1}{y}\,dy = \int x\,dx$$

Integrating both sides yields

$$\log|y| = \frac{x^2}{2} + C$$

That is

$$|y| = Ae^{\frac{x^2}{2}} \quad \text{for } A = e^C > 0 \quad \text{or} \quad y = Ke^{\frac{x^2}{2}} \quad \text{for any real number } K$$

This approach is called solving by *separation of variables.*

Example 1.18.

(i) The solution to the differential equation $\frac{dy}{dx} = x^4$ *is simply the integral* $y = \frac{x^5}{5} + C.$

(ii) The differential equation $\frac{dy}{dx} = y$ *can be solved by separation of variables*

$$\int \frac{1}{y} dy = \int 1 dx \implies \log|y| = x + C \implies y = Ke^x \quad \textit{for any realnumber } K$$

(iii) Solving $\frac{dy}{dx} = \frac{y}{x}$

$$\int \frac{1}{y} dy = \int \frac{1}{x} dx \implies \log|y| = \log|x| + C$$
$$\implies y = Kx \quad \textit{for any real number } K$$

(iv) Solving $\frac{dy}{dx} = 2xy^2$

$$\int \frac{1}{y^2} dy = \int 2x dx \implies \frac{-1}{y} = x^2 + C \implies y = \frac{-1}{x^2 + C}$$

□

Not all differential equations are separable. Another approach is to multiply by an *integrating factor*. When the differential equation is of the form below, the differential equation becomes an exact derivative which is easy to integrate.

$$\frac{dy}{dx} + P(x)y = Q(x) \tag{1.9}$$

The integrating factor is then

$$I = e^{\int P(x)dx}$$

Multiplying Equation (1.9) by I

$$I\frac{dy}{dx} + IPy = IQ$$

$$\implies \frac{d(Iy)}{dx} = IQ \qquad \text{by product rule in Section 1.3}$$

$$\implies Iy = \int IQdx \qquad \text{integrating both sides w.r.t. } x$$

Exercise 1.4. *Solve the differential equation*

$$\frac{dy}{dx} + \frac{3}{x}y = x^2 \text{ where } x > 0$$

Proof. Here $P(x) = \frac{3}{x}$ and $Q(x) = x^2$.
Then

$$I = e^{\int \frac{3}{x}dx} = e^{3\log x} = x^3$$

Multiplying by I and integrating:

$$\underbrace{x^3}_{I} y = \int \underbrace{x^3}_{I} \underbrace{x^2}_{Q} dx = \int x^5 dx = \frac{x^6}{6} + C$$

Dividing across by x^3

$$y = \frac{x^3}{6} + \frac{C}{x^3}$$

□

Exercise 1.5. *Solve*

$$\frac{dy}{dx} + y = e^{-x}$$

where $y(0) = 1$.

Proof. Here $P(x) = 1$ and $Q(x) = e^{-x}$. So $I = e^{\int P(x)dx} = e^{\int 1 dx} = e^x$.

The solution is then:

$$\underbrace{e^x}_{I} y = \int \underbrace{e^x}_{I} \underbrace{e^{-x}}_{Q} dx = \int 1 dx = x + C$$

Multiplying across by e^{-x}

$$y = e^{-x}(x + C)$$

Using the initial condition $y(0) = 1$, we can find the constant of integration C. Letting $x = 0$ and $y = 1$ then

$$1 = C$$

Therefore, the solution is

$$y = e^{-x}(x + 1)$$

□

1.17 Summary

The ideas and techniques covered in this chapter will either directly feature in later topics or be used to facilitate understanding when we move on to stochastic calculus.

Understanding how limits work in calculus aids understanding of how stochastic processes evolve and how integrals of stochastic processes are defined. While many stochastic processes (like Brownian motion) are nowhere differentiable in the classical sense, the intuition of differentials still draws from calculus. Classical integration is key to understanding its stochastic counterpart Itô integrals. Itô's formula which is a stochastic analogue of the chain rule uses ideas similar to Taylor expansions. Stochastic differential equations are central to stochastic calculus. These are like regular differential equations but with random terms (often involving Brownian motion). A background in solving ordinary differential equations helps interpret and solve stochastic differential equations.

Once the basics of calculus are mastered, the student is then well-positioned to address the probabilistic tools that stochastic calculus also requires.

Further reading

A high school calculus textbook will provide a useful refresher for the computational side of calculus – derivatives, integrals, Taylor series, etc. But stochastic calculus rests on a much deeper understanding of calculus.

There are many excellent texts to strengthen foundations in ordinary calculus. Simmons (1995), Strang (1991) or Spivak (2006) are all great for building intuition and rigor. These texts provide a sufficient level of detail required to progress through the material presented here.

To go beyond this level and think about these topics like a mathematician, Abbott (2015) or Apostol (1974) are a useful place to start.

Notes

1. Partial derivatives are assumed to exist at and near (x_0, y_0) and be continuous at (x_0, y_0) (Simmons, 1995).
2. The limit when it exists is known as the *Riemann integral*. Integrals can be defined in other ways also.

2 Statistical background

Overview

This chapter provides a review of the concepts in probability and statistics necessary for studying stochastic processes. It begins with foundational material on random variables, distributions and expectations. The next sections on conditional expectations, moment generating functions and the tower rule are important for a deeper understanding of the mathematical statistics seen in stochastic calculus. The final material on convergence theorems will become relevant when discussing stochastic integration.

Having a strong statistical background is important for understanding stochastic calculus because stochastic calculus is fundamentally about understanding and working with randomness that evolves over time.

Once again many topics are described with minimal detail and only a few examples. It is expected that most students will have encountered these ideas in previous courses. The references at the end of the chapter provide any extra context the student may require from fundamentals to more advanced topics.

2.1 Discrete random variables

When two dice are rolled, the result could be that the first die lands on 6 and the second die lands on 1 or in shorthand (6,1). The set of all possible outcomes is known as the *sample space* (denoted Ω). In this case, there are 36 possible outcomes and the sample space is:

$$\begin{aligned}\Omega = \{&(1,1),(1,2),(1,3),(1,4),(1,5),(1,6),(2,1),\ldots,\\ &(6,3),(6,4),(6,5),(6,6)\}\end{aligned}$$

The sample space is *discrete* if there are a finite or countably infinite[1] number of possible outcomes.

We may be interested in computing totals. For example, the total for the outcome (6,1) is 7. The elements of the sample space that produce a total of 7 are

$$A = \{(1,6),(2,5),(3,4).(4.3),(5,2),(6,1)\}$$

DOI: 10.4324/9781003477624-3

A total of 7 produced by rolling two dice is called an *event*.

Definition 2.1. *An event is a subset of a sample space.*

The next step is to assign a probability to events. We would like the assignment to follow some simple rules. A function P defined on a discrete sample space Ω is a *probability measure* if

1 For any event A, $P(A) \geq 0$. In other words, the probability of any event happening is non-negative.
2 For mutually exclusive events A and B, the probability of either event happening is the sum of the probability of each event happening. In symbols $P(A \cup B) = P(A) + P(B)$.
3 The probability of some outcome in the sample space happening is one, that is, $P(\Omega) = 1$.

Since all 36 possible outcomes in the sample space for rolling two dice are equally likely, the intuitive probability measure for this experiment would assign a probability of $1/36$ to each outcome.

For the sample space when two dice are rolled, each outcome is associated with a total. And the probability measure assigns a probability to each of these outcomes. The total when two dice are rolled is then a type of *random variable*.

Definition 2.2. *Given a sample space Ω and a probability measure P a random variable X is a function that assigns a numerical value to each outcome.*

Let the random variable X be the total when rolling two dice, then every outcome is associated with a value of X and a probability. The probability of a total of 7 is the sum of the probabilities of the outcomes producing this event $P(X = 7) = 6/36$. We then have the following probability distribution:

x	$P(X = x)$
2	1/36
3	2/36
4	3/36
5	4/36
6	5/36
7	6/36
8	5/36
9	4/36
10	3/36
11	2/36
12	1/36

Definition 2.3. *Given X a discrete random variable, a probability distribution describes the probability $P(X = x)$ of each value of x occurring.*

From the rules of probability measures above, a function $f(x)$ is valid as a probability distribution for a discrete random variable if it obeys the following conditions.

Theorem 2.1. *A function $f(x) = P(X = x)$ is valid as a probability distribution for a discrete random variable X if*

1 $f(x) \geq 0$
2 $\sum_x f(x) = 1$

2.2 Continuous random variables

A sample space is *continuous* if there are a infinite number of possible outcomes such as the height of an adult male randomly selected from the US population. The probability of a continuous random variable taking on a specific value is zero. It is more meaningful to consider the probability the continuous random variable falls into a range of values. We define a *probability density function* instead of a probability distribution function.

Definition 2.4. *A probability density function $f(x)$ of the random variable X is defined as*

$$P(a \leq X \leq b) := \int_a^b f(x)dx$$

for any real numbers a and b.

The rules of probability measures then imply the following theorem.

Theorem 2.2. *A function $f(x) = P(X = x)$ is valid as a probability density function for a continuous random variable X if*

1 $f(x) \geq 0$ *for all* x
2 $\int_{-\infty}^{\infty} f(x) = 1$

2.3 Moments of random variables

The *expectation* (or *expected value*) of a random variable is its average over many repeated experiments. If we repeatedly toss two dice recording the total each time, then the expectation is the average value we get after an infinite number of repetitions.

Given $f(x)$ as in the discrete or continuous case, the *expectation* of a random variable X is

$$\mathbb{E}X := \sum_x xf(x) \text{ or } \mathbb{E}X := \int_{-\infty}^{\infty} xf(x)\,dx \tag{2.1}$$

These properties follow from the definition:

(a) $\mathbb{E}(aX + b) = a\mathbb{E}X + b$ for any real numbers a and b.
(b) $\mathbb{E}(aX + bY) = a\mathbb{E}X + b\mathbb{E}Y$ for any random variables X,Y and real numbers a,b.

For a function g of a random variable X, it can be shown that

$$\mathbb{E}\,g(X) = \sum_x g(x)f(x) \text{ or } \mathbb{E}(X) = \int_{-\infty}^{\infty} g(x)f(x)\,dx \tag{2.2}$$

The kth *moment* of a random variable X is $\mathbb{E}X^k$. The first moment $\mathbb{E}X$ is the expectation of the random variable (Equation 2.1). Sometimes the symbol μ will be used for the first moment. Higher moments will be important when we consider moment-generating functions in the next section.

The kth *central moment* of a random variable X is $\mathbb{E}(X - \mu)^k$. The second central moment $\mathbb{E}(X - \mu)^2$ gives the spread of the distribution of X around the expectation. It is more commonly known as the *variance*.

The *variance* of a random variable X with expectation μ can be written as

$$\begin{aligned} varX &:= \mathbb{E}(X - \mu)^2 \\ &= \mathbb{E}(X^2 - 2\mu X + \mu^2) \\ &= \mathbb{E}X^2 - 2\mu\mathbb{E}X + \mu^2 \\ &= \mathbb{E}X^2 - \mu^2 \end{aligned}$$

Sometimes the symbol σ^2 will be used for the variance. The square root of the variance σ is the *standard deviation*.

Given a random variable X with variance σ^2, it is easily shown that:

$$var(aX + b) = a^2 varX \quad \text{for real numbers } a, b$$

Note that adding a constant to a random variable does not change its variance.

2.4 Multivariate distributions

The example of rolling two dice above generated a sample space. A random variable X, the total of the two dice, was then associated with this sample space. Other random variables could be associated with the same sample space. We could calculate a second random variable Y which is the product of the two dice.

The probability of getting a total of 4 and a product of 3 can then be determined. The elements of the sample space that produce a total of 4 and a product of 3 are

$$\{(1,3),(3,1)\}$$

and therefore

$$P(X = 4, Y = 3) = \frac{2}{36} = \frac{1}{18}$$

Similarly $P(X = x, Y = y)$ can be calculated for any x and y that can occur. The function $f(x, y) = P(X = x, Y = y)$ is called a *joint probability distribution.*

The equivalent in the continuous case is a *joint probability density function.* The probability that X and Y are in the region R in the xy-plane is

$$\iint_R f(x, y)\,dx\,dy$$

where $f(x, y)$ is the joint probability density function.

Given $f(x, y)$ as in the discrete or continuous case, the expectation $\mathbb{E}(XY)$ is

$$\mathbb{E}(XY) = \sum_x \sum_y xyf(x, y) \text{ or } \mathbb{E}(XY) = \int_{-\infty}^{\infty} \int_{-\infty}^{\infty} xyf(x, y)\,dx\,dy$$

The *covariance* of random variables X and Y is defined as

$$cov(X, Y) := \mathbb{E}\Big((X - \mathbb{E}X)(Y - \mathbb{E}Y)\Big)$$

It tells us how X and Y vary in relation to each other.

The covariance of random variables X with expectation μ_X and Y with expectation μ_Y can be written as

$$\begin{aligned} cov(X, Y) &= \mathbb{E}\Big((X - \mu_X)(Y - \mu_Y)\Big) \\ &= \mathbb{E}(XY - \mu_X Y - \mu_Y X + \mu_X \mu_Y) \\ &= \mathbb{E}(XY) - \mu_X \mathbb{E}Y - \mu_Y \mathbb{E}X + \mu_X \mu_Y \\ &= \mathbb{E}(XY) - \mu_X \mu_Y \end{aligned} \tag{2.3}$$

Sometimes the symbol σ_{XY} will be used for the covariance.

In the previous section, the expectation of a linear combination $aX + bY$ of two random variables X and Y was given. Here we derive an expression for the variance of $aX + bY$ where random variables X and Y have expectation μ_X

Table 2.1 Linear combinations of random variables X, Y for real numbers a, b

Combination	*Expectation*	*Variance*
X	μ_X	σ_X^2
Y	μ_Y	σ_Y^2
$aX + b$	$a\mu_X + b$	$a^2\sigma_X^2$
$aX + bY$	$a\mu_X + b\mu_Y$	$a^2\sigma_X^2 + b^2\sigma_Y^2 + 2ab\,\sigma_{XY}$

and μ_Y and variance σ_X^2 and σ_Y^2 respectively.

$$
\begin{aligned}
var(aX+bY) &= \mathbb{E}(aX+bY)^2 - \Big(\mathbb{E}(aX+bY)\Big)^2 \\
&= \mathbb{E}(a^2X^2 + 2abXY + b^2Y^2) - (a\mu_X + b\mu_Y)^2 \\
&= a^2\mathbb{E}X^2 + 2ab\mathbb{E}(XY) + b^2\mathbb{E}Y^2 - (a^2\mu_X^2 + 2ab\mu_X\mu_Y + b^2\mu_Y^2) \\
&= a^2\Big(\mathbb{E}X^2 - \mu_X^2\Big) + b^2\Big(\mathbb{E}Y^2 - \mu_Y^2\Big) + 2ab\Big(\mathbb{E}(XY) - \mu_X\mu_Y\Big) \\
&= a^2\sigma_X^2 + b^2\sigma_Y^2 + 2ab\,\sigma_{XY}
\end{aligned}
$$

We will refer back to the properties of linear combinations of random variables many times in this text. They are summarized in Table 2.1 for convenience.

The *correlation coefficient* also tells us how X and Y vary in relation to each other. It is standardized to take values in the interval $[-1, 1]$.

The correlation coefficient is defined as

$$\rho_{XY} := \frac{\sigma_{XY}}{\sigma_X\sigma_Y}$$

2.5 Conditional probability

The *conditional probability* of A given B is the probability that A occurs given that B has occurred (written $P(A|B)$). It is defined as

$$P(A|B) := \frac{P(A \cap B)}{P(B)}$$

where $A \cap B$ is the event that A and B both happen.

Two events A and B are *independent* if the occurrence of one does not affect the probability of the other occurring. Therefore the probability that A occurs does not change if B occurs

$$P(A|B) = P(A)$$

This implies

$$\frac{P(A \cap B)}{P(B)} = P(A) \implies P(A \cap B) = P(A)P(B)$$

The disjoint events $B_1, ..., B_n$ are a *partition* of the sample space Ω if $B_1 \cup B_2 \cup .... \cup B_n = \Omega$.

The following result states that the probability of an event A happening can be found by summing the probabilities $P(A \cap B_i) = P(A|B_i)P(B_i)$ over the partition.

Theorem 2.3. *The 'law of total probability' states that given an event A and partition $B_1, ..., B_n$*

$$P(A) = \sum_{i=1}^{n} P(A|B_i)\, P(B_i)$$

Next we will define conditional expectations as they feature prominently in stochastic calculus. But there are a few preliminary definitions that need to be put in place first.

If X and Y are discrete random variables, the *marginal distribution* of Y is

$$f(y) := \sum_{x} f(x, y)$$

If X and Y are continuous, the *marginal density* of Y is

$$f(y) := \int_{-\infty}^{\infty} f(x, y)\, dx$$

In keeping with the definition of conditional probabilities, the *conditional distribution* of X given Y for X and Y discrete or the *conditional density* of X given Y for X and Y continuous is then the function

$$f(x|y) := \frac{f(x, y)}{f(y)}$$

where $f(y)$ is defined as above and $f(x, y)$ is defined as in the previous section.

The *conditional expectation* is the expected value of a random variable when additional information is known. More formally, the conditional expectation of X given $Y = y$ written as $\mathbb{E}(X | Y = y)$ is

$$\mathbb{E}(X | Y = y) := \sum_x x f(x|y) \text{ or } \mathbb{E}(X | Y = y) := \int_{-\infty}^{\infty} x f(x|y)\, dx$$

for discrete X and Y and continuous X and Y, respectively.

The next result will be useful when dealing with martingales.

Theorem 2.4. *The 'law of total expectation' or 'law of iterated expectations' states that the expected value of a random variable X is equal to the expected value of the conditional expectation of X given Y, that is,*

$$\mathbb{E}X = \mathbb{E}(\mathbb{E}(X | Y)) \tag{2.4}$$

Proof. The proof largely follows from the law of total probability above. The result is proven here for the discrete case while the continuous case is omitted but follows the same logic.

$$\begin{aligned}
\mathbb{E}X &= \sum_x x\, P(X = x) \\
&= \sum_x x \left(\sum_y P(X = x | Y = y)\, P(Y = y) \right) \\
&\quad \text{rearranging the sums} \\
&= \sum_y P(Y = y) \left(\sum_x x\, P(X = x | Y = y) \right) \\
&= \sum_y P(Y = y)\, \mathbb{E}(X | Y = y) \\
&= \mathbb{E}(\mathbb{E}(X | Y))
\end{aligned}$$

□

2.6 Examples of probability distributions

2.6.1 Binomial distribution

The binomial distribution is a discrete probability distribution that models the number of successes in a fixed number of independent trials, each with the same probability of success. In each trial there are only two possible outcomes: success or failure. Flipping a coin 20 times and counting the number of heads can be modeled by the binomial distribution. Or rolling a die 10 times and counting the number of 6's.

The probability of x successes in n trials is given by the probability distribution

$$f(x) = \frac{n!}{k!(n-k)!} p^k (1-p)^{n-k}$$

where p is the probability of success.

The expectation and variance of a binomially distributed variable X are

$$\mathbb{E}X = np \quad \text{and} \quad varX = np(1-p)$$

This distribution underpins the binomial model for pricing derivatives (Chapter 8).

2.6.2 *Poisson distribution*

The Poisson distribution is a discrete probability distribution that models the number of events occurring in a fixed interval of time or space. For example, the number of insurance claims arriving in a month or the number of car accidents on a section of road. It is assumed that:

- Events occur independently of each other
- The average rate λ of events is constant over time (space)
- Two events cannot occur at the exact same instant (location)

The probability of x events occurring is given by the probability distribution

$$f(x) = \frac{\lambda^x e^{-\lambda}}{x!}$$

where λ is the average rate of events per interval.

The expectation and variance of a Poisson distributed variable X are

$$\mathbb{E}X = \lambda \quad \text{and} \quad varX = \lambda$$

This distribution underpins models of insurance claims processes.

2.6.3 *Uniform distribution*

The continuous uniform distribution describes a situation where a random variable can take any value within an interval with equal probability.

The continuous uniform distribution has the density function

$$f(x) = \begin{cases} \frac{1}{b-a}, & \text{for } a \leq x \leq b \\ 0, & \text{otherwise} \end{cases} \tag{2.5}$$

for some real numbers a and b.

The expectation and variance of a random variable X with this distribution are

$$\mathbb{E}X = \frac{a+b}{2} \quad \text{and} \quad varX = \frac{(b-a)^2}{12}$$

2.6.4 Normal distribution

The normal distribution is a continuous probability distribution with density function

$$f(x) = \frac{1}{\sigma\sqrt{2\pi}} e^{-\frac{(x-\mu)^2}{2\sigma^2}} \tag{2.6}$$

for some real numbers μ and σ, where $\sigma > 0$.

The function $f(x)$ is graphed in Figure 2.1 and seen to be symmetric around $x = \mu$.

The expectation and variance of a normally distributed variable X are

$$\mathbb{E}X = \mu \quad \text{and} \quad varX = \sigma^2$$

and in symbols this will be represented by $X \sim N(\mu, \sigma^2)$.

The *standard normal distribution* has mean $\mu = 0$ and variance $\sigma^2 = 1$. In this case, the graph of the density function will be centered on the y-axis. We will use the symbol $\phi(z)$ to denote the density function of the standard normal distribution, that is,

$$\phi(z) = \frac{1}{\sqrt{2\pi}} e^{-\frac{z^2}{2}}$$

and the symbol Z to represent a random variable with this distribution.

From Definition 2.4, the probability that Z is between a and b is given by the area under the curve

$$P(a \le Z \le b) = \int_a^b \phi(z)dz$$

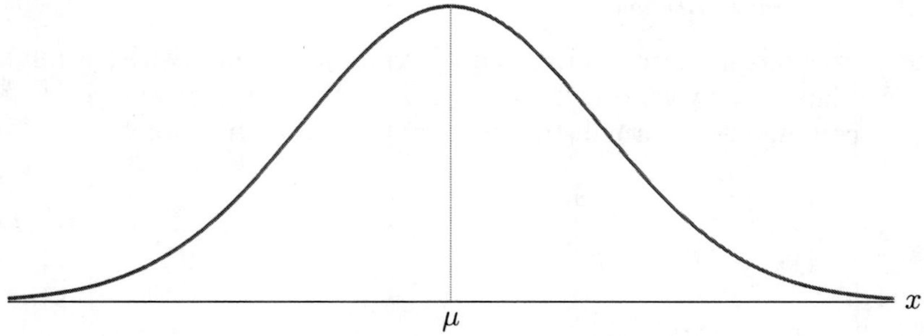

Figure 2.1 Density function of the normal distribution

We will use the symbol $\Phi(x)$ to denote the *cumulative distribution function* of the standard normal distribution, that is,

$$\Phi(x) = \int_{-\infty}^{x} \phi(z)dz = \frac{1}{\sqrt{2\pi}} \int_{-\infty}^{x} e^{-\frac{z^2}{2}} dz$$

From the first fundamental theorem of calculus $\Phi'(x) = \phi(x)$.

The cumulative distribution function can be evaluated using

- NORM.S.DIST(x,1) function in Excel[2]
- pnorm(x,0,1) in *R*

Example 2.1.

(i) Using

- *NORM.S.DIST(−1.5,1) in Excel*
- *pnorm(−1.5,0,1) in R*

the probability that Z is less than −1.5 is

$$P(Z < -1.5) = 0.067$$

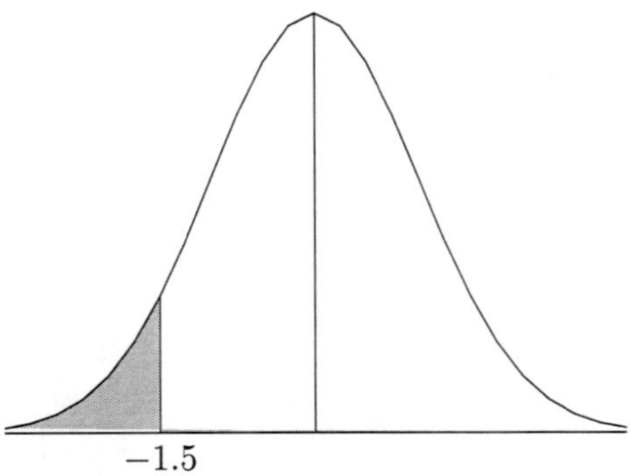

(ii) The probability that Z is greater than −1.5, $P(Z > -1.5)$, *is* $1 - 0.067 = 0.933$ *since the total area under the curve is 1 from Theorem 2.2.*

(iii) The probability that Z is greater than 1.5, $P(Z > 1.5)$, *is also 0.067 from the symmetry of the density function about the y-axis.*

(iv) Using

- *NORM.S.DIST(0.8,1) - NORM.S.DIST(−0.9,1) in Excel.*
- *pnorm(0.8,0,1) − pnorm(−0.9,0,1) in R*

the probability that Z is between −0.9 and 0.8 is

$$P(-0.9 < Z < 0.8) = 0.604$$

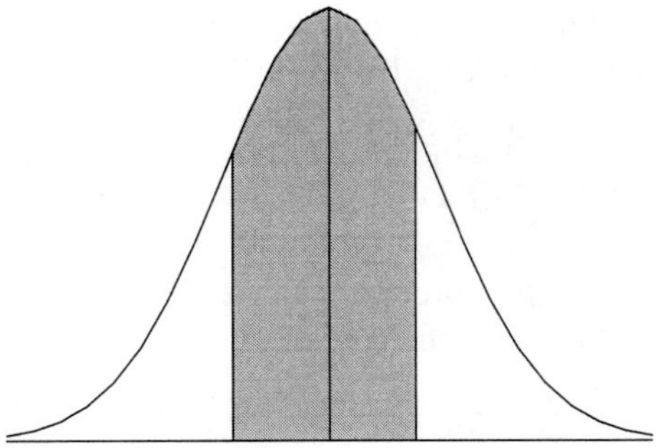

The following result allows us to use the standard normal distribution to calculate probabilities for any normally distributed variable.

Theorem 2.5. *Given $X \sim N(\mu, \sigma^2)$, then*

$$Z = \frac{X - \mu}{\sigma}$$

has the standard normal distribution.

Proof. We will use integration by substitution (Section 1.14).
Letting $z = \frac{x-\mu}{\sigma}$ and $z_1 = \frac{x_1-\mu}{\sigma}; z_2 = \frac{x_2-\mu}{\sigma}$

$$\begin{aligned} P(x_1 < X < x_2) &= \frac{1}{\sigma\sqrt{2\pi}} \int_{x_1}^{x_2} e^{-\frac{(x-\mu)^2}{2\sigma^2}} dz \\ &= \frac{1}{\sigma\sqrt{2\pi}} \int_{z_1}^{z_2} e^{-\frac{z^2}{2}} dx \\ &= P(z_1 < Z < z_2) \end{aligned}$$

□

Example 2.2. *The area under a normal curve with mean 2 and variance 9 between 1 and 5 is given by*[3]

$$P(1 < X < 5) = P\left(\frac{1-2}{3} < Z < \frac{5-2}{3}\right) = P(-1/3 < Z < 1) = 0.472$$

Stock market returns are not quite normally distributed. Their distribution is *leptokurtic* meaning that extremely negative and extremely positive returns happen more often than the normal distribution would suggest. A leptokurtic distribution has fatter tails and a pointier peak than the normal distribution.[4]

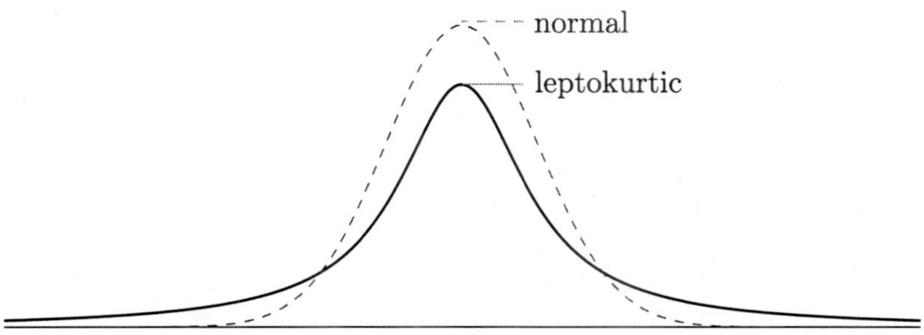

Figure 2.2 Leptokurtic distribution

The normal distribution is extremely important in statistics due to the *central limit theorem* which states that the average of a sample will be approximately normally distributed regardless of the population distribution. The intuition is that the sample average will tend toward the population mean. Any extreme sample values should be balanced out in the sample average as long as the sample is large enough. Therefore, deviations of the sample average from the population mean should be symmetric and large deviations unlikely.

The normal distribution will feature heavily in derivative pricing as it is the basis of Brownian motion processes. Also, a lognormal model of asset prices is the basis of option pricing. A lognormally distributed random variable is a random variable whose logarithm is normally distributed. That is, if $X = e^Y$ is lognormally distributed, then Y is normally distributed. The properties of lognormal distributions are discussed in Chapter 7.

2.7 Moment-generating functions

Moment-generating functions (MGF) allow us to deal with all the moments of a random variable at the one time. The MGF of a random variable X is

$$M_X(r) := \mathbb{E}e^{rX} \tag{2.7}$$

for some real number r.

To see how this function generates moments, we refer back to the Taylor expansion of e^x around $x = 0$ (Example 1.12).

$$\begin{aligned} e^x &= 1 + x + \frac{x^2}{2!} + \frac{x^3}{3!} + \dots \\ &= \sum_{k=0}^{\infty} \frac{x^k}{k!} \end{aligned}$$

$$
\begin{aligned}
\implies e^{rX} &= \sum_{k=0}^{\infty} \frac{(rX)^k}{k!} \\
&= \sum_{k=0}^{\infty} \frac{X^k r^k}{k!} \\
\implies M_X(r) = \mathbb{E}e^{rX} &= \sum_{k=0}^{\infty} \frac{\mathbb{E}X^k\, r^k}{k!}
\end{aligned}
$$

Taking the kth derivative of $M_X(r)$ with respect to r and evaluating it at $r = 0$, we get the kth moment of X

$$
\mathbb{E}X^k = \frac{d^k}{dr^k} M_X(r)\bigg|_{r=0} \tag{2.8}
$$

If two random variables have the same MGF, then they have the same moments for all k. In which case it follows that they must have the same probability distribution.

The MGF of normally distributed random variables will be of special interest in later chapters.

Theorem 2.6. *The MGF of a normally distributed random variable $X \sim N(\mu, \sigma^2)$ is $e^{r\mu + \frac{1}{2}r^2\sigma^2}$.*

Proof. From Theorem 2.5, X can be written as $X = \mu + \sigma Z$, where Z has the standard normal distribution.

Using the density function of the standard normal distribution, the MGF of σZ is:

$$
\begin{aligned}
\mathbb{E}e^{r\sigma Z} &= \int_{-\infty}^{\infty} e^{r\sigma z} \frac{1}{\sqrt{2\pi}} e^{-\frac{1}{2}z^2}\, dz \\
&= \int_{-\infty}^{\infty} \frac{1}{\sqrt{2\pi}} e^{-\frac{1}{2}(z^2 - 2r\sigma z)}\, dz \\
&= \int_{-\infty}^{\infty} \frac{1}{\sqrt{2\pi}} e^{-\frac{1}{2}((z - r\sigma)^2 - r^2\sigma^2)}\, dz \\
&= e^{\frac{1}{2}r^2\sigma^2} \int_{-\infty}^{\infty} \frac{1}{\sqrt{2\pi}} e^{-\frac{1}{2}(z - r\sigma)^2}\, dz \\
&= e^{\frac{1}{2}r^2\sigma^2}
\end{aligned}
$$

The last line follows since the integrand is the probability density function of a normal random variable with mean $r\sigma$ and variance 1. By the definition of a probability density function, the integral should therefore equal one (Theorem 2.2).

Then

$$\mathbb{E}e^{rX} = \mathbb{E}e^{r(\mu+\sigma Z)} = e^{r\mu}\mathbb{E}e^{r\sigma Z} = e^{r\mu}e^{\frac{1}{2}r^2\sigma^2} = e^{r\mu+\frac{1}{2}r^2\sigma^2}$$

□

Setting $r = 1$ we have a result that will prove to be very useful in a number of applications in this book.

$$\mathbb{E}e^{X} = e^{\mu+\frac{\sigma^2}{2}} \tag{2.9}$$

Another useful result follows from this theorem. If $Z \sim N(0,1)$, then $\sigma Z \sim N(0,\sigma^2)$. Setting μ to zero in Equation (2.9) and using the Taylor expansion of the exponential function

$$\begin{aligned}\mathbb{E}e^{\sigma Z} &= e^{\frac{\sigma^2}{2}} \\ &= 1 + \underline{\frac{\sigma^2}{2}} + \underline{\frac{1}{2!}\left(\frac{\sigma^2}{2}\right)^2} + \frac{1}{3!}\left(\frac{\sigma^2}{3}\right)^3 + \ldots\end{aligned}$$

We will now do the Taylor expansion before finding the expectation:

$$\begin{aligned}\mathbb{E}e^{\sigma Z} &= \mathbb{E}\left(1 + \sigma Z + \frac{1}{2!}(\sigma Z)^2 + \frac{1}{3!}(\sigma Z)^3 + \frac{1}{4!}(\sigma Z)^4 + \ldots\right) \\ &= 1 + \sigma\mathbb{E}Z + \underline{\frac{\sigma^2}{2!}\mathbb{E}Z^2} + \frac{\sigma^3}{3!}\mathbb{E}Z^3 + \underline{\frac{\sigma^4}{4!}\mathbb{E}Z^4} + \ldots\end{aligned}$$

These expansions can only be equal only if they are equal term by term. Comparing the coefficients of powers of σ in the underlined terms, we have

$$\mathbb{E}Z^2 = 1, \mathbb{E}Z^4 = 3 \text{ etc.} \tag{2.10}$$

and the expectation of all odd powers of Z must be zero.

We can also use MGFs to prove that linear combinations of independent normally distributed random variables are normally distributed.

First we have to show the MGF of a sum of independent random variables is the product of their MGFs.

Theorem 2.7. *Given X_1 and X_2 independent random variables and $Y = X_1 + X_2$, then*

$$M_Y(r) = M_{X_1}(r)M_{X_2}(r)$$

Proof. A proof is provided for the discrete case.

If X_1 and X_2 independent, then the joint probability distribution $f(x_1, x_2) = f(x_1)f(x_2)$

$$\begin{aligned}
M_Y(r) &= \mathbb{E}e^{r(X_1+X_2)} \\
&= \sum_{x_1}\sum_{x_2} e^{r(x_1+x_2)} f(x_1, x_2) \\
&= \sum_{x_1}\sum_{x_2} e^{rx_1} e^{rx_2} f(x_1)f(x_2) \\
&= \sum_{x_1} e^{rx_1} f(x_1) \sum_{x_2} e^{rx_2} f(x_2) \\
&= \mathbb{E}e^{rX_1}\mathbb{E}e^{rX_2} \\
&= M_{X_1}(r)M_{X_2}(r)
\end{aligned}$$

Proof for the continuous case follows the same logic. □

Theorem 2.8. *Given X_1 and X_2 independent and normally distributed, then $Y = aX_1 + bX_2$ is also normally distributed for real numbers a, b.*

Proof. Let X_1 and X_2 have means μ_1, μ_2 and variances σ_1^2, σ_2^2 .

From Theorem 2.6, the MGF of X_i for $i = 1, 2$ is

$$e^{r\mu_i + \frac{1}{2}r^2\sigma_i^2}$$

Then Y has MGF

$$\begin{aligned}
M_Y(r) &= M_{aX_1}(r)M_{bX_2}(r) \\
&= \mathbb{E}e^{raX_1}\mathbb{E}e^{rbX_2} \\
&= e^{ra\mu_1 + \frac{1}{2}r^2a^2\sigma_1^2} e^{rb\mu_2 + \frac{1}{2}r^2b^2\sigma_2^2} \\
&= e^{r(a\mu_1 + b\mu_2) + \frac{1}{2}r^2(a^2\sigma_1^2 + b^2\sigma_2^2)}
\end{aligned}$$

This is the MGF of a normally distributed random variable with mean $a\mu_1 + b\mu_2$ and variance $a^2\sigma_1^2 + b^2\sigma_2^2$. Therefore, $Y = aX_1 + bX_2$ must be normally distributed. □

2.8 Convergence of random variables

In stochastic calculus, many of the proofs rely on finding the value or random variable to which a sequence of random variables converges. The situation is more complicated than the concept of limits for real-valued functions described in Section 1.1 of the previous chapter as it depends on what we mean by a sequence of random variables approaching a limit.

Here are some different ways a sequence of random variables $X_1, X_2, \ldots, X_n$, ... can converge to X:

- **Convergence in probability** X_n converges in probability to X if the probability that X_n is far from X becomes small as n becomes large.

  $$\lim_{n\to\infty} P(|X_n - X| > \epsilon) = 0 \text{ for any } \epsilon > 0$$

  That is, no matter how small is our choice of ϵ, the probability that X_n is further than ϵ away from X eventually becomes zero as the sequence continues. It is written as $X_n \stackrel{p}{\to} X$.
- **Mean-square convergence** means that X_n is approaching X since the average squared error between them is becoming zero.

  $$\lim_{n\to\infty} \mathbb{E}(X_n - X)^2 = 0$$

  It is written as $X_n \stackrel{m.s.}{\to} X$. This type of convergence becomes useful when discussing Itô integrals. Mean-square convergence can be shown to imply convergence in probability.
- **Almost sure convergence** means that $X_n(\omega)$ approaches $X(\omega)$ for almost all ω in the sample space. There may be some ω for which this is not true but the subset of these elements has probability zero.

  More formally, the set of $w \in \Omega$ for which the following is true is a set with probability one

  $$\lim_{n\to\infty} X_n(\omega) = X(\omega)$$

  It is written as $X_n \stackrel{a.s.}{\to} X$. Almost sure convergence can be shown to imply convergence in probability.

Theorem 2.9. *Given a constant k and a sequence of random variables X_n such that $\mathbb{E}X_n \to k$ and $varX_n \to 0$ as $n \to \infty$, then $X_n \stackrel{m.s.}{\to} k$.*

Proof.

$$\begin{aligned}
\mathbb{E}(X_n - k)^2 &= \mathbb{E}(X_n^2 - 2kX_n + k^2) \\
&= \mathbb{E}X_n^2 - 2k\mathbb{E}X_n + k^2 \\
&= \left(\mathbb{E}X_n^2 - (\mathbb{E}X_n)^2\right) + \left((\mathbb{E}X_n)^2 - 2k\mathbb{E}X_n + k^2\right) \\
&= varX_n + (\mathbb{E}X_n - k)^2
\end{aligned}$$

If $varX_n \to 0$ and $\mathbb{E}X_n \to k$ as $n \to \infty$ the right-hand side becomes zero and by definition $X_n \overset{m.s.}{\to} k$. □

There are many situations in which we will exchange limits with expectations and integrals but this is not always possible. The following theorems specify when the limit of expectations for a sequence of random variables is the expectation of the limit and when the limit of an integral of a sequence of functions is the integral of the limit. The results for expectations and integrals are bundled together as the conditions for exchanging limits are very similar in both cases.

First we clarify what we mean by functions converging. A sequence of real-valued functions[5] $f_1, f_2, \ldots\ldots, f_n, \ldots$ converges to a function f *almost everywhere* if $f_n(x)$ approaches $f(x)$ for almost all real numbers x.

In the following theorem, the condition for exchanging limits is that the random variable (or function) must be steadily rising to the limit.

Theorem 2.10. ***Monotone convergence***

If $X_n \overset{a.s.}{\to} X$ *and*

$$0 \leq X_1(\omega) \leq X_2(\omega) \leq \ldots \text{ for almost all } \omega$$

then

$$\lim_{n\to\infty} \mathbb{E}X_n = \mathbb{E}X$$

If $\lim_{n\to\infty} f_n = f$ *almost everywhere and*

$$0 \leq f_1(x) \leq f_2(x) \leq \ldots \text{ for almost all } x$$

then

$$\lim_{n\to\infty} \int_{-\infty}^{\infty} f_n(x)\, dx = \int_{-\infty}^{\infty} f(x)\, dx$$

In the next theorem, the condition for exchanging limits is that the random variable (or function) must have an well-behaved upper bound.

Theorem 2.11. ***Dominated convergence***

If $X_n \overset{a.s.}{\to} X$ *and there is another random variable* Y *such that*

$$\mathbb{E}Y < \infty \text{ and } |X_n(\omega)| \leq Y(\omega) \text{ for almost all } \omega \text{ and every } n$$

then

$$\lim_{n\to\infty} \mathbb{E}X_n = \mathbb{E}X$$

If $\lim_{n\to\infty} f_n = f$ *almost everywhere and there is another function g such that*

$$\int_{-\infty}^{\infty} g(x)\,dx < \infty \text{ and } |f_n(x)| \leq g(x) \text{ for almost all } x \text{ and every } n$$

then

$$\lim_{n\to\infty} \int_{-\infty}^{\infty} f_n(x)\,dx = \int_{-\infty}^{\infty} f(x)\,dx$$

If either one of the conditions stipulated in the theorems is true, then it is safe to proceed to exchange limits.

Proofs of these theorems are left to more advanced texts.

2.9 Summary

A solid grounding in probability and statistics is required to understand random processes and stochastic calculus. Probability distributions, expectations and variances will feature in the sections on Brownian motion; conditional expectations are part of the definition of a martingale and Itô integrals require an understanding of convergence of random variables. The concepts of independence, moments and MGFs will also feature in many places throughout the text.

Further reading

Miller and Miller (2012) provides a calculus-based refresher on core probability and statistics. Ross (2020) is another classic introductory text.

Klenke (2020) progresses from introductory statistics to topics useful for studying stochastic processes. Calin (2015); Klebaner (2012) and Shreve (2004) all have introductory chapters covering the mathematical and statistical tools matching the needs of stochastic calculus.

Notes

1. Outcomes can be put into a one-to-one correspondence with the numbers $\{1, 2, 3, \ldots\}$.
2. NORM.S.DIST(x,1) gives the cumulative distribution function at x. NORM.S.DIST (x,0) gives the density function evaluated at x

3. This can also be evaluated using
   - NORM.DIST(5,2,3,1) - NORM.DIST(1,2,3,1) in Excel.
   - pnorm(5,2,3) – pnorm(1,2,3) in R.
4. A *platykurtic* distribution has flatter tails and a lower, wider peak than a normal distribution.
5. Technically we should restrict ourselves to only considering what are known as *Borel-measurable* functions in these convergence theorems but most real-valued functions are Borel-measurable.

3 Introduction to derivative securities

Overview

In the rest of this course, we will describe the evolution of asset prices in order to derive option pricing results. In this chapter, we will infer the characteristics of derivative securities simply from the concept of no arbitrage. On completion, you will be able to

- state what is meant by arbitrage and the law of one price.
- outline the factors that affect option prices. This will help clarify the elements that should be included in a pricing model and the sign of their partial derivatives (*Greeks*).
- derive specific results which are not model dependent:
  - show how to value a forward contract.
  - explain what is meant by put-call parity.

3.1 Terminology

First we introduce some basic ideas.

Definition 3.1. *A* ***derivative*** *is a contract which promises to make a payment in the future, the amount of which derives from the behavior of some asset.*

The underlying asset could be a share, bond, interest rate, currency, commodity (e.g. gold, wheat) or some other type of asset.

There are many types of derivative securities. The most common types are the following:

- An **option** gives an investor the right but not the obligation to buy or sell a specified asset for a specified price at or within a specified time.
- **Forwards** are contracts directly negotiated between two parties to buy or sell an asset for a specified price on a set future date.
- **Futures** are similar to forwards except the trade is executed through an exchange. As a result, the contract is standardized and the exchange bears

DOI: 10.4324/9781003477624-4

the risk of default. To mitigate this risk, the exchange requires daily settlement as prices change during the time of the contract.

- **Swaps** are agreements to exchange cash flows between two parties.

There are two main types of options: calls and puts.

A **call** option is the right to buy the underlying asset for a set price at or within a specified time. A **put** option is the right to sell an asset for a set price at or within a specified time. The holder of an option is not obliged to trade but does so if it is profitable. The writer is obliged to trade if the holder wants to.

There are a few key terms we will use. The future price agreed for transfer of the asset is called the **strike price**. The **expiration date** is the final day the option can be exercised. The price paid for the option is the **premium**.

Options vary as to when they can be exercised. The two most common variants are **European** options, where exercise can only occur at expiry, and **American** options, where exercise can occur on any date before expiry.

Here are examples of how calls and puts work.

Example 3.1. *Apple stock is currently $170 per share. An investor buys a one-month European call option on this stock at strike price $180 for a premium of $5.*

During the month, the stock price can go up or down.

Let's say it goes up to $200. The investor exercises the option buying the stock at $180 and immediately selling it on in the market for $200. Those transactions make the investor $20.

What would happen if the price went down? Let's say the stock price after one month is $165. The investor does not want to buy at $180 and lose money by then selling in the market for a lower price. The option in this scenario is worthless.

A call option is therefore a bet that the price of the underlying asset will go higher than the strike price.

The payoff from the option can be described mathematically. Let S_t be the stock price at time t, the strike price is K and the option expires at T.

If the stock price is above the strike price at maturity ($S_T > K$), the investor can buy the share at K and sell immediately on the market for S_T. The payoff is then $S_T - K$. If $S_T \leq K$, the payoff is 0. That is

$$\text{Payoff} = \begin{cases} S_T - K, & \text{if } S_T > K \\ 0, & \text{if } S_T \leq K \end{cases}$$

or written in mathematical shorthand

$$\text{Payoff} = [S_T - K]^+$$

We say the call option at time t is **in the money** if $S_t > K$; **out of the money** if $S_t < K$ and **at the money** if $S_t = K$.

Example 3.2. *An investor buys a one-month European put option on Tesla stock at strike price \$250 for a premium of \$10.*

The stock price after one month is \$210. The investor buys at this price and sells the stock on to the option holder for \$250. Those transactions make the investor \$40.

If instead the price rises to \$300, then it is not worthwhile for the investor to buy at this price and sell in the market at \$250. The option then expires worthless.

A put option is therefore a bet that the price of the underlying asset will go lower than the strike price.

Summarizing the payoff mathematically, when $S_T < K$ the investor buys the stock in the market at S_T and sells to the option holder at K. The payoff is then $K - S_T$. If $S_T \geq K$ the payoff is 0. This is written in mathematical shorthand as

$$\text{Payoff} = [K - S_T]^+$$

We say the call option at time t is **in the money** if $S_t < K$; **out of the money** if $S_t > K$ and **at the money** if $S_t = K$.

Short selling an asset means investing in such a way that the investor gains if the asset value falls. This can be achieved by borrowing the asset, selling it then repurchasing it when the price falls and returning the asset to the original lender. Being **long** in an asset implies a position where the investor gains if the asset value rises.

3.2 Factors affecting equity option prices

A number of mathematical models have been used to value options. One of the more widely used is the Black–Scholes model. This model is the main focus of the later part of this text.

In the Black–Scholes model, option prices on a dividend-paying stock at time t will be seen to depend on six parameters

S_t	current underlying share price
K	strike price
T	expiration date
σ	volatility of underlying share
r	risk-free interest rate
q	the dividend rate

The volatility is yet to be defined but let's say for the moment it relates to how much the stock price varies.

Why should these factors affect option prices? Each factor is considered in turn below keeping all other factors fixed.

3.2.1 *Current share price, S_t*

As an American call can be exercised at any time within the duration of the option contract, its value at time t is $[S_t - K]^+$. If S_t increases, then the value of an American call obviously rises.

If S_t increases, the probability of a European call being in-the-money at maturity is now higher so the value of a European call should also rise.

For an American put, the value at time t is $[K - S_t]^+$ so if S_t increases, then the option price decreases. For a European put, the probability of the option being in-the-money at maturity is lower as S_t rises so the value of a European put should also decline.

3.2.2 *Strike price, K*

The strike price has the opposite effect on the option to the current share price. If K increases, then the value of an American call $[S_t - K]^+$ decreases. And the probability of a European call being in-the-money at maturity is now lower so the European call price also decreases.

Reverse arguments to those above also apply to put options.

3.2.3 *Expiration date, T*

For American options, a longer time to maturity nests exercise opportunities of shorter terms so the option becomes more valuable as T increases. For European options, usually the longer the time to expiry the greater the chance that the underlying share can move in favor of the holder of the option before expiry. So we would expect the European option price to increase with time to maturity.[1]

3.2.4 *Volatility, σ*

The same argument applies to volatility. Volatility relates to how much the share price varies. The higher the volatility, the greater the chance that the share price can move in favor of the holder of the option before expiry. Higher volatility should therefore increase call and put option prices.

3.2.5 *Risk-free rate, r*

When the risk-free rate is high, the present value of the strike price is low. The holder of a call option then requires less cash now to exchange for the share in the future. We would then expect higher risk-free rates to make calls more attractive. Conversely, the discounted strike price is worth less to the holder of a put option for a high risk-free rate making puts less attractive.

Table 3.1 Effect of increasing factor on the price of stock options

Variable	*European*		*American*	
	call	*put*	*call*	*put*
S_t	+	−	+	−
K	−	+	−	+
T	+	+	+	+
σ	+	+	+	+
r	+	−	+	−
q	−	+	−	+

3.2.6 *Dividend rate, q*

When dividends are issued the value of the firm is reduced by the sum of all dividends paid. Each share in the firm's stock is also reduced in value proportionately. Dividends therefore drive down the share price when issued.

Th effect of an increase in the dividend rate is then the same as the effect of decreasing the share price as described above. Call prices go down and put prices go up.

The effects of changes in the six parameters are summarized in Table 3.1.

In addition to knowing the direction of the change in stock option prices we can also quantify the size of this change. Chapter 11 is devoted to this topic. Given a pricing formula, the partial derivative of the price with respect to a factor quantifies the price change holding all other factors constant. These partial derivatives are known as the *Greeks*.

Pricing formulas for European options in the Black–Scholes model are considered next.

3.3 Pricing European options

The Black–Scholes pricing formula for a European call option c_t at time t takes the form:

$$c_t = S_t e^{-q(T-t)}\Phi(d_1) - e^{-r(T-t)}K\Phi(d_2) \tag{3.1}$$

where

$$d_1 = \frac{\log(S_t/K) + (r - q + \frac{1}{2}\sigma^2)(T-t)}{\sigma\sqrt{T-t}}, d_2 = d_1 - \sigma\sqrt{T-t}$$

The parameters are as described above while $\Phi(\cdot)$ is the cumulative distribution function of the standard normal distribution as described in Section 2.6.4.

The derivation of this formula will be explained later. The following exercise explains how the formula is applied to a pricing problem.

Exercise 3.1. *You are considering the purchase of 100 units of a 3-month European call option with strike price \$25 on a stock. You are given:*

(i) The Black–Scholes framework holds.

(ii) The stock is currently selling for \$20.

(iii) The stockâŁ™s volatility is 24%.

(iv) The stock pays dividends continuously at a rate proportional to its price. The dividend yield is 3%.

(v) The continuously compounded risk-free interest rate is 5%.

Calculate the price of the block of 100 options.

[MFE - Sample 2010 Q6]

Proof. Here

$$T - t = 3/12, K = 25, S_t = 20, \sigma = 0.24, q = 0.03, r = 0.05$$

Then

$$\begin{aligned} d_1 &= \frac{\log(S_t/K) + (r - q + \frac{1}{2}\sigma^2)(T-t)}{\sigma\sqrt{T-t}} \\ &= \frac{\log(20/25) + (0.05 - 0.03 + \frac{1}{2}0.24^2) \times 0.25}{0.24\sqrt{0.25}} = -1.7579 \end{aligned}$$

and

$$d_2 = d_1 - \sigma\sqrt{T-t} = -1.7579 - 0.24\sqrt{0.25} = -1.8779$$

From Equation (3.1), the call price is

$$\begin{aligned} c_t &= S_t e^{-q(T-t)}\Phi(d_1) - e^{-r(T-t)}K\Phi(d_2) \\ &= 20e^{-0.03\times0.25}\Phi(-1.7579) - 25e^{-0.05\times0.25}\Phi(-1.8779) = 0.0362 \end{aligned}$$

The price of the block of 100 options is then \$3.62. □

The Black–Scholes pricing formula for a European put option p_t at time t looks similar:

$$p_t = Ke^{-r(T-t)}\Phi(-d_2) - S_t e^{-q(T-t)}\Phi(-d_1)$$

where d_1 and d_2 are as defined before. The next exercise prices a European put option with this formula.

> **Exercise 3.2.** *You are asked to determine the price of a European put option on a stock. Assuming the Black–Scholes framework holds, you are given:*
>
> (i) *The stock price is \$100.*
>
> (ii) *The put option will expire in 6 months.*
>
> (iii) *The strike price is \$98.*
>
> (iv) *The continuously compounded risk-free interest rate is* $r = 0.055$.
>
> (v) $q = 0.01$
>
> (vi) $\sigma = 0.50$
>
> *Calculate the price of this put option.*
>
> *[MFE - Spring 2007 Q3]*

Proof. Here

$$S_t = 100, T - t = 6/12, K = 98, r = 0.055, q = 0.01, \sigma = 0.50$$

Then

$$\begin{aligned} d_1 &= \frac{\log(S_t/K) + (r - q + \frac{1}{2}\sigma^2)(T - t)}{\sigma\sqrt{T - t}} \\ &= \frac{\log(100/98) + (0.055 - 0.01 + \frac{1}{2}0.50^2) \times 0.5}{0.50\sqrt{0.5}} = 0.2976 \end{aligned}$$

and

$$d_2 = d_1 - \sigma\sqrt{T - t} = 0.2976 - 0.50\sqrt{0.5} = -0.0560$$

Then the put price is

$$\begin{aligned} p_t &= Ke^{-r(T-t)}\Phi(-d_2) - S_t e^{-q(T-t)}\Phi(-d_1) \\ &= 98e^{-0.055\times 0.5}\Phi(0.0560) - 100e^{-0.01\times 0.5}\Phi(-0.2976) = 11.69 \end{aligned}$$

The price of this put option is \$11.69. □

In order to derive these formulas, the real world has to be greatly simplified by making a number of assumptions. The aim is to develop a mathematical model of investor behavior that retains key features of financial markets and provides a reasonable approximation in pricing and risk management applications.

The first set of assumptions concern the evolution of asset prices. The stock price is assumed to follow a particular statistical process over time known as **geometric Brownian motion**. And the parameters of this stock price process are assumed constant over the duration of the option. Chapter 7 examines these assumptions in more detail.

The second set of assumptions concerns the operation of financial markets. The market is presumed to operate under certain ideal conditions. Financial markets are **efficient** in that security prices reflect all available information. There are no assets that are undervalued or overvalued. Markets are also **arbitrage-free** meaning that there are no opportunities to make a profit without risk. These important ideas are explored next.

3.4 Efficient markets

The principle of efficient markets rests on the assumption that asset prices fully reflect all available information relevant to the value of these assets. Price changes are then unpredictable. Studying past price behavior to predict future prices known as *technical analysis* or analyzing company financial information such as earnings and asset values known as *fundamental analysis* is no better than choosing stocks at random.[2] Active fund management or building portfolios based on trading strategies may produce above-market returns in the short run but in the longer run, it is impossible to outperform the market without taking above-market risk.

If new information arrives in the market that indicates that earnings from this asset will be higher, this news spreads quickly and rational investors attempting to profit from this news purchase the asset bidding the price up. This new information could be anything not anticipated by the market, for example, a better-than-expected earnings announcement, a successful tender or a new innovative product. Information arrival is unpredictable so price changes are random and changes from today to tomorrow should be independent of price changes from yesterday. This random behavior is reflected in the use of *random walks* and *martingales* to represent the evolution of asset prices mathematically. These concepts are developed in the next chapters.

The existence of market anomalies such as systematic rules for identifying undervalued stocks or temporal effects on returns are significant challenges to market efficiency. Many examples of investor irrationality recorded by Shiller (2001) and others also mean that the efficiency of financial markets is still contested (Lo, 2018*b*).

In practice, markets vary in their degree of efficiency according to ease of trading, costs of exploiting inefficiencies and the resources available to investors to exploit them (Damodaran, 2018). For example, the market for publicly traded equities is much more efficient than the market for real estate where assets are heterogeneous and unit prices are high.

3.5 Arbitrage

The principle of no arbitrage is a simple but powerful idea that motivates our approach to pricing derivatives throughout this book. Pricing forwards using no-arbitrage arguments will be developed in the next sections of this chapter.

Put very simply, the principle of no arbitrage says that two securities that give the same payments must have the same price. Otherwise, there is an opportunity to make a profit without any risk.

As an example of an arbitrage opportunity, imagine a share in Shell PLC is priced at \$70 on the New York Stock Exchange and \$65 on the London Stock Exchange. An arbitrageur could buy the share in London and sell it in New York pocketing the \$5 difference. If an arbitrage opportunity such as this exists, then all active market participants would follow the same strategy and the price in London would be bid up as investors purchase the stock while the price in New York would fall due to sales. The arbitrage opportunity would instantaneously disappear.

As a second example, consider a forward contract to purchase a non-dividend share currently trading at \$5 for \$10 in one year's time. If the risk-free interest rate is 2% p.a. with continuous compounding the forward writer could borrow \$5 to purchase the share now then sell it for \$10 in one year's time repaying the loan at $\$5 \times e^{0.02} = \$5.10$. A profit of $\$10 - \$5.10 = \$4.90$ has been made with no risk.

Once again, market participants would recognize this arbitrage opportunity bidding up the share price to S_t and bidding down the forward price to K. At market equilibrium, the forward price should equal the loan repayment so that there is no-arbitrage profit. That is,

$$K - S_t \times e^{0.02} = 0$$

A similar mechanism applies when the forward price is too low. If the forward contract was to purchase the share for \$4 in one year's time, the forward holder could short the share and invest the proceeds at the risk-free rate. In one year's time, the investment is worth $\$5 \times e^{0.02} = \$5.10$ and the share is repurchased for \$4 yielding a risk-free profit of \$1.10. This arbitrage opportunity would disappear quickly and market equilibrium restored.

This logic then gives us a way of determining the price of a forward. Choose K so that profit is zero in one year's time based on the current price and the current risk-free interest rate. Also, if profit after one year is now zero with certainty, then the value of this contract to both holder and writer must be zero when initially negotiated.

It is not so straightforward when pricing options but the assumption of no arbitrage will be critical in risk-neutral pricing also. Option pricing in the Black–Scholes model relies on being able to set up a portfolio that exactly matches the option payoff. Then by no-arbitrage arguments, the option price must equal the value of that portfolio.

A slightly more formal definition of arbitrage will be useful in other applications.

Definition 3.2. *An arbitrage opportunity means that a portfolio currently with a net value of zero at some future time yields a non-zero probability of strictly positive profit and a zero probability of loss.*

In the second example above with high forward price, the arbitrageur initially has a portfolio of a forward contract and a share worth \$5 funded by borrowings of \$5. Taken together, the portfolio has a net value of zero. After one year, the portfolio yields a certain profit of \$4.90 with a zero probability of loss.

We will use the expressions *the principle of no arbitrage* and *law of one price* interchangeably to signify that there are no-arbitrage opportunities in the market. Two securities that give the same payoff must have the same price otherwise there is an arbitrage opportunity. Arbitrageurs will set up a zero-cost portfolio by shorting the expensive one and going long on the cheap one. Any such opportunities are eliminated by rational investors acting immediately to make profits.

3.6 Pricing forward contracts

Forwards are priced using the principle of no arbitrage as described in the previous section.

A **forward contract** is an agreement made at time t between two parties under which the holder agrees to buy from the writer a specified amount of an asset at a set price (K) on a set date (T). The underlying asset could be a commodity or financial security.

The payoff from the forward is easily demonstrated. Let S_T be the asset price at time T. At maturity, the holder must pay K for the asset. The value of this transaction is $S_T - K$. If the asset price is above the forward price at maturity then the payoff is positive. Otherwise, it is negative and the holder makes a loss.

The payoff for the writer is the opposite. She buys the asset at S_T and receives K from the holder of the forward contract in exchange for the asset. Her payoff is then $K - S_T$.

The next result is a more formal expression of the no-arbitrage argument of the previous section for an asset with no income. We will construct a zero-cost portfolio at time t. By the principle of no arbitrage, it cannot yield a profit at T.

Theorem 3.1. *The forward price for an asset with no income is* $K = S_t e^{r(T-t)}$. *A zero-coupon bond or a non-dividend share provides no income for example.*

Proof. Do the following

(a) Negotiate a forward contract to sell one unit of the asset with price K maturing at T.
(b) Borrow to buy one unit of the asset for S_t.

Remember the value of the forward contract when initially negotiated is zero. The transactions then have zero net cost. The cash flows are

$$0(\text{forward}) - S_t(\text{asset}) + S_t(\text{borrowing})$$

At T, the asset is sold to the holder of the forward contract for K and the loan is repaid. Given risk-free interest rate r, the cashflows at maturity are

$$K - S_t e^{r(T-t)}$$

The cashflows at maturity must be zero to avoid arbitrage profits. Then

$$K = S_t e^{r(T-t)}$$

In other words, the forward price is the future value of the underlying asset. □

If the asset generates an income during the duration of the forward contract such as a coupon bond or a share paying a known dividend at discrete times, then the argument can be easily amended.

Theorem 3.2. *The forward price is* $K = S_t e^{r(T-t)} - c e^{r(T-\tau)}$ *where the asset provides a fixed amount* c *to the holder at time* τ *before maturity at time* T.

Proof. As before

(a) Negotiate a forward contract to sell one unit of the asset with price K maturing at T.
(b) Borrow to buy the asset for S_t.

The transactions have zero net cost.

At τ, the income c is received and invested at the risk-free interest rate r.

At T, the asset is sold to the holder of the forward contract for K, the loan is repaid and the investment matures. The cashflows are

$$K - S_t e^{r(T-t)} + c e^{r(T-\tau)}$$

To avoid arbitrage profits

$$K = S_t e^{r(T-t)} - c e^{r(T-\tau)}$$

The forward price is now the future value of the underlying asset minus the future value of the income received. □

This formula is sometimes given in a different form.

$$K = S_t e^{r(T-t)} - ce^{r(T-\tau)} = (S_t - ce^{-r(\tau-t)})e^{r(T-t)} = (S_t - I)e^{r(T-t)}$$

where I is the present value of the fixed income.

Exercise 3.3. *A fixed-interest security with par value €100 pays coupons of 6% pa half-yearly. That is, the bond holder receives €3 every six months.*

Three months before the next coupon, an investor negotiates a forward contract to buy the security in six months' time. The current price of the security is €103.11, and the risk-free interest rate is 5.8% pa.

(i) Calculate the forward price.

(ii) Calculate the forward price of a contract to buy the security in ten months' time.

Proof.
(i) $K = S_t e^{r(T-t)} - ce^{r(T-\tau)} = 103.11 \times e^{0.058\times\frac{6}{12}} - 3 \times e^{0.058*(\frac{6}{12}-\frac{3}{12})}$
$= 103.099 = €103.10$

(ii) There is a second coupon at nine months within the duration of this longer forward contract. The future value of this coupon must also be subtracted from the forward price.

$$\begin{aligned} K &= S_t e^{r(T-t)} - c_1 e^{r(T-\tau_1)} - c_2 e^{r(T-\tau_2)} \\ &= 103.11 \times e^{0.058\times\frac{10}{12}} - 3 \times e^{0.058\times(\frac{10}{12}-\frac{3}{12})} - 3 \times e^{0.058\times(\frac{10}{12}-\frac{9}{12})} \\ &= 102.097 = €102.10 \end{aligned}$$

□

The last case we will consider is an asset paying income at a known yield such as a dividend-paying share. The amount of dividend is not known but the dividend yield is.

The **dividend yield** of an asset is the annual amount of dividends paid relative to the asset price.

$$q := \frac{\text{Annual dividends per asset}}{\text{Price per asset}}$$

To use our no-arbitrage model, we assume income is received continuously and immediately reinvested in assets. One asset then becomes $e^{q(T-t)}$ assets at time T.[3]

Theorem 3.3. *The forward price is* $K = S_t e^{(r-q)(T-t)}$ *where the asset has a known dividend yield* q.

Proof. Do the following

(a) Negotiate a forward contract to sell $e^{q(T-t)}$ units of the asset.
(b) Borrow to buy one asset for S_t.

The transactions have zero net cost.

At time T, one asset has become $e^{q(T-t)}$ assets. These assets are sold to the holder of the forward contract and the loan is repaid. The cashflows are

$$Ke^{q(T-t)} - S_t e^{r(T-t)}$$

To avoid arbitrage profits

$$K = S_t e^{(r-q)(T-t)}$$

The forward price is the future value of $e^{-q(T-t)}$ underlying assets. □

Although the value of a forward contract at the time it is first negotiated is zero, it has a value as time progresses.

Theorem 3.4. *The value of a forward contract* f_t *at time* t *after initially negotiated is*

$$f_t = (K_t - K_0)e^{-r(T-t)}$$

where K_0 *is the original forward price;* K_t *is the forward price if negotiated now and* $T - t$ *is the time from now to delivery.*

Proof. Form a portfolio consisting of

(a) A forward contract already negotiated at time zero to buy an asset for K_0 at T.
(b) A forward contract negotiated now to sell the asset for K_t at T.

The payoff at T is $K_t - K_0$ and this is risk-free. By the law of one price, the portfolio must have the same current value as an investment at the risk-free rate that yields $K_t - K_0$ at T. Therefore, the current value of the portfolio is $(K_t - K_0)e^{-r(T-t)}$.

The current value of the forward contract negotiated now is 0. Let the forward contract negotiated at time zero have value f_t now.

Therefore

$$f_t = (K_t - K_0)e^{-r(T-t)}$$

□

3.7 Put-call parity

Put-call parity defines a relationship between the prices of European call and European put options with the same strike price and expiration date. It ensures that no-arbitrage opportunities exist between these related securities.

This relationship is useful in this book as once the value of one option is known the value of the other option follows.[4]

Assume the underlying asset pays no income. Let K be the strike price of the options and r be the continuously compounded risk-free interest rate. Now we set up two portfolios that give equivalent payoffs.

Consider portfolio A consisting of one European call option on an asset plus a risk-free zero-coupon bond with par value K. At maturity, the asset price S_T is either below or above the strike price K.

- If $S_T > K$, the call option is exercised with payoff $S_T - K$. The bond is worth K. The portfolio is then worth

$$(S_T - K) + K = S_T$$

- If $S_T \leq K$, the call option is worthless. The bond is worth K. The portfolio is then worth K.

A second portfolio can be created to replicate this payoff. Consider portfolio B consisting of one European put option on the asset plus one unit of the asset. At maturity, the payoffs are

- If $S_T > K$, the put option is worthless and the asset value is S_T. The portfolio is then worth S_T.
- If $S_T \leq K$, the put option is exercised with payoff $K - S_T$. The asset is worth S_T. The portfolio is then worth

$$(K - S_T) + S_T = K$$

The payoffs from the two portfolios are identical. By the law of one price, these two portfolios must have the same value at time t.

$$c_t + Ke^{-r(T-t)} = p_t + S_t \tag{3.2}$$

Therefore, once we have completed the derivation of the price of a European call option on a non-dividend paying share in the Black–Scholes model, the price of the corresponding European put option can be found from put-call parity.

What is the arbitrage opportunity if put-call parity does not apply? If the right-hand side of Equation (3.2) is less than the left-hand side, then the arbitrageur could sell a call and a bond and use the proceeds to buy a put and an asset. The amount left over is invested at the risk-free rate. At expiration, the contracts are settled at no cost and the risk-free investment matures. A risk-free profit has been made at zero cost.

It is important to remember the conditions when put-call parity applies. The options must have the same strike price and maturity. And the options must be European. When the options are American in style, it does not suffice to compare the portfolio values at maturity alone since one or both options may be exercised prior to maturity. Indeed, it becomes impossible to derive a parity relationship between the prices of calls and puts.[5]

We also assumed the underlying asset generates no income. If the asset has a known dividend yield q, the parity relationship is easily amended.

Portfolio A consists of one European call plus a zero-coupon bond as before. Portfolio B now consists of one European put but only $e^{-q(T-t)}$ units of the asset. With continuous reinvesting we have one share at maturity and the payoffs match across portfolios as before. Since the payoffs are the same, the current value of the two portfolios must also equate giving the amended put-call parity relationship:

$$c_t + Ke^{-r(T-t)} = p_t + S_t e^{-q(T-t)} \tag{3.3}$$

Exercise 3.4. *You are given the following:*

- *The current price to buy one share of XYZ stock is 500.*
- *The stock does not pay dividends.*
- *The continuously compounded risk-free interest rate is 6%.*
- *A European call option on one share of XYZ stock with a strike price of K that expires in one year costs 66.59.*
- *A European put option on one share of XYZ stock with a strike price of K that expires in one year costs 18.64.*

Using put-call parity, calculate the strike price, K.

[IFM - Sample Q2]

Proof.

$$\begin{aligned} c_t + Ke^{-r(T-t)} &= p_t + S_t \\ \implies 66.59 + Ke^{-0.06\times 1} &= 18.64 + 500 \\ \implies K &= 480 \end{aligned}$$

□

3.8 Early exercise of non-dividend paying American call

American options can be exercised on any date before maturity but in general it is never optimal to exercise a non-dividend paying American call option early.

To see why this is true first compare American and European call options on the same stock with the same strike price K. The American call has all the rights of a European call with the additional benefit of early exercise. The American call should therefore be more valuable at every point in time.

Consider again portfolio A from the previous section consisting of one European call option on an asset with price c_t plus a risk-free zero-coupon bond with par value K. At maturity this portfolio always has a value of at least the asset price S_T. Since we are certain that this will be true, then the value of the portfolio at any time t must be at least the asset price S_t. Hence we have an expression for the lower bound of a European call.

$$c_t + Ke^{-r(T-t)} \geq S_t \implies c_t \geq S_t - Ke^{-r(T-t)}$$

Then the American call has a value C_t greater than this lower bound. But the lower bound is higher than the payoff of the American option if exercised early which is $S_t - K$.

$$C_t \geq c_t \geq S_t - Ke^{-r(T-t)} > S_t - K$$

Therefore the American call will always have a value higher than its early exercise payoff.

The payment of dividends on the other hand may or may not motivate early exercise.

3.9 Summary

This chapter introduces the basic definitions of derivative securities. Also, a number of important results are established without any reference to the dynamics of underlying asset prices.

The factors that affect option prices are considered and a rationale for how option prices move in response to changes in these factors is provided.

The idea of no arbitrage or law of one price is the foundation of how we think about valuing derivatives. This is the rule that there are no 'free lunches' in financial markets. If two assets or strategies give the same payoff, they must have the same price.

Two important results come straight from no arbitrage. First, we can work out the value of a forward contract. Second, we learn about put-call parity, which shows the link between European call and European put options with the same strike and maturity.

To price options we require models that describe how asset prices evolve over time. But it will be seen that the principle of no arbitrage is still fundamental to pricing and managing the risks of these securities.

Key things to remember

- The application of the law of one price to pricing a forward contract: Section 3.6.
- The value of a forward contract after initially negotiated: Section 3.6.
- Put-call parity for dividend and non-dividend paying share options: Section 3.7.

Further reading

Hull (2021) is the classic starting point for understanding the fundamentals of financial derivatives and is widely used in universities and industry. Pirie (2021) is less mathematical and geared more toward professional practice giving a broad and fairly deep view of derivatives from basic instruments through to risk management.

Notes

1. However, this does not always hold. An investor may know some future event will drive down the underlying share price at time $\tau < T$ (e.g. share dividends) then the price will be higher for European call options with maturity less than τ. In this case, the option price decreases with maturity (Hull, 2021). Also, a European put holder receives a payment in the future when the option is exercised and the share is sold. The benefit of this payment is reduced as the time to expiry increases. This effect may occasionally outweigh the positive effect especially if the discount rate is high and the option is deep-in-the-money (Pirie, 2021).
2. In Malkiel's memorable phrase, a blindfolded chimpanzee throwing darts at the *Wall Street Journal* could select a portfolio as well as financial experts (Malkiel, 2023).
3. If the dividend yield is q per annum then the dividend yield in the time interval Δt is $q\Delta t$. If we own ϕ assets and the current asset price is S_t, then we receive $\phi q \Delta t\, S_t$ in income. Immediately reinvesting this amount we buy $\phi q \Delta t$ assets. The change in the number of assets owned $\Delta\phi$ is $\phi q \Delta t$. In the limit, this becomes $d\phi = \phi q dt$.

   Solving this differential equation $\frac{d\phi}{\phi} = qdt \implies \log\frac{\phi_T}{\phi_t} = q(T-t) \implies \phi_T = \phi_t e^{q(T-t)}$, that is, one share becomes $e^{q(T-t)}$ shares at time T.
4. More generally, it demonstrates how synthetic positions can be created using the different combinations of the call and put options along with the underlying asset to produce equivalent payoffs.
5. Though a put-call inequality relationship can be derived. See p222, Sundaram and Das (2011)

Questions

1 The market price of Stock A is 50. A customer buys a put contract at strike price 50 on Stock A for 500. The put contract is for 100 shares of A. Assume an interest rate of zero. Calculate the customer's maximum possible loss.

> **Solution** When $S_T < 50$, the payoff from the option is $50 - S_T$.
> If $S_T \geq K$, the payoff is 0. So the most the investor loses is the 500 premium.

2 For a certain stock, investor A purchases a call option at strike price 45 while investor B purchases a 135-strike put option. Both options are European with the same expiration date. Assume that there are no transaction costs.

If the final stock price at expiration is S, investor A's payoff will be 12. Calculate investor B's payoff at expiration, if the final stock price is S.

[IFM - Sample Q48]

> **Solution** Payoff from call is $[S - K]^+$ so $S - 45 = 12 \implies S = 57$
> Payoff from put is $[K - S]^+$ so investor B's payoff at expiration is 135-57 = 78.

3 Determine which, if any, of the following positions has or have an unlimited loss potential from adverse price movement in the underlying asset, regardless of the initial premium received.

I. Short 1 forward contract
II. Short 1 call option
III. Short 1 put option

[IFM - Sample Q26]

> **Solution**
>
> I. The seller is obliged to buy the asset in the market and deliver it at the agreed price. There are no limits to potential losses as the asset price rises.
> II. If the price rises above the strike, the holder will exercise the option. The seller is then obliged to buy the asset in the market and deliver it at the strike price. There are no limits to how high the asset price may rise; therefore, there are no limits to the size of potential losses.
> III. The seller is obliged to buy the asset at the strike price. The asset could be worthless. The investor has therefore a loss limited to the strike price. Loss potential is limited in this case.

4 You are given the following information about two options, A and B:

- i Option A is a one-year European put with exercise price 45.
- ii Option B is a one-year American call with exercise price 55.
- iii Both options are based on the same underlying asset, a stock that pays no dividends.
- iv Both options go into effect at the same time and expire at $t = 1$.

You are also given the following information about the stock price:

- i The initial stock price is 50.
- ii The stock price at expiration is also 50.
- iii The minimum stock price (from $t = 0$ to $t = 1$) is 46.
- iv The maximum stock price (from $t = 0$ to $t = 1$) is 58.

Determine which of the following statements is true.

- (a) Both options A and B are 'at-the-money' at expiration.
- (b) Both options A and B are 'in-the-money' at expiration.
- (c) Both options A and B are 'out-of-the-money' throughout each option's term.
- (d) Only option A is ever 'in-the-money' at some time during its term.
- (e) Only option B is ever 'in-the-money' at some time during its term.

[IFM - Sample Q44]

Solution At expiration the price is 50 and both options are 'out-of-the-money' eliminating answers (a) and (b).

With a strike price of 45 and a minimum stock price of 46, option A is never in the money, eliminating answer (d).

With a strike price of 55, option B will be in the money at the time the stock price is 58, eliminating answer (c) and verifying answer (e).

5 A share has price S_t at time t, dividend yield q and volatility σ. The risk-free rate of interest is r per annum. Which of the events below move the price of a call option on this share in the same direction as the price of a put option? The strike for both options is K.

- a. An increase in the share price S_t.
- b. An increase in the strike price K.
- c. A decrease in the dividend yield q.
- d. A reduction in the risk-free rate r.
- e. An increase in the volatility σ.

Solution An increase in the volatility σ. See Table 3.1.

6 A three-month forward contract exists on a zero-coupon corporate bond. The bond's current price is \$ 950. The continuously compounded risk-free rate is 6% p.a. Calculate the forward price.

Solution

$$K = S_0 e^{r(T-t)}$$
$$= 950 \times e^{\frac{3}{12}}$$
$$= \$964.36$$

7 You are given the following information about Stock XYZ:

(i) The current price of the stock is 35 per share.

(ii) The expected continuously compounded rate of return is 8%.

(iii) The stock pays semi-annual dividends of 0.32 per share, with the next dividend to be paid two months from now.

The continuously compounded risk-free interest rate is 4%. Calculate the current one-year forward price for stock XYZ.

[IFM - Sample Q51]

Solution

$$K = S_t e^{r(T-t)} - c_1 e^{r(T-\tau_1)} - c_2 e^{r(T-\tau_2)}$$
$$= 35 \times e^{0.04\times 1} - 0.32 \times e^{0.04\times(1-\frac{2}{12})} - 0.32 \times e^{0.04\times(1-\frac{8}{12})}$$
$$= 35.77$$

The expected continuously compounded rate of return is irrelevant.

8 The current price of a share is \$ 150. The share pays a dividend yield of 2.5%. Calculate the forward price of a six-month contract on one share if the continuously compounded risk-free rate is 4% p.a.

Solution

$$K = S_t e^{(r-q)(T-t)}$$
$$= 150e^{(0.04-0.025)\frac{6}{12}}$$
$$= \$151.13$$

9 You are analyzing a bond. The bond has:

Par value	\$1,000
Current price	\$975
Annual coupon rate	12%
Coupon payment frequency	semiannual
Term to maturity	10.75 years

Current continuously compounded risk-free interest rates per annum are:

Term	*Interest rate*
3 month	8.0%
6 month	8.5%
9 month	9.0%
11 month	12.0%

Calculate the eleven-month forward price of this bond.

[FETE - Fall 2011 Q6(b)]

Solution Discount rates are given not forward rates. This indicates that we use the expression for a forward with fixed income using present value.

$$K = (S_t - I)e^{r(T-t)}$$

The present value of fixed income (I) is

$$I = 60 \times e^{-0.08\times\frac{3}{12}} + 60 \times e^{-0.09\times\frac{9}{12}} = 114.90$$

Then

$$K = (975 - 114.90)e^{0.12\times\frac{11}{12}} = \$960.12$$

10 A market maker in stock index forward contracts observes a 6-month forward price of 112 on the index. The index spot price is 110 and the continuously compounded dividend yield on the index is 2%. The continuously compounded risk-free interest rate is 5%.

Describe actions the market maker could take to exploit an arbitrage opportunity.

[IFM – Sample Q21]

Solution The no-arbitrage price is $K = S_t e^{(r-q)(T-t)} = 110e^{(0.05-0.02)\frac{6}{12}} = 111.66$. The forward price is therefore too high.

Following the second example in Section 3.5, the market maker could negotiate a contract to sell $e^{0.02\times\frac{6}{12}}$ units of the index in six months' time. She borrows \$110 to purchase the index now and sells the units in six months' time repaying the loan at $\$110 \times e^{0.05\times\frac{6}{12}} = \$112.78$. A profit of $112e^{0.02\times\frac{6}{12}} - \$112.78 = \$0.34$ has been made with no risk.

11 You have invested an amount in stock at time zero.

A colleague suggests the following strategy: sell a call and then purchase a put with the proceeds in order to remove all risk from the portfolio.

Critique the suggestion from your colleague.

[FETE – Fall 2012 Q10(c)]

Solution Put-call parity states

$$c_t + Ke^{-r(T-t)} = p_t + S_t$$

which rearranged gives

$$Ke^{-r(T-t)} = p_t + S_t - c_t$$

The colleague is correct. A combination of buying a put, owning stock and selling a call is equivalent to a risk-free investment.

12 The PS index has the following characteristics:

- One share of the PS index currently sells for 1,000.
- The PS index does not pay dividends.

Sam wants to lock in the ability to buy this index in one year for a price of 1,025. He can do this by buying or selling European put and call options with a strike price of 1,025.

The annual effective risk-free interest rate is 5%.

Determine the hedging strategy that will achieve SamâŁ™s objective and also give the cost today of establishing this position.

[IFM - Sample Q5]

Solution Sam aims to create a forward using calls and puts. This is known as a *synthetic forward.*

Rearranging the put-call parity relationship

$$c_t - p_t = S_t - Ke^{-r(T-t)}$$

At maturity, the right-hand side becomes $S_T - K$, which is the payoff from buying the asset at T for a price of K. This indicates the hedging strategy is to buy a call and sell a put.

Note that the *effective* interest rate (i) is calculated as if compounded annually so $1+i = e^r$, where r is the continuously compounded rate. The cost is $S_t - Ke^{-r} = 1000 - 1025\left(\frac{1}{1+0.05}\right) = 23.81$.

13 Derive the put-call parity relationship for a share that has a dividend of known amount d payable prior to the expiration date.

Solution Assume the dividend is paid at τ before expiration at T.

Portfolio A consists of one European call plus a zero-coupon bond. Portfolio B consists of one European put and one share. We need to subtract off the dividend from the right-hand side of Equation (3.2) so that the payoffs match across portfolios as before. Since the payoffs are the same, the current value of the two portfolios must also equate giving the put-call parity relationship:

$$c_t + Ke^{-r(T-t)} = p_t + S_t - de^{-r(\tau-t)}$$

Part II

Stochastic calculus

4 Standard Brownian motion

Introduction

In the previous chapter, we developed a number of results for forwards and options without knowing how the underlying asset evolves over time. But we need models of financial assets to determine a reasonable price for an option.

To price derivative contracts we need to combine a model of asset behavior with a rule driving investor behavior in the market. The principle of no arbitrage (section sec:arb) provides such a rule. We will then construct a portfolio to hedge the derivative and use the absence of arbitrage to find its price.

Many models of financial assets are available to us and differ by settings and assumptions. There could be finitely many outcomes and a finite number of time periods considered or our model could have infinitely many scenarios with continuous time. We will deal with the former situation in Chapter 8. Our main focus in this book is the latter situation.

Our asset models will combine both time and uncertainty in a *stochastic process*. As in the case of a stochastic variable, the process can have a range of different outcomes at any time described by a probability distribution. In addition, this probability distribution also evolves over time.

The most important stochastic process studied here is standard Brownian motion. This process drives the randomness in our models of equity prices and interest rates. This is defined in Section 4.1 and its non-differentiability is shown in Section 4.2. Standard Brownian motion has the important property that it is a martingale. Martingales are defined in Section 4.4 with a number of examples where processes are tested for this property. The chapter concludes with Section 4.5 where standard Brownian motion is simulated in R.

4.1 Definition

In 1828, the English botanist Robert Brown discovered the erratic motion of grains of pollen suspended in water. The grains were observed to move by random amounts in constantly changing directions. Independently and much later, the French mathematician, Louis Bachelier, developed a theory of financial speculation where changes in stock prices occur randomly. Based on

DOI: 10.4324/9781003477624-6

his observations of the Paris stock exchange, Bachelier theorized that these random stock price movements followed a normal distribution and also that they were unaffected by any previous price changes. The stock price then evolves over time as an accumulation of these random stock price movements in a process now called Brownian motion. A firm mathematical foundation for this process was then provided by MIT mathematics professor Norbert Wiener in 1923.

Brownian motion is defined formally below.

Definition 4.1. *A stochastic process $W = \{W_t : t \geq 0\}$ is said to be a* ***standard Brownian motion***[1] *if it is continuous and it satisfies the following properties:*

(i) $W_0 = 0$
(ii) *If $t_1 < t_2 \leq t_3 < t_4$, then the increments $W_{t_2} - W_{t_1}$ and $W_{t_4} - W_{t_3}$ are independent.*
(iii) *For any given t_1 and t_2 with $t_2 > t_1$, the distribution of the increment $W_{t_2} - W_{t_1}$ is a normal distribution with mean 0 and variance $t_2 - t_1$.*

We will also use the simpler notation $W = \{W_t\}_{t \geq 0}$. The first property is known as the *normalization* property. The second is the *independent increments* property. We note that the future of the process only depends on the past through the current value so the process evolves by successively adding independent random increments which are normally distributed.

The evolution of the process is uncertain and sample paths or *realizations* will vary. We will see in Section 4.5 how to simulate the process to generate such realizations. One particular realization is given Figure 4.1.

We notice the process takes on the value zero at $t = 0$ and its value may increase or decrease over time. It may also become negative. Also, since $W_t = W_t - W_0$ then from the third property W_t is normally distributed with mean 0 and variance t. The value of W_t then varies more as t increases.

Exercise 4.1. *Let W be a standard Brownian motion starting with $W_0 = 0$.*

1 *What is the probability that W_5 takes a positive value?*

2 *What is the probability that W_1 takes a value between −2 and 2?*

3 *What is the probability that W_5 takes a value between −2 and 2?*

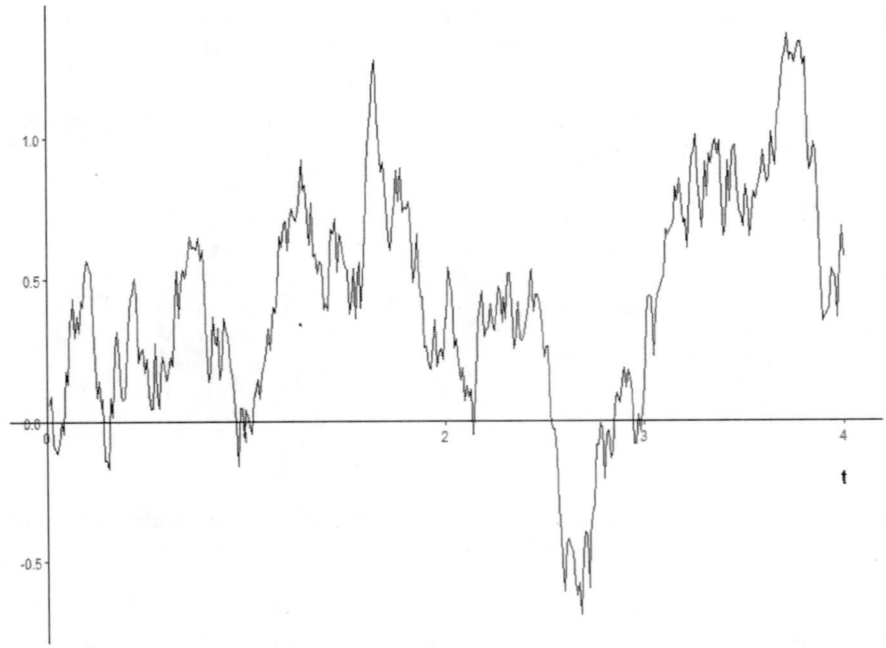

Figure 4.1 A realization of standard Brownian motion

Proof.

1 We have that W_t is normally distributed with mean 0 and variance t or in notation $W_t \sim N(0, t)$.

$$W_5 \sim N(0, 5) \implies P(W_5 > 0) = 0.5$$

In fact, $P(W_t > 0) = 0.5$ for all t.

2 W_1 is normally distributed with mean 0 and variance 1, that is, the standard normal distribution $Z \sim N(0, 1)$.

$$P(-2 < W_1 < 2) = P(-2 < Z < 2) = 0.9545$$

3 W_5 has the distribution $N(0, 5)$ therefore

$$P(-2 < W_5 < 2) = P\left(\frac{-2}{\sqrt{5}} < Z < \frac{2}{\sqrt{5}}\right) = 0.6289$$

So the probability that W_t is in the interval (-2,2) has declined as t increased from $t = 1$ to $t = 5$. This is due to the variance of W_t increasing with t. □

By referring back to Definition 4.1, it can be shown that the following are also the standard Brownian motion:

- $\{-W_t\}_{t\geq 0}$
- $Y_t = \begin{cases} W_t, & 0 \leq t \leq T \\ 2W_T - W_t, & t \geq T \end{cases}$
- $\left\{\frac{1}{\sqrt{\alpha}} W_{\alpha t}\right\}_{t\geq 0}$ where $\alpha > 0$
- $Z_t = \begin{cases} 0, & t = 0 \\ tW_{1/t}, & t > 0 \end{cases}$

In addition, for a given $T > 0$, $\{W_{T-t} - W_T\}$ behaves like a standard Brownian motion over the interval $0 \leq t \leq T$.

Exercise 4.2. *Prove that if* W *is a standard Brownian motion, then so is* $\left\{\frac{1}{\sqrt{\alpha}} W_{\alpha t}\right\}_{t\geq 0}$ *where* $\alpha > 0$.

Proof. We prove that this process satisfies the three properties of Definition 4.1.

(i) *Normalization* property: Let $t = 0$ in the expression given in the question and check that the result is zero.

$$\frac{1}{\sqrt{\alpha}} W_{\alpha\times 0} = \frac{1}{\sqrt{\alpha}} W_0 = \frac{1}{\sqrt{\alpha}} \times 0 = 0$$

(ii) *Independent increments* property: Check that increments over non-overlapping intervals are independent.
If $t_1 < t_2 \leq t_3 < t_4$, then we have

$$\frac{1}{\sqrt{\alpha}} W_{\alpha t_2} - \frac{1}{\sqrt{\alpha}} W_{\alpha t_1} = \frac{1}{\sqrt{\alpha}} (W_{\alpha t_2} - W_{\alpha t_1})$$

and

$$\frac{1}{\sqrt{\alpha}} W_{\alpha t_4} - \frac{1}{\sqrt{\alpha}} W_{\alpha t_3} = \frac{1}{\sqrt{\alpha}} (W_{\alpha t_4} - W_{\alpha t_3}).$$

These must be independent since increments of the standard Brownian motion are independent for any non-overlapping intervals. We know $[\alpha t_1, \alpha t_2]$ and $[\alpha t_3, \alpha t_4]$ are non-overlapping since $[t_1, t_2]$ and $[t_3, t_4]$ are non-overlapping.[2]

(iii) Lastly, check that the distribution of the process increment is normally distributed with mean 0 and variance given by the time interval.
Let $t_1 < t_2$ then

$$\begin{aligned}\frac{1}{\sqrt{\alpha}}W_{\alpha t_2} - \frac{1}{\sqrt{\alpha}}W_{\alpha t_1} &= \frac{1}{\sqrt{\alpha}}(W_{\alpha t_2} - W_{\alpha t_1}) \\ &\sim \frac{1}{\sqrt{\alpha}}N(0, \alpha t_2 - \alpha t_1) \quad \text{from (iii) in Definition 4.1} \\ &= \frac{1}{\sqrt{\alpha}}\sqrt{\alpha}N(0, t_2 - t_1) \qquad \text{from Table 2.1} \\ &= N(0, t_2 - t_1)\end{aligned}$$

Therefore the process increment has the correct distribution.

In conclusion, then $\left\{\frac{1}{\sqrt{\alpha}}W_{\alpha t}\right\}_{t \geq 0}$ is a standard Brownian motion.

□

The independent increments property of standard Brownian motion is intuitively attractive for modeling financial assets as it fits with the theory of efficient markets. If securities are priced efficiently then current prices reflect all available information and any price changes, reflecting new information, should be independent of what financial markets already know. But standard Brownian motion is not a good model for asset prices for reasons that have already become clear. In Figure 4.1, W_t was seen to become negative in the interval over which the process was simulated. It can be proven mathematically that standard Brownian motion is certain to become negative eventually for any realization.[3] A model where asset prices can possibly take negative values is not useful. Another problem is that, in practice, the movements in asset prices are not found to be independent of the current asset price. Larger movements are more frequent if the current asset price is high. Another process derived from this model (*geometric Brownian motion*) which we will see in Section 6.5 addresses these shortcomings.

4.2 Non-differentiability of standard Brownian motion

The two most important properties of standard Brownian motion are that it is not differentiable and that it is a martingale.

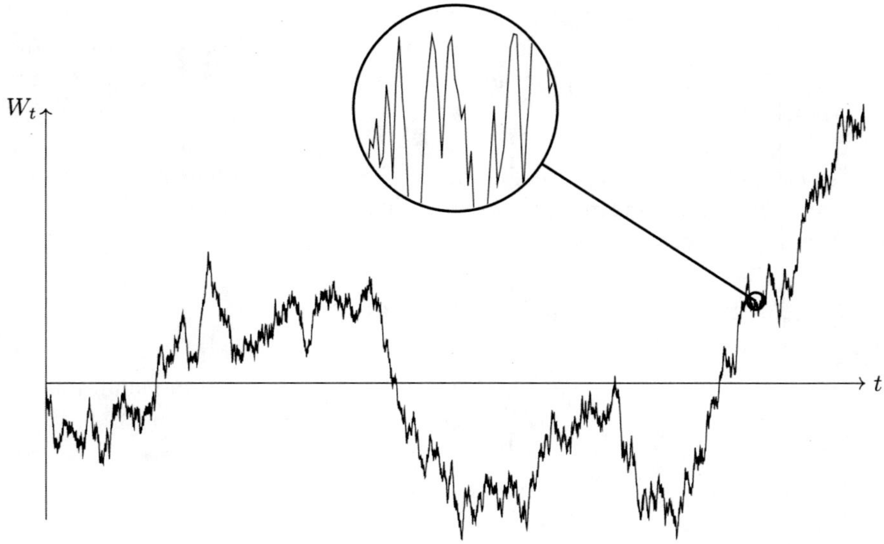

Figure 4.2 A realization of standard Brownian motion

In Section 1.3, differentiable functions were said to be smooth and continuous. Examples of non-differentiable functions with abrupt changes in direction were given.

Zooming in on Brownian motion at any point, we will observe such abrupt changes in direction. Since the function is not smooth, it is not differentiable.

We can also see this a little more formally.

We have from Definition 1.2 that the derivative of a function $y = f(t)$ is

$$\frac{dy}{dt} = \lim_{\Delta t \to 0} \frac{f(t + \Delta t) - f(t)}{\Delta t}$$

$f(t)$ is then non-differentiable if this limit does not exist for any value of t.

Theorem 4.1. *Standard Brownian motion is non-differentiable.*

A formal proof is provided by Durrett (2012). The intuition is given below.

Proof. From the definition

$$\frac{dW}{dt} = \lim_{\Delta t \to 0} \frac{W_{t+\Delta t} - W_t}{\Delta t}$$

But $W_{t+\Delta t} - W_t \sim N(0, \Delta t)$ and so $\frac{W_{t+\Delta t} - W_t}{\Delta t} \sim \frac{1}{\Delta t} N(0, \Delta t) = N(0, \frac{1}{\Delta t})$ from Table 2.1

As $\Delta t \to 0$, this does not converge to any particular random variable in the sense explained in Section 2.8 since $\frac{1}{\Delta t}$ grows without limit. □

Most stochastic processes are not differentiable. The calculus of stochastic processes then will not follow the rules of ordinary calculus in Chapter 1 we are familiar with.

First, we cannot evaluate the derivative of functions of standard Brownian motion using the chain rule (Equation 1.1). Integrals in Chapter 1 were found by moving between integration and differentiation using the fundamental theorems of calculus. If we cannot evaluate derivatives of functions of standard Brownian motion how can we evaluate integrals of functions of standard Brownian motion? Second, we have a bigger problem as integration with respect to standard Brownian motion is ill-defined. In later chapters, we are required to evaluate integrals such as

$$\int_0^T X_t \, dW$$

where X_t may be a function of W_t as well as time. Theorem 1.3 indicates that we could integrate by substitution

$$\int_0^T X_t \, dW = \int_0^T X_t \frac{dW}{dt} dt$$

but this will not work because the derivative is undefined.[4] This is why we need to use stochastic calculus not ordinary calculus to deal with integrals of standard Brownian motion.

4.3 Filtrations

Before we consider the definition of a martingale, we need to explore the idea of conditional expectation introduced in Chapter 2. The conditional probability of A given B is the probability that A occurs given that B has occurred. Similarly the conditional expectation of X given $Y = y$ is the expected value of random variable X when random variable Y has taken the value y.

We will need to evaluate expectations of processes conditional on all relevant information known currently. At each time t, all information relevant to the process up to that point is known. The future is still uncertain – only the past and present is known. The concept of *filtrations* is a formalization of the idea of information relevant to the process up to time t.

Let the filtration $\mathcal{F}$ represent all information known about a process $X = \{X_t\}_{t \geq 0}$. Very little information is known initially and information grows with time. Therefore, all the information $\mathcal{F}_s$ known at time s is included in all the information $\mathcal{F}_t$ known at a later time t. The term 'filtration' is used because uncertainty about the evolution of the process is getting filtered out as time passes.

In financial markets, there are many factors that determine stock prices such as the state of the company's finances, sector competition and macroeconomic cycles. Our models are much simpler than this. Whether we are modeling stock prices or interest rates the only source of uncertainty we consider comes from the randomness of standard Brownian motion. Therefore our filtration $\mathcal{F}$ only depends on the values that the standard Brownian process has taken up to a point in time.

We assume that we cannot be certain about the value of the process X at some point T in the future. Instead we will estimate it imprecisely using the information currently available. The conditional expectation of X_T given $\mathcal{F}_t$ written in shorthand as $\mathbb{E}(X_T|\mathcal{F}_t)$ or more simply $\mathbb{E}_t X_T$ provides such an estimate. If we decide not to use the information available to us to calculate this expectation, then we have the *unconditional* expectation which we will write as $\mathbb{E}X_T$.

The process X could be the amount of a particular stock a financial trader decides to hold in a portfolio over time. She is uncertain about her future holdings X_T at time T since her investment strategy may lead her to sell or buy more of this stock depending on future circumstances but she can have a guess $\mathbb{E}_t X_T$ based on current available information.

On the other hand, she does not need to guess her current holding X_t of the stock. This is known with certainty and X_t is not a random variable. It is the known current number of shares in the portfolio.

$$\mathbb{E}_t X_t = X_t \tag{4.1}$$

She makes her decision purely on the basis of information available up to the present $\mathcal{F}_t$ and we say that X_t is *adapted* to $\mathcal{F}$. That is, at time t, the value X_t depends only on information up to t not information from the future. In modeling financial processes, it makes sense that all our variables and decisions should depend only on the past and present not on future knowledge. All stochastic processes encountered from now on can be presumed adapted and Equation (4.1) can be presumed to always hold.

4.4 The martingale property

Recall the law of iterated expectations (Theorem 2.4):

$$\mathbb{E}X = \mathbb{E}(\mathbb{E}(X|Y))$$

In words, the law states that finding the expectation of X is the same as finding the expectation of X conditional on $Y = y$ then taking the average across all possible y.

This idea can be extended to expectations conditional on a filtration. You are given a time s and a later time t and an even later time T. Then the *tower rule* states that predicting X_T based on $\mathcal{F}_s$ is the same as first predicting based on $\mathcal{F}_t$ then taking the average of these predictions across all possible $\mathcal{F}_t$ based on information at s

$$\mathbb{E}(X_T|\mathcal{F}_s) = \mathbb{E}\Big(\ \mathbb{E}(X_T|\mathcal{F}_t)\ |\mathcal{F}_s\Big) \quad \text{where } 0 \leq s \leq t \leq T$$

or more succinctly

$$\mathbb{E}_s X_T = \mathbb{E}_s(\mathbb{E}_t X_T) \quad \text{where } 0 \leq s \leq t \leq T \tag{4.2}$$

Exercise 4.3. *Given a standard Brownian motion $W = \{W_t\}_{t \geq 0}$ compute*

(i) $\mathbb{E}W_T^2$

(ii) $\mathbb{E}_t W_T^2$

where $0 \leq t \leq T$

Proof.

First derive a simple result that follows from the properties of normal distributions.

If $X \sim N(0, \sigma^2)$, then $X = \sigma Z$, where Z is the standard normal distribution.

Then $\mathbb{E}X^2 = \mathbb{E}(\sigma Z)^2 = \sigma^2 \mathbb{E}Z^2 = \sigma^2$ since $\mathbb{E}Z^2 = 1$ from Section 2.7.

(i) For all t, $W_t \sim N(0, t)$. In particular, $W_T \sim N(0, T)$ so $\mathbb{E}W_T^2 = T$ from the result in the box.

(ii)
$$\mathbb{E}_t W_T^2 = \mathbb{E}_t\Big(\ W_t + (W_T - W_t)\Big)^2$$
$$= \mathbb{E}_t\Big(\ W_t^2 + 2(W_T - W_t)W_t + (W_T - W_t)^2\Big)$$

Since the value of W_t is known at t we have

$$= W_t^2 + 2W_t\mathbb{E}_t(W_T - W_t) + \mathbb{E}_t(W_T - W_t)^2$$
$$= W_t^2 + (T - t)$$

The last line follows from the fact that the increment of W over the time interval from t to T is independent of current or past values of W by definition. Therefore, the expectation of $W_T - W_t$ conditional on $\mathcal{F}_t$ is the same as the unconditional expectation. We have from Definition 4.1(iii) that $W_T - W_t \sim N(0, T-t)$. Hence

$$\mathbb{E}_t(W_T - W_t) = \mathbb{E}(W_T - W_t) = 0$$

and

$$\mathbb{E}_t(W_T - W_t)^2 = \mathbb{E}(W_T - W_t)^2 = T - t$$

from the result in the box above. □

Definition 4.2. *A stochastic process X is said to be a martingale if $\mathbb{E}X_t$ finite for all t and $\mathbb{E}_t X_T = X_t$ for all $0 \leq t \leq T$.*

The martingale property means that the expected future value of the process is its current value. That is, the process is not expected to drift upwards or downwards away from where it is now.

The condition that $\mathbb{E}X_t$ is finite for all t equates to $\mathbb{E}|X_t|$ finite (see p.15 Shreve (2004)). This can be easier in practice to prove for stochastic processes.[5]

This is an important definition. The theory of pricing and hedging of financial derivatives is formulated in terms of martingales. The name originates in an 18th-century gambling strategy of doubling bets repeatedly (Mansuy, 2022). By coincidence, it is also the name given to the straps in a horse's harness that discourages the horse from raising its head too high. This provides a useful analogy in that a martingale process is expected to stay at the same level over time.

Theorem 4.2. *Standard Brownian motion is a martingale.*

Proof. We will first show $\mathbb{E}_t W_T = W_t$ for all $0 \leq t \leq T$.

The procedure to prove this part of the martingale property for any process will always be the same. Divide the process into that part that is already known at time t and the increment from t to T that is unknown.

$$\begin{aligned} \mathbb{E}_t W_T &= \mathbb{E}_t(W_t + (W_T - W_t)) \\ &= W_t + \mathbb{E}_t(W_T - W_t) && \text{since } W_t \text{ known at } t \\ &= W_t \end{aligned}$$

The alternative representation for finite expectations will be demonstrated, that is, $\mathbb{E}|W_t|$ finite. Here we can evaluate this expression as a function of time t. First we derive a result for the expectation of the absolute value of powers of Z a random variable with standard normal distribution.

$\mathbb{E}|Z| = \sqrt{\frac{2}{\pi}}$ for $Z \sim N(0,1)$ since

$$\begin{aligned}
\mathbb{E}|Z| &= \int_{-\infty}^{\infty} |z| \frac{1}{\sqrt{2\pi}} e^{-\frac{z^2}{2}}\, dz \\
&\qquad \text{using the p.d.f. of the standard normal distribution} \\
&= 2\int_{0}^{\infty} z \frac{1}{\sqrt{2\pi}} e^{-\frac{z^2}{2}}\, dz \\
&\qquad \text{since the integrand has same values for } -x \text{ and } x \\
&= \frac{2}{\sqrt{2\pi}} \int_{0}^{\infty} z e^{-\frac{z^2}{2}}\, dz \\
&= \frac{2}{\sqrt{2\pi}} \left(-e^{-z^2/2} \Big|_0^{\infty} \right) \qquad \text{see Example 1.15 (vii)} \\
&= \frac{2}{\sqrt{2\pi}} = \sqrt{\frac{2}{\pi}}
\end{aligned}$$

All odd powers of Z can be dealt with recursively in the same way giving $\mathbb{E}|Z^{2n+1}| = 2^n n! \sqrt{\frac{2}{\pi}}$

The absolute value can be simply dropped for even powers of Z. Then use the result in Equation (2.10).

Recall $W_t \sim N(0,t) = \sqrt{t}Z$. Then $\mathbb{E}|W_t|$ finite since

$$\mathbb{E}|W_t| = \mathbb{E}|\sqrt{t}Z| = \sqrt{t}\mathbb{E}|Z| = \sqrt{t}\sqrt{\frac{2}{\pi}} < \infty \quad \text{for all } t \geq 0$$

In other words, $\mathbb{E}|W_t|$ is never infinitely big for any t.

Since standard Brownian motion satisfies both conditions of the definition then it is a martingale. □

Exercise 4.4. *Let W be a standard Brownian motion. Show that the stochastic process defined by $\{e^{\lambda W_t - \frac{1}{2}\lambda^2 t}\}_{t \geq 0}$ is a martingale using the property that $\mathbb{E}e^X = e^{\mu + \frac{\sigma^2}{2}}$ when $X \sim N(\mu, \sigma^2)$.*

Proof. Recall that $\mathbb{E}e^X = e^{\mu + \frac{\sigma^2}{2}}$ follows from setting $r = 1$ in the MGF of the normal distribution (Equation 2.9).

We will first show

$$\mathbb{E}_t e^{\lambda W_T - \frac{1}{2}\lambda^2 T} = e^{\lambda W_t - \frac{1}{2}\lambda^2 t}$$

for all $0 \le t \le T$.

Once again, the first step is to divide the process into the part known at t and the increment from t to T that is as yet unknown.

We have $W_T - W_t \sim N(0, T-t)$ which implies that $\lambda(W_T - W_t) \sim N(0, \lambda^2(T-t))$. Hence

$$\begin{aligned}
\mathbb{E}_t e^{\lambda W_T - \frac{1}{2}\lambda^2 T} &= \mathbb{E}_t e^{\lambda(W_t + (W_T - W_t)) - \frac{1}{2}\lambda^2 T} \\
&= e^{\lambda W_t - \frac{1}{2}\lambda^2 T}\mathbb{E}_t e^{\lambda(W_T - W_t)} && \text{since } W_t \text{ known at } t \\
&= e^{\lambda W_t - \frac{1}{2}\lambda^2 T}\mathbb{E} e^{\lambda(W_T - W_t)} && \text{same argument as in Exercise 4.3(ii)} \\
&= e^{\lambda W_t - \frac{1}{2}\lambda^2 T} e^{\frac{1}{2}\lambda^2 (T-t)} && \text{using the property given} \\
&= e^{\lambda W_t - \frac{1}{2}\lambda^2 t}
\end{aligned}$$

We now demonstrate that this process has finite expectations, that is, $\mathbb{E} e^{\lambda W_t - \frac{1}{2}\lambda^2 t}$ finite. We can evaluate this expression at time t.

We have $W_t \sim N(0, t) \implies \lambda W_t \sim N(0, \lambda^2 t)$ so for all t

$$\begin{aligned}
\mathbb{E} e^{\lambda W_t - \frac{1}{2}\lambda^2 t} &= e^{-\frac{1}{2}\lambda^2 t}\mathbb{E} e^{\lambda W_t} && \text{removing non-stochastic elements} \\
&= e^{-\frac{1}{2}\lambda^2 t} e^{\frac{1}{2}\lambda^2 t} && \text{using the property given} \\
&= 1
\end{aligned}$$

Therefore $\mathbb{E} e^{\lambda W_t - \frac{1}{2}\lambda^2 t}$ is finite for all t.

Since $\{e^{\lambda W_t - \frac{1}{2}\lambda^2 t}\}_{t \ge 0}$ satisfies both conditions of the definition, then it is a martingale. □

In most cases, we use a simple result called the *triangle inequality* to demonstrate the finite expectations condition for the alternative representation.

The triangle inequality simply states that the absolute value of a sum of real numbers is always less than or equal to the sum of the absolute values.

$$|a + b| \le |a| + |b| \quad \text{for all real numbers } a, b$$

So for example:

$$\begin{aligned}
|2 + 3| &= |2| + |3|, && \text{that is, equality applies when } a, b \ge 0 \\
|-2 + 3| &< |-2| + |3|, && \text{that is, inequality applies when } a \text{ or } b < 0
\end{aligned}$$

Exercise 4.5. *Let W be a standard Brownian motion. Prove that the process $\{W_t^2 - t\}_{t \geq 0}$ is a martingale.*

Proof. First show $\mathbb{E}_t(W_T^2 - T) = W_t^2 - t$ for all $0 \leq t \leq T$.

$$\begin{aligned} \mathbb{E}_t(W_T^2 - T) &= \mathbb{E}_t W_T^2 - T && \text{since } T \text{ non-stochastic} \\ &= W_t^2 + (T - t) - T && \text{from Exercise 4.3(ii)} \\ &= W_t^2 - t \end{aligned}$$

Next show that $\{W_t^2 - t\}_{t \geq 0}$ has finite expectations, that is, $\mathbb{E}|W_t^2 - t|$ finite for all $t \geq 0$.

$$\begin{aligned} \mathbb{E}|W_t^2 - t| &\leq \mathbb{E}(|W_t^2| + |t|) && \text{from triangle inequality} \\ &= \mathbb{E}W_t^2 + t && \text{since } t \text{ non-stochastic} \\ &= 2t && \text{from Exercise 4.3(i)} \\ &< \infty \end{aligned}$$

Since the process satisfies both conditions of the definition, it is a martingale. □

4.5 Simulation

We can simulate standard Brownian motion in R. First, break the time interval from $t = 0$ to $t = 4$ into small steps of $dt = 0.01$.

```
t  <-  c(1 : 400)
t  <-  t/100
```

We then model $dW \sim N(0, 0.01)$. Generate 400 normally distributed random numbers from a distribution with mean zero and standard deviation $\sqrt{0.01}$, putting the result in the 400-by-1 column vector dW.

```
dW  <-  rnorm(400, 0, sqrt(0.01))
```

Calculate the cumulative sum of this vector working downwards from the first element. This will generate each of the W_t terms.

```
Wt  <-  cumsum(dW)
```

Now plot the results.

```
plot(t,Wt,type = "l", xlab = "Time", ylab = "W_t")
```

```
Simulation of standard Brownian motion
t   <-  c(1 : 400)
t   <-  t/100
dW   <-  rnorm(400, 0, sqrt(0.01))
Wt   <-  cumsum(dW)
plot(t,Wt, type = "l", xlab = "Time", ylab = "W_t")
```

4.6 Summary

Standard Brownian motion has been introduced in this chapter. It is the foundation of all the stochastic processes used to model financial assets that are presented in this book. Standard Brownian motion has two important properties.

The first is non-differentiability. Most stochastic processes are not differentiable. As a result, the calculus of stochastic processes will not follow the rules of ordinary calculus. The rules of stochastic calculus are developed in the next chapter.

The second is the martingale property. This important property is at the core of the theory of pricing and hedging of financial derivatives.

Key things to remember

- Definition of standard Brownian motion (Definition 4.1)
- How to check that a process is a standard Brownian motion (Exercise 4.2).
- Definition of martingale (Definition 4.2).
- How to check that a process is a martingale (property 4.2 and Exercise 4.4).

The property that $\mathbb{E}e^X = e^{\mu+\frac{1}{2}\sigma^2}$ when $X \sim N(\mu, \sigma^2)$ first introduced in Chapter 2 will come up again repeatedly.

Further reading

A historical perspective on standard Brownian motion can be found in Sullivan and Weithers (1991) who trace the origins of option pricing theory to Louis Bachelier's pioneering work on stochastic processes. For a rigorous yet accessible introduction to the mathematics underpinning modern financial models, see Hirsa and Neftci (2013). Readers looking for a more intuitive and less formal approach to stochastic concepts may benefit from Calin (2015). For a mathematically detailed exploration of continuous-time finance models, Shreve (2004) remains an essential text complementing the more concise exposition in Durrett (2012). Finally, Klebaner (2012) presents a balanced introduction to stochastic calculus, emphasizing applications and examples that bridge theory and financial practice.

Notes

1. The terms *Brownian motion* and *Wiener process* are generally regarded as synonyms.
2. Multiplication by scalar is monotonic (i.e. preserves order).
3. See corollary 3.3.7 in Calin (2015)
4. More advanced calculus outside the scope of this text would indicate we could use the definition of the RiemannâŁ"Stieltjes integral. However, the RiemannâŁ"Stieltjes integral is not defined for functions of unbounded variation where variation was defined in Section 1.11. W_t has unbounded variation since

$$\sum_{0}^{n-1}(W_{t_{i+1}} - W_{t_i})^2 \leq \delta \sum_{0}^{n-1} |W_{t_{i+1}} - W_{t_i}|$$

where δ is the biggest absolute increment $|W_{t_{i+1}} - W_{t_i}|$ over all n intervals. If the variation of W_t (the sum on the right-hand side) is finite, then, in the limit with δ getting infinitesimally small, the sum on the left-hand side known as the quadratic variation must become zero. The quadratic variation of W_t is shown in Section 6.3 to be non-zero.
5. A random variable with probability density function given by the Cauchy distribution $f(x) = \frac{1}{\pi}\left[\frac{\gamma}{(x-x_0)^2+\gamma^2}\right]$ where $\gamma > 0$ is an example of a random variable with non-finite expectation.

Questions

$W = \{W_t\}_{t \geq 0}$ is standard Brownian motion in the following problems.

1 What is the probability that W_2 takes a value in the interval $(-1,1)$?

Solution W_2 is normally distributed with mean 0 and variance 2.

$$P(-1 < W_2 < 1) = P(-\frac{1}{\sqrt{2}} < Z < \frac{1}{\sqrt{2}}) = 0.521$$

2 Show that $Z_t = \begin{cases} 0, & t = 0 \\ tW_{1/t}, & t > 0 \end{cases}$ is a standard Brownian motion using the result that

$$t_2 W_{1/t_2} - t_1 W_{1/t_1} \text{ and } t_4 W_{1/t_4} - t_3 W_{1/t_3}$$

are independent for $t_1 < t_2 \leq t_3 < t_4$.

Solution

(i) $Z_0 = 0$ by definition

(ii) You are given the second property. $Z_{t_2} - Z_{t_1}$ and $Z_{t_4} - Z_{t_3}$ are independent.

(iii) Any linear combination $aW_s + bW_t$ for $s < t$ is normally distributed since $aW_s + bW_t = (a + b)W_s + b(W_t - W_s)$ which is a linear combination of independent normal variables and therefore normal from Theorem 2.8. $Z_{t_2} - Z_{t_1} = t_2 W_{1/t_2} - t_1 W_{1/t_1}$ is then normal.

$$\begin{aligned}
\mathbb{E}\left(Z_{t_2} - Z_{t_1}\right) &= \mathbb{E}\left(t_2 W_{1/t_2} - t_1 W_{1/t_1}\right) \\
&= t_2 \mathbb{E} W_{1/t_2} - t_1 \mathbb{E} W_{1/t_1} = 0 - 0 = 0 \\
var\left(Z_{t_2} - Z_{t_1}\right) &= var\left(t_2 W_{1/t_2} - t_1 W_{1/t_1}\right) \\
&= t_2^2 var W_{1/t_2} + t_1^2 var W_{1/t_1} - 2t_2 t_1 cov(W_{1/t_2}, W_{1/t_1}) \\
&= t_2^2(1/t_2) + t_1^2(1/t_1) - 2t_2 t_1(1/t_2) \\
&= t_2 - t_1
\end{aligned}$$

$Cov(W_s, W_t) = s$ where $s < t$ from Question 5(v).

3 Two standard Brownian motions W_1 and W_2 are correlated if the increments $W_1, t_2 - W_1, t_1$ and $W_2, t_2 - W_2, t_1$ are correlated.

Let W_1 and W_2 be standard Brownian motions then show

$$aW_1 + bW_2$$

is a standard Brownian motion where

$$a = \frac{\alpha}{\sqrt{\alpha^2 + \beta^2 + 2\rho\alpha\beta}} \text{ and } b = \frac{\beta}{\sqrt{\alpha^2 + \beta^2 + 2\rho\alpha\beta}}$$

for any real numbers α, β, and ρ is the correlation coefficient of W_1 and W_2. You can assume that $aW_1 + bW_2$ is normally distributed.

Solution

(i) $aW_{1,0} + bW_{2,0} = 0$

(ii) If $t_1 < t_2 \leq t_3 < t_4$, then

$$\begin{aligned}&(aW_{1,t_2} + bW_{2,t_2}) - (aW_{1,t_1} + bW_{2,t_1})\\&\quad = a(W_{1,t_2} - W_{1,t_1}) + b(W_{2,t_2} - W_{2,t_1})\end{aligned}$$

which is independent of

$$\begin{aligned}&(aW_{1,t_4} + bW_{2,t_4}) - (aW_{1,t_3} + aW_{2,t_3})\\&\quad = a(W_{1,t_4} - W_{1,t_3}) + b(W_{2,t_4} - W_{2,t_3})\end{aligned}$$

(iii)

$$\begin{aligned}&(aW_{1,t_2} + bW_{2,t_2}) - (aW_{1,t_1} + bW_{2,t_1})\\&\quad = a(W_{1,t_2} - W_{1,t_1}) + b(W_{2,t_2} - W_{2,t_1})\\&\quad \sim aN(0, t_2 - t_1) + bN(0, t_2 - t_1)\\&\quad = N(0, (a^2 + b^2 + 2\rho ab)(t_2 - t_1))\\&\quad = N(0, t_2 - t_1)\end{aligned}$$

from properties of random variables in Table 2.1

4 For what value of the correlation coefficient between two standard Brownian motions W_1 and W_2 is $W_1 + W_2$ also a standard Brownian motion?

Solution Setting $a = b = 1$ in expressions in the previous question, we get:

$$\alpha = \sqrt{\alpha^2 + \beta^2 + 2\rho\alpha\beta} \implies 0 = \beta^2 + 2\rho\alpha\beta$$
$$\beta = \sqrt{\alpha^2 + \beta^2 + 2\rho\alpha\beta} \implies 0 = \alpha^2 + 2\rho\alpha\beta$$

So $\alpha = \beta$ and $\alpha^2 = -2\rho\alpha^2 \implies \rho = -0.5$

5 Where $0 < s < t$ find
(i) $\mathbb{E}\big((W_t^2 - t)(W_s^2 - s)\big)$
(ii) $\mathbb{E}_s W_t^3$.
(iii) $var(W_s + W_t)$.
(iv) $\mathbb{E}(W_s W_t)$.
(v) $\rho(W_s, W_t)$
(vi) $cov(e^{W_s}, e^{W_t})$

Solution

(i)

$$\begin{aligned}
\mathbb{E}\big((W_t^2 - t)(W_s^2 - s)\big) &= \mathbb{E}(W_t^2 W_s^2 - sW_t^2 - tW_s^2 + ts) \\
&= \mathbb{E}W_t^2 W_s^2 - s\mathbb{E}W_t^2 - t\mathbb{E}W_s^2 + ts \\
&= \mathbb{E}(W_s + W_t - W_s)^2 W_s^2 - st - ts + ts \\
&= \mathbb{E}W_s^4 + 2\mathbb{E}(W_t - W_s)\mathbb{E}W_s^3 \\
&\quad + \mathbb{E}(W_t - W_s)^2 \mathbb{E}W_s^2 - st \\
&= 3s^2 + (t - s)s - st = 2s^2
\end{aligned}$$

(ii)

$$\begin{aligned}
\mathbb{E}_s W_t^3 &= \mathbb{E}_s(W_s + W_t - W_s)^3 \\
&= \mathbb{E}_s\Big(W_s^3 + 3W_s^2(W_t - W_s) + 3W_s(W_t - W_s)^2 \\
&\quad + (W_t - W_s)^3\Big) \\
&= W_s^3 + 3W_s\mathbb{E}_s(W_t - W_s)^2 \\
&= W_s^3 + 3W_s(t - s)
\end{aligned}$$

(iii)

$$\begin{aligned}
var(W_s + W_t) &= var(W_s + (W_s + W_t - W_s)) \\
&= var(2W_s + W_t - W_s) \\
&= 4var(W_s) + var(W_t - W_s) \text{ since independent} \\
&= 4s + (t - s) \\
&= 3s + t
\end{aligned}$$

(iv)

$$\begin{aligned}
\mathbb{E}(W_s W_t) &= \mathbb{E}\big(W_s(W_s + W_t - W_s)\big) \\
&= \mathbb{E}W_s^2 + \mathbb{E}(W_s(W_t - W_s)) \\
&= \mathbb{E}W_s^2 + \mathbb{E}W_s\,\mathbb{E}(W_t - W_s) \\
&= s
\end{aligned}$$

(v)

$$\begin{aligned}
\rho(W_s, W_t) &= \frac{cov(W_s, W_t)}{\sqrt{var\,W_s\;var\,W_t}} \\
&= \frac{\mathbb{E}(W_s W_t) - \mathbb{E}W_s\,\mathbb{E}W_t}{\sqrt{var\,W_s\;var\,W_t}} \\
&= \frac{s - 0 \times 0}{\sqrt{st}} \\
&= \sqrt{\frac{s}{t}}
\end{aligned}$$

(vi) $cov(e^{W_s}, e^{W_t}) = \mathbb{E}(e^{W_s}e^{W_t}) - \mathbb{E}e^{W_s}\mathbb{E}e^{W_t} = \mathbb{E}e^{W_s+W_t} - e^{s/2}e^{t/2} = e^{(3s+t)/2} - e^{s/2}e^{t/2}$
Using Equation (2.9) and the result for $var(W_s + W_t)$ from the above question.

6 Evaluate the following expressions for $0 < s < t < u$:
(i) $\mathbb{E}(W_s W_t W_u)$
(ii) $\mathbb{E}_s(W_t W_u)$

[QFIQF–Fall 2023 Q2(a)]

Solution

(i)

$$\begin{aligned}\mathbb{E}(W_s W_t W_u) &= \mathbb{E}\big(W_s(W_s + W_t - W_s) \\ &\quad (W_s + W_t - W_s + W_u - W_t)\big) \\ &= \mathbb{E}W_s^3 + 2\mathbb{E}(W_s^2(W_t - W_s)) \\ &\quad + \mathbb{E}(W_s(W_t - W_s)^2) + \mathbb{E}(W_s^2(W_u - W_t)) \\ &\quad + \mathbb{E}(W_s(W_t - W_s)(W_u - W_t)) \\ &= 2\mathbb{E}W_s^2\,\mathbb{E}(W_t - W_s) + \mathbb{E}W_s\,\mathbb{E}(W_t - W_s)^2 \\ &\quad + \mathbb{E}W_s^2\,\mathbb{E}(W_u - W_t) \\ &\quad + \mathbb{E}W_s\,\mathbb{E}(W_t - W_s)\,\mathbb{E}(W_u - W_t) \\ &= 0\end{aligned}$$

(ii)

$$\begin{aligned}\mathbb{E}_s(W_t W_u) &= \mathbb{E}_s\big((W_s + W_t - W_s)(W_s + W_t - W_s + W_u - W_t)\big) \\ &= W_s^2 + 2W_s\mathbb{E}_s(W_t - W_s) + \mathbb{E}_s(W_t - W_s)^2 \\ &\quad + W_s\mathbb{E}_s(W_u - W_t) + \mathbb{E}_s(W_t - W_s)(W_u - W_t) \\ &= W_s^2 + (t - s) + \mathbb{E}_s(W_t - W_s)\,\mathbb{E}_s(W_u - W_t) \\ &= W_s^2 + (t - s)\end{aligned}$$

7 Show that the following are martingales:

(i) $W_t^3 - 3tW_t$

(ii) $aX_t + bY_t + c$, where X_t and Y_t martingales and $\{a, b, c\}$ are real numbers.

Solution

(i) First show $\mathbb{E}_t(W_T^3 - 3TW_T) = W_t^3 - 3tW_t$ for all $0 \le t \le T$.

$$\begin{aligned}\mathbb{E}_t(W_T^3 - 3TW_T) &= \mathbb{E}_t W_T^3 - 3T\mathbb{E}_t W_T \\ &= (W_t^3 + 3W_t(T - t)) - 3TW_t\end{aligned}$$

from the previous question and W_t a martingale

$$= W_t^3 - 3tW_t$$

Next show that $\{W_t^3 - 3tW_t\}_{t\geq 0}$ has finite expectations, that is, $\mathbb{E}|W_t^3 - 3tW_t|$ finite for all $t \geq 0$.

$$\begin{aligned}
\mathbb{E}|W_t^3 - 3tW_t| &\leq \mathbb{E}(|W_t^3| + 3t|W_t|) \\
&= \mathbb{E}|W_t^3| + 3t\mathbb{E}|W_t| \\
&= \mathbb{E}|W_t^3| + 3t\sqrt{t}\sqrt{\frac{2}{\pi}} \\
&= 2t\sqrt{t}\sqrt{\frac{2}{\pi}} + 3t\sqrt{t}\sqrt{\frac{2}{\pi}} \\
&< \infty
\end{aligned}$$

Since the process satisfies both conditions of the definition, it is a martingale.

(ii) $\mathbb{E}_t Z_T = \mathbb{E}_t(aX_T + bY_T + c) = a\mathbb{E}_t(X_T) + b\mathbb{E}_t(Y_T) + c = aX_t + bY_t + c$ since X_t and Y_t are martingales.

$\mathbb{E}|Z_t|$ finite since $\mathbb{E}|Z_t| = \mathbb{E}|aX_t + bY_t + c| \leq E(|aX_t| + |bY_t| + |c|) = |a|\mathbb{E}|X_t| + |b|\mathbb{E}|Y_t| + |c| < \infty$

8 Show that e^{W_t} is not a martingale.

Solution

$$\mathbb{E}_t e^{W_T} = \mathbb{E}_t e^{W_t + (W_T - W_t)} = e^{W_t}\mathbb{E}_t e^{W_T - W_t} = e^{W_t} e^{\frac{1}{2}(T-t)} \neq e^{W_t}$$

9 Find the constant c such that $(a + bW_t)^2 + ct$ is a martingale for any real numbers a, b.

Solution

$$\begin{aligned}
\mathbb{E}_t((a + bW_T)^2 + cT) &= a^2 + 2ab\,\mathbb{E}_t W_T + b^2\mathbb{E}_t W_T^2 + cT \\
&= a^2 + 2abW_t + b^2(W_t^2 + (T - t)) + cT \\
&= (a + bW_t)^2 + b^2(T - t) + cT
\end{aligned}$$

Therefore, $\mathbb{E}_t((a + bW_T)^2 + cT) = (a + bW_t)^2 + ct$ if $b^2(T - t) + cT = ct \implies c = -b^2$

Check that expectations are finite for this c:

$$\begin{aligned}
\mathbb{E}|(a + bW_t)^2 - b^2 t| &\leq \mathbb{E}\big(|(a + bW_t)^2| + |b^2 t|\big) \\
&= \mathbb{E}(a + bW_t)^2 + b^2 t
\end{aligned}$$

$$= a^2 + 2ab\mathbb{E}W_t + b^2\mathbb{E}W_t^2 + b^2 t$$
$$= a^2 + 2b^2 t$$

So $\mathbb{E}|(a + bB_t)^2 - b^2 t|$ finite.

10 An asset follows a stochastic process $\{A_t\}_{t \geq 0}$ given by the relationship

$$A_t = A_0 e^{W_t}$$

Given that the asset is initially worth \$10,

(i) calculate the asset's expected value at $t = 4$ and

(ii) calculate the probability that the asset value is greater than \$30 at $t = 4$.

Solution

(i) From Equation (2.9), given $X \sim N(\mu, \sigma^2)$, then

$$\mathbb{E}e^X = e^{\mu + \frac{1}{2}\sigma^2}$$

So $W_t \sim N(0, t) \implies \mathbb{E}10e^{W_t} = 10e^{\frac{t}{2}}$
The asset's expected value at $t = 4$ is then $10e^2 = 73.9$.

(ii) $P(10e^{W_4} > 30) = P(W_4 > \log 3) = P(Z > \frac{\log 3}{\sqrt{4}}) = 0.291$

11 An asset follows a stochastic process $\{S_t\}_{t \geq 0}$ given by the relationship

$$S_t = S_0 e^{\mu t + \sigma W_t}$$

Given that μ and σ are constants and the asset is initially worth \$1, derive an expression for $P(S_t \leq x)$.

Solution

$P(S_t \leq x) = P(e^{\mu t + \sigma W_t} \leq x) = P(\mu t + \sigma W_t \leq \log x) = P(W_t \leq \frac{\log x - \mu t}{\sigma}) = P(z \leq \frac{\log x - \mu t}{\sigma\sqrt{t}})$ where $Z \sim N(0, 1)$

12 Why is the following R code an incorrect simulation of standard Brownian motion?

```
t  <-  c(1 : 100)
Wt  <-  rnorm(100, 0, sqrt(t))
plot(t, Wt, type = "l", xlab = "Time", ylab = "Wt")
```

Solution

This is a common student mistake. The simulation correctly captures that $W_t \sim N(0, t)$. However, it ignores how the process is formed from successively adding independent random increments.

5 Itô formula

Introduction

The calculus of stochastic processes does not follow the rules of ordinary calculus. We saw in the previous chapter that standard Brownian motion is not differentiable. This prevents us from differentiating functions of standard Brownian motion using the chain rule, and integration by substitution is also impossible. Without integration by substitution, it is not clear how we evaluate integrals with respect to W.

In the next chapter, we will derive a mathematically meaningful expression for the integral

$$\int_0^T X_t \, dW$$

where X_t is a stochastic process adapted to W but there is no mathematical basis for stochastic differentiation.

However, throughout this chapter, we will use differential notation in relation to stochastic processes. This should be understood as mathematical shorthand for the integral form. It is easier to work with this differential notation in practical applications.

The total differential of a function was introduced in Section 1.8 as the change in the function's value due to changes in its variables. In this chapter, Itô's formula in differential form is introduced. For practical purposes, it can be considered as a stochastic extension of the total differential.

5.1 Itô's formula for Brownian motion

Using an argument based on the Taylor series expansion from Section 1.10, it will be shown in the next chapter that given f a continuous and differentiable function of a variable:

$$f(W_T) - f(W_0) = \int_0^T f'(W_t) dW + \int_0^T \frac{f''(W_t)}{2} dt$$

DOI: 10.4324/9781003477624-7

This is the Itô formula in integral form. The derivatives are not with respect to W. They are the derivatives of $f(x)$ with respect to x then evaluated at W_t.

The Itô formula can be written in differential form as

$$df(W_t) = f'(W_t)dW + \frac{f''(W_t)}{2}dt \tag{5.1}$$

although this has no clearly defined mathematical meaning. It is simply a convenient shorthand.

Comparing this result to the expression for the total differential in Equation (1.2), we can see that the second term is the new element introduced by stochastic calculus. All our ordinary calculus intuition then has to be amended for this extra term.

Exercise 5.1. *Using Itô's formula find* $d(W_t^2)$.

Proof. Itô's formula tells us that

$$df(W_t) = f'(W_t)dW_t + \frac{f''(W_t)}{2}dt$$

We first find the derivatives of the function $f(W_t) = W_t^2$ given in the question.

$$f'(W_t) = 2W_t$$
$$f''(W_t) = 2$$

Then substituting into the formula:

$$\begin{aligned} d(W_t^2) &= 2W_t dW + \frac{2}{2}dt \\ &= 2W_t dW + dt \end{aligned}$$

The first term follows the rules of ordinary calculus, while the second term is the element particular to stochastic calculus. □

Exercise 5.2. *Using Itô's formula find* $d(W_t^4)$.

Proof. Here we have $f(W_t) = W_t^4$ so

$$f'(W_t) = 4W_t^3$$
$$f''(W_t) = 12W_t^2$$

Using Itô's formula:

$$\begin{aligned} d(W_t^4) &= 4W_t^3 dW + \frac{12W_t^2}{2} dt \\ &= 4W_t^3 dW + 6W_t^2 dt \end{aligned}$$

Once again the first term is what we would expect from ordinary calculus, while the second term is the additional stochastic calculus term. □

Exercise 5.3. *Find*

(a) $d(e^{kW_t})$, *where* k *is a constant*
(b) $d(\log W_t)$
(c) $d(5W_t^3 + 2W_t)$

Proof.

(a) $d(e^{kW_t}) = ke^{kW_t} dW + \dfrac{k^2 e^{kW_t}}{2} dt$

(b) $d(\log W_t) = \dfrac{1}{W_t} dW - \dfrac{1}{2W_t^2} dt$

(c) $d(5W_t^3 + 2W_t) = (15W_t^2 + 2)dW + 15W_t dt$ □

5.2 Formula for Itô processes

An *Itô process* is a type of stochastic process written as the sum of a term involving changes in time and a term involving changes in standard Brownian motion. Most of the stochastic processes describing the evolution of financial variables in this book will take this form:

$$dX = \mu(X_t, t)dt + \sigma(X_t, t)dW$$

where $\mu(X_t, t)$ is known as the *drift coefficient*, and $\sigma(X_t, t)$ is the *diffusion coefficient*.

Repeating the derivation using a Taylor series expansion gives the Itô formula in differential form for Itô processes as

$$df(X_t) = f'(X_t)dX + \frac{f''(X_t)}{2}(dX)^2 \tag{5.2}$$

The only difficulty here is to evaluate $(dX)^2$. Dropping the dependence of μ and σ on (X_t, t) temporarily for clarity:

$$(dX)^2 = (\mu dt + \sigma dW)^2 = \mu^2 (dt)^2 + 2\mu\sigma\, dt\, dW + \sigma^2 (dW)^2$$

In the next chapter, it will be seen that

$$\begin{aligned} (dt)^2 &= 0 \\ dt\, dW &= 0 \\ (dW)^2 &= dt \end{aligned}$$

or in tabular form:

	dt	dW
dt	0	0
dW	0	dt

(5.3)

Therefore,

$$(dX)^2 = \sigma(X_t, t)^2 dt$$

Exercise 5.4. *Given the stochastic process*

$$dX = \mu X_t dt + \sigma X_t dW$$

where μ and σ are constants. Find $d(\log X_t)$.

Proof. Itô's formula is

$$df(X_t) = f'(X_t)dX + \frac{f''(X_t)}{2}(dX)^2$$

Find the derivatives of the function $f(X_t) = \log X_t$ given in the question.

$$\begin{aligned} f'(X_t) &= \frac{1}{X_t} \\ f''(X_t) &= -\frac{1}{X_t^2} \end{aligned}$$

And $(dX)^2 = (\sigma X_t)^2 dt$

Then substituting into the formula:

$$
\begin{aligned}
d(\log X_t) &= \frac{1}{X_t}dX - \frac{1}{2X_t^2}(dX)^2 \\
&= \frac{1}{X_t}(\mu X_t dt + \sigma X_t dW) - \frac{1}{2X_t^2}(\sigma^2 X_t^2 dt) \\
&= (\mu - \frac{1}{2}\sigma^2)dt + \sigma dW
\end{aligned}
$$

□

Exercise 5.5. *Given the stochastic process*

$$dX = -\beta X_t dt + dW$$

where β is a constant. Find $d(X_t^2)$.

Proof.

$$
\begin{aligned}
d(X_t^2) &= 2X_t dX + \frac{1}{2}(2)(dX)^2 \\
&= 2X_t(-\beta X_t dt + dW) + dt \\
&= (1 - 2\beta X_t^2)dt + 2X_t dW
\end{aligned}
$$

□

5.3 Itô formula in two variables

The Itô formula can be extended to two variables by using the Taylor series expansion in two variables (Equation 1.5). Given f a continuous and differentiable function of Itô processes X and Y, the Itô formula becomes:

$$df(X_t, Y_t) = \frac{\partial f}{\partial x}dX + \frac{\partial f}{\partial x}dY + \frac{1}{2}\left(\frac{\partial^2 f}{\partial x^2}(dX)^2 + 2\frac{\partial^2 f}{\partial x \partial y}\, dX\, dY + \frac{\partial^2 f}{\partial y^2}(dY)^2\right) \tag{5.4}$$

Exercise 5.6. *Use the two-dimensional Itô formula to derive:*

1 The stochastic product rule:

$$d(X_t Y_t) = Y_t dX + X_t dY + dXdY$$

2 The stochastic quotient rule:

$$d\frac{X_t}{Y_t} = \frac{1}{Y_t}dX - \frac{X_t}{Y_t^2}dY + \frac{X_t}{Y_t^3}(dY)^2 - \frac{1}{Y_t^2}dXdY$$

Note the similarities between these expressions and the ordinary calculus product and quotient rules given in Section 1.3.

Proof.

1. Itô's formula in two variables is

$$df(X_t, Y_t) = \frac{\partial f}{\partial x}dX + \frac{\partial f}{\partial x}dY + \frac{1}{2}\left(\frac{\partial^2 f}{\partial x^2}(dX)^2 + 2\frac{\partial^2 f}{\partial x \partial y}\, dX\, dY + \frac{\partial^2 f}{\partial y^2}(dY)^2\right)$$

Find the derivatives of the function $f(X_t, Y_t) = X_t Y_t$.

$\frac{\partial f}{\partial x} = Y_t \quad \frac{\partial f}{\partial y} = X_t$

$\frac{\partial^2 f}{\partial x^2} = 0 \quad \frac{\partial^2 f}{\partial x \partial y} = 1 \quad \frac{\partial^2 f}{\partial y^2} = 0$

Then substituting into the formula:

$$\begin{aligned} d(X_t Y_t) &= Y_t dX + X_t dY + \frac{1}{2}\left(0 \times (dX)^2 + 2 \times 1 \times dX\, dY + 0 \times (dY)^2\right) \\ &= Y_t dX + X_t dY + dX\, dY \end{aligned}$$

2. Finding the derivatives of the function $f(X_t, Y_t) = \frac{X_t}{Y_t}$.

$\frac{\partial f}{\partial x} = \frac{1}{Y_t} \quad \frac{\partial f}{\partial y} = -\frac{X_t}{Y_t^2}$

$\frac{\partial^2 f}{\partial x^2} = 0 \quad \frac{\partial^2 f}{\partial x \partial y} = -\frac{1}{Y_t^2} \quad \frac{\partial^2 f}{\partial y^2} = 2\frac{X_t}{Y_t^3}$

Then substituting into the formula:

$$\begin{aligned} d\left(\frac{X_t}{Y_t}\right) &= \frac{1}{Y_t}dX - \frac{X_t}{Y_t^2}dY + \frac{1}{2}\left(0 \times (dX)^2 - 2\frac{1}{Y_t^2}\, dX\, dY + 2\frac{X_t}{Y_t^3}(dY)^2\right) \\ &= \frac{1}{Y_t}dX - \frac{X_t}{Y_t^2}dY - \frac{1}{Y_t^2}\, dX\, dY + \frac{X_t}{Y_t^3}(dY)^2 \end{aligned} \tag{5.5}$$

□

If X and Y are driven by the same standard Brownian motion, then the covariation term $dXdY$ is straightforward to calculate:

Let

$dX = \mu^x(X_t, t)dt + \sigma^x(X_t, t)dW$ and
$dY = \mu^y(X_t, t)dt + \sigma^y(X_t, t)dW$

Then using multiplication Table 5.3 and dropping the dependence on (X_t, t) temporarily for clarity:

$$\begin{aligned} dXdY &= (\mu^x dt + \sigma^x dW)(\mu^y dt + \sigma^y dW) \\ &= \sigma^x \sigma^y dt \end{aligned}$$

It can be shown that $dW_1\,dW_2 = 0$, where $W_{1,t}$ and $W_{2,t}$ are independent Brownian processes.[1]

So if X_t and Y_t are driven by two different standard Brownian motions correlated with coefficient ρ, then we can model the processes flexibly as[2]

$$\begin{aligned} &dX = \mu^x dt + \sigma^x dW_1 \text{ and} \\ &dY = \mu^y dt + \sigma^y(\rho dW_1 + \sqrt{1-\rho^2} dW_2) \quad \text{for } -1 \le \rho \le 1 \end{aligned}$$

The covariation term becomes

$$\begin{aligned} dXdY &= (\mu^x dt + \sigma^x dW_1)\left(\mu^y dt + \sigma^y(\rho dW_1 + \sqrt{1-\rho^2} dW_2)\right) \\ &= \sigma^x \sigma^y \rho\, dt \end{aligned}$$

As an important special case we can let one of the two processes be a function of time only. The covariation term is then zero and the Itô formula (5.4) simplifies as

$$\begin{aligned} df(t, X_t) &= \frac{\partial f}{\partial t} dt + \frac{\partial f}{\partial x} dX + \frac{1}{2}\left(\frac{\partial^2 f}{\partial t^2}(dt)^2 + 2\frac{\partial^2 f}{\partial t \partial x} dt\, dX + \frac{\partial^2 f}{\partial x^2}(dX)^2\right) \\ &= \frac{\partial f}{\partial t} dt + \frac{\partial f}{\partial x} dX + \frac{1}{2}\frac{\partial^2 f}{\partial x^2}(dX)^2 \qquad \text{using the table in Table 5.3} \end{aligned}$$

We will refer to this version as the *time-dependent* Itô formula.

If the Itô process takes the form $dX = \mu(X_t, t)dt + \sigma(X_t, t)dW$, then dropping the dependence on (X_t, t) temporarily

$$\begin{aligned} df(t, X_t) &= \frac{\partial f}{\partial t} dt + \frac{\partial f}{\partial x} dX + \frac{1}{2}\frac{\partial^2 f}{\partial x^2}(dX)^2 \\ &= \frac{\partial f}{\partial t} dt + \frac{\partial f}{\partial x}(\mu dt + \sigma dW) + \frac{1}{2}\frac{\partial^2 f}{\partial x^2}\sigma^2 dt \\ &= \left(\frac{\partial f}{\partial t} + \frac{\partial f}{\partial x}\mu(X_t, t) + \frac{1}{2}\frac{\partial^2 f}{\partial x^2}\sigma^2(X_t, t)\right) dt + \frac{\partial f}{\partial x}\sigma(X_t, t) dW \end{aligned} \tag{5.6}$$

Exercise 5.7. *Given the stochastic process*

$$dS = \mu S_t dt + \sigma S_t dW$$

Find

$$d\left(\frac{e^{r(T-t)}}{S_t}\right)$$

where μ, σ, r, T *are all constants.*

Proof. Finding the derivatives of the function $f(t, S_t) = \frac{e^{r(T-t)}}{S_t}$.

$$\frac{\partial f}{\partial t} = \frac{-re^{r(T-t)}}{S_t} \quad \frac{\partial f}{\partial s} = -\frac{e^{r(T-t)}}{S_t^2} \quad \frac{\partial^2 f}{\partial s^2} = 2\frac{e^{r(T-t)}}{S_t^3}$$

and $(dS)^2 = \sigma^2 S_t^2 dt$.
Then substituting into the time-dependent Itô formula:

$$\begin{aligned} d\left(\frac{e^{r(T-t)}}{S_t}\right) &= \frac{-re^{r(T-t)}}{S_t}dt - \frac{e^{r(T-t)}}{S^2}dS + \frac{1}{2} \times 2\frac{e^{r(T-t)}}{S_t^3} \times \sigma^2 S_t^2 dt \\ &= \frac{-re^{r(T-t)}}{S_t}dt - \frac{e^{r(T-t)}}{S^2}(\mu S_t dt + \sigma S_t dW) + \frac{e^{r(T-t)}}{S_t}\sigma^2 dt \\ &= \frac{e^{r(T-t)}}{S_t}\Big((-r - \mu + \sigma^2)dt - \sigma dW\Big) \end{aligned}$$

Or using Equation (5.6):

$$\begin{aligned} d\left(\frac{e^{r(T-t)}}{S_t}\right) &= \left(\frac{\partial f}{\partial t} + \frac{\partial f}{\partial s}\mu(X_t, t) + \frac{1}{2}\frac{\partial^2 f}{\partial s^2}\sigma^2(X_t, t)\right)dt + \frac{\partial f}{\partial s}\sigma(X_t, t)dW \\ &= \left(\frac{-re^{r(T-t)}}{S_t} - \frac{e^{r(T-t)}}{S_t^2}\mu S_t + \frac{e^{r(T-t)}}{S_t^3}\sigma^2 S_t^2\right)dt - \frac{e^{r(T-t)}}{S_t^2}\sigma S_t dW \\ &= \frac{e^{r(T-t)}}{S_t}\Big((-r - \mu + \sigma^2)dt - \sigma dW\Big) \end{aligned}$$

□

Exercise 5.8. *Given the stochastic process*

$$dX = -\gamma X_t dt + \sigma dW$$

use the time-dependent Itô formula to find $d(e^{\gamma t} X_t)$, *where* γ, σ *are constants.*

Proof.

$$
\begin{aligned}
d(e^{\gamma t}X_t) &= \gamma e^{\gamma t}X_t dt + e^{\gamma t}dX + \frac{1}{2} \times 0 \times (dX)^2 \\
&= \gamma e^{\gamma t}X_t dt + e^{\gamma s}(-\gamma X_t dt + \sigma dW) \\
&= \sigma e^{\gamma t}dW
\end{aligned}
$$

□

5.4 Summary

This material is quite challenging.

A suggested pathway is to

- first learn to apply the Itô formula to functions of W_t: Section 5.1
- then apply the Itô formula to functions of Itô processes: Section 5.2
- then recognize when to apply the time-dependent Itô formula: Section 5.3

Why do the normal rules of calculus not apply to stochastic processes? This is fundamental to understanding the entire chapter.

Once these elements are mastered, then the rest of the chapter can be tackled.

Key things to remember

- The Itô formula
  - for standard Brownian motion (Equation 5.1)
  - for Itô processes (Equation 5.2)
  - for two variables (Equation 5.4)
  - time-dependent formula (Equation 5.6)
- The multiplication table (Table 5.3).

Further reading

Klebaner (2012) provides a clear and intuitive foundation for stochastic calculus, combining rigorous exposition with numerous real-world examples and applications across science and finance. For a more advanced and finance-oriented treatment, Shreve (2004) offers a detailed account of Itô integrals, stochastic differential equations, and continuous-time financial modeling. For greater depth, Oksendal (2013) remains a classic reference, balancing theoretical rigor with clear exposition and numerous examples. Finally, Calin (2015) provides a more conversational and intuitive approach, ideal for readers new to the topic or seeking conceptual clarity before tackling more advanced treatments.

Notes

1. Proof given in Shreve (2004) p.165.
2. Letting $\sigma_1 = \rho$ and $\sigma_2 = \sqrt{1-\rho^2}$ then $U = \rho W_1 + \sqrt{1-\rho^2} W_2$ is standard Brownian motion from Chapter 4 Exercise 3. And

$$\rho(W_{1,t}, U_t) = \frac{\mathbb{E}(W_{1,t} U_t) - \mathbb{E} W_{1,t} \mathbb{E} U_t}{\sqrt{var W_{1,t}\, var U_t}} = \frac{\rho \mathbb{E}(W_{1,t}^2)}{\sqrt{t \times t}} = \frac{\rho t}{t} = \rho \text{ for all } t$$

Questions

$W = \{W_t\}_{t\geq 0}$ is standard Brownian motion in the following problems.

1 Find
   (a) $d(W_t^3)$
   (b) $d(W_t^\theta)$, where θ is a constant.
   (c) $d(W_t e^{W_t})$
   (d) $d(tW_t^2)$
   (e) $d(e^{at+bW_t})$, where a, b are constants.
   (f) $d(e^{t+W_t^2})$

Solution

(a) $d(W_t^3) = 3W_t^2 dW + 3W_t dt$

(b)
$$d(W_t^\theta) = \theta W_t^{\theta-1} dW + \frac{1}{2}\theta(\theta-1)W_t^{\theta-2}dt$$
$$= \theta W_t^{\theta-2}(W_t dW + \frac{1}{2}(\theta-1)dt)$$

(c) Treating $W_t e^{W_t}$ as $f(W_t)$ in the Itô formula

$$d(W_t e^{W_t}) = (e^{W_t} + W_t e^{W_t})dW + \frac{1}{2}(e^{W_t} + e^{W_t} + W_t e^{W_t})(dW)^2$$
$$= e^{W_t}\left((1+W_t)dW + (1+\frac{1}{2}W_t)dt\right)$$

You could also use the stochastic product rule.

$$d(W_t e^{W_t}) = e^{W_t}dW + W_t d(e^{W_t}) + dWd(e^{W_t})$$
$$= e^{W_t}dW + W_t(e^{W_t}dW + \frac{e^{W_t}}{2}dt) + e^{W_t}dt$$

from Exercise 5.3(a)

$$= e^{W_t}\left((1+W_t)dW + (1+\frac{1}{2}W_t)dt\right)$$

(d) Using the time-dependent Itô formula

$$d(tW_t^2) = W_t^2 dt + 2tW_t dW + \frac{1}{2}(2t)(dW)^2$$
$$= (W_t^2 + t)dt + 2tW_t dW$$

(e) Using the time-dependent Itô formula

$$d(e^{at+bW_t}) = ae^{at+bW_t}dt + be^{at+bW_t}dW + \frac{1}{2}b^2e^{at+bW_t}dt$$
$$= e^{at+bW_t}\left((a + \frac{b^2}{2})dt + bdW\right)$$

(f) Using the time-dependent Itô formula

$$d(e^{t+W_t^2}) = e^{t+W_t^2}dt + (2W_te^{t+W_t^2})dW + \frac{1}{2}(2e^{t+W_t^2}$$
$$+ 4W_t^2e^{t+W_t^2})(dW)^2$$
$$= 2e^{t+W_t^2}\left((1 + W_t^2)dt + W_tdW\right)$$

2 If $dX = Y_tdt + Z_tdW$, then find $(dX)^2$.

Solution $(dX)^2 = Z_t^2dt$

3 Given the stochastic process

$$dX = \mu X_tdt + \sigma X_tdW$$

where μ and σ are constants. Find
(a) $d(2X_t)$
(b) $d(X_t^2)$
(c) $d(e^{X_t})$
(d) $d(1 + X_t)^{-1}$
(e) $d(X_t^ne^{mt})$, where n, m constants

Solution
(a) $d(2X_t) = 2dX = 2\mu X_tdt + 2\sigma X_tdW$
(b)
$$d(X_t^2) = 2X_tdX + \frac{1}{2}2\,(dX)^2$$
$$= 2X_t(\mu X_tdt + \sigma X_tdW) + \sigma^2X_t^2dt$$
$$= X_t^2\left((2\mu + \sigma^2)dt + 2\sigma dW\right)$$

(c)
$$d(e^{X_t}) = e^{X_t}dX + \frac{1}{2}\,e^{X_t}(dX)^2$$
$$= e^{X_t}(\mu X_tdt + \sigma X_tdW) + \frac{1}{2}e_t^X(\sigma^2X_t^2dt)$$

$$= X_t e^{X_t}\left((\mu + \frac{1}{2}\sigma^2 X_t)dt + \sigma dW\right)$$

(d)

$$d(1+X_t)^{-1} = -(1+X_t)^{-2}dX + \frac{1}{2}2(1+X_t)^{-3}(dX)^2$$
$$= -(1+X_t)^{-2}(\mu X_t dt + \sigma X_t dW)$$
$$+ (1+X_t)^{-3}\sigma^2 X_t^2 dt$$
$$= \frac{X_t}{(1+X_t)^2}\left((-\mu + \frac{\sigma^2 X_t}{1+X_t})dt - \sigma dW\right)$$

(e) Using the time-dependent Itô formula applied to Itô processes (Equation 5.6)

$$d(X_t^n e^{mt}) = \left(mX_t^n e^{mt} + nX_t^{n-1}e^{mt}(\mu X_t) + \frac{1}{2}n(n-1)\right.$$
$$\left. X_t^{n-2}e^{mt}(\sigma^2 X_t^2)\right)dt + nX_t^{n-1}e^{mt}(\sigma X_t)dW$$
$$= X_t^n e^{mt}\left((m + n\mu + \frac{1}{2}n(n-1)\sigma^2)dt + n\sigma dW\right)$$

4 Given the stochastic process

$$dX = a dW$$

where a is a constant. Find $d(e^{X_t})$.

Solution

$$d(e^{X_t}) = e^{X_t}dX + \frac{1}{2}e^{X_t}(dX)^2$$
$$= ae^{X_t}\left(\frac{a}{2}dt + dW\right)$$

5 Given the stochastic process

$$dX = Y_t dW + Y_t^2 dt$$

and $Z_t = e^{-4X_t}$ find dZ in terms of Z_t and Y_t.

Solution

$$\begin{aligned} dZ = de^{-4X_t} &= -4e^{-4X_t}dX + \frac{16}{2}e^{-4X_t}(dX)^2 \\ &= -4Z_t(Y_t dW + Y_t^2 dt) + 8Z_t Y_t^2 dt \\ &= 4Z_t Y_t\left(Y_t dt - dW\right) \end{aligned}$$

6 Given the stochastic process

$$dX = Y_t dW + Z_t dt$$

determine $d(e^{3t^2 X_t})$.

Solution Using the time-dependent Itô formula applied to Itô process (Equation 5.6).

$$\begin{aligned} d(e^{3t^2 X_t}) &= (6tX_t e^{3t^2 X_t} + 3t^2 e^{3t^2 X_t}(Z_t) \\ &\quad + \frac{1}{2}(9t^4 e^{3t^2 X_t})(Y_t^2))dt + (3t^2 e^{3t^2 X_t})(Y_t)dW \\ &= te^{3t^2 X_t}\left(\left(6X_t + 3tZ_t + \frac{9}{2}t^3 Y_t^2\right)dt + 3tY_t dW\right) \end{aligned}$$

7 Verify that the process X_t given by

$$X_t = X_0 e^{(\mu - \frac{1}{2}\sigma^2)t + \sigma W_t}$$

where X_0, μ, σ are constants that satisfies the stochastic differential equation

$$dX = \mu X_t dt + \sigma X_t dW$$

Solution
Using result from Question 1(e) with $a = \mu - 1/2\sigma^2$ and $b = \sigma$:

$$\begin{aligned} dX &= X_0 e^{(\mu - \frac{1}{2}\sigma^2)t + \sigma W_t}(\mu dt + \sigma dW) \\ &= \mu X_t dt + \sigma X_t dW \end{aligned}$$

8 Given the stochastic processes

$$dB = rB_t dt$$
$$dS = \mu S_t dt + \sigma S_t dW$$

where r, μ, σ are all constants find $d(\frac{S_t}{B_t})$.

Solution
Use the stochastic quotient rule.

$$\begin{aligned} d(\frac{S_t}{B_t}) &= \frac{1}{B_t}dS - \frac{S_t}{B_t^2}dB \\ &= \frac{1}{B_t}(\mu S_t dt + \sigma S_t dW) - \frac{S_t}{B_t^2}(rB_t dt) \\ &= (\mu - r)\frac{S_t}{B_t}dt + \sigma\frac{S_t}{B_t}dW \end{aligned}$$

9 Assume that X_t and Y_t are stochastic processes satisfying the following stochastic differential equations:

$$dX = \mu_1 X_t dt + X_t(\alpha\, dW_1 + \beta\, dW_2), \quad X_0 = 1$$
$$dY = \mu_2 Y_t dt + Y_t(\beta\, dW_1 + \gamma\, dW_2), \quad Y_0 = 1$$

where $\mu_1, \mu_2, \alpha, \beta, \gamma$ are positive constant parameters, W_1 and W_2 are two independent standard Wiener processes.

Let

$$S_t = \sqrt{X_t Y_t}$$

Derive the stochastic differential equation of S_t.

[QFIQF spring 2020 q1(a)]

Solution
The covariation term is

$$\begin{aligned} dXdY &= \left(\mu_1 X_t dt + X_t(\alpha\, dW_1 + \beta\, dW_2)\right) \\ &\quad \left(\mu_2 Y_t dt + Y_t(\beta\, dW_1 + \gamma\, dW_2)\right) \\ &= X_t Y_t(\alpha\beta + \beta\gamma)dt \end{aligned}$$

Use the Itô formula in two variables.

$$
\begin{aligned}
d\sqrt{X_tY_t} &= \frac{1}{2}(X_tY_t)^{-\frac{1}{2}}Y_t dX + \frac{1}{2}(X_tY_t)^{-\frac{1}{2}}X_t dY \\
&\quad + \frac{1}{2}\Big(-\frac{1}{4}(X_tY_t)^{-\frac{3}{2}}Y_t^2(dX)^2 + 2(-\frac{1}{4}(X_tY_t)^{-\frac{3}{2}}X_tY_t \\
&\quad + \frac{1}{2}(X_tY_t)^{-\frac{1}{2}})\, dX\, dY - \frac{1}{4}(X_tY_t)^{-\frac{3}{2}}X_t^2(dY)^2\Big) \\
&= \frac{1}{2}S_t(\mu_1 dt + (\alpha\, dW_1 + \beta\, dW_2)) + \frac{1}{2}S_t(\mu_2 dt + (\beta\, dW_1 + \gamma\, dW_2)) \\
&\quad + \frac{1}{2}\Big(-\frac{1}{4}S_t(\alpha^2 + \beta^2))dt + \frac{1}{2}S_t(\alpha\beta + \beta\gamma)dt - \frac{1}{4}S_t(\beta^2 + \gamma^2)dt\Big) \\
&= \frac{1}{2}S_t\Big(\mu_1 + \mu_2 - \frac{1}{4}((\alpha-\beta)^2 + (\beta-\gamma)^2)\Big)dt \\
&\quad + \frac{1}{2}S_t\Big((\alpha+\beta)dW_1 + (\beta+\gamma)dW_2\Big)
\end{aligned}
$$

10 Given the stochastic processes

$$
\begin{aligned}
dS_1 &= \mu_1 S_{1,t}dt + \sigma_1 S_{1,t}dW_1 \\
dS_2 &= \mu_2 S_{2,t}dt + \sigma_2 S_{2,t}dW_2
\end{aligned}
$$

where W_1 and W_2 are standard Brownian processes correlated with coefficient ρ. Find $df(S_{1,t}, S_{2,t})$ for function f.

Solution

Use the Itô formula in two variables.

$$
\begin{aligned}
df\,(S_{1,t}, S_{2,t}) &= \left(\frac{\partial f}{\partial s_1}dS_1 + \frac{\partial f}{\partial s_2}dS_2\right) \\
&\quad + \frac{1}{2}\left(\frac{\partial^2 f}{\partial s_1^2}(dS_1)^2 + 2\frac{\partial^2 f}{\partial s_1 \partial s_2}dS_1 dS_2 + \frac{\partial^2 f}{\partial s_2^2}(dS_2)^2\right) \\
&= \left(\frac{\partial f}{\partial s_1}dS_1 + \frac{\partial f}{\partial s_2}dS_2\right) + \left(\frac{\partial^2 f}{\partial s_1^2}\left(\sigma_1^2 S_{1,t}^2 dt\right)\right. \\
&\quad \left. + 2\frac{\partial^2 f}{\partial s_1 \partial s_2}(\rho\sigma_1\sigma_2 S_{1,t}S_{2,t}dt) + \frac{\partial^2 f}{\partial s_2^2}\left(\sigma_2^2 S_{2,t}^2 dt\right)\right) \\
&= \left(\frac{\partial f}{\partial s_1}\mu_1 S_{1,t} + \frac{\partial f}{\partial s_2}\mu_2 S_{2,t} + \frac{1}{2}\frac{\partial^2 f}{\partial s_1^2}\sigma_1^2 S_{1,t}^2\right. \\
&\quad \left. + \frac{\partial^2 f}{\partial s_1 \partial s_2}\rho\sigma_1\sigma_2 S_{1,t}S_{2,t} + \frac{1}{2}\frac{\partial^2 f}{\partial s_2^2}\sigma_2^2 S_{2,t}^2\right) dt \\
&\quad + \left(\frac{\partial f}{\partial s_1}\sigma_1 S_{1,t}dW_1 + \frac{\partial f}{\partial s_2}\sigma_2 S_{2,t}dW_2\right)
\end{aligned}
$$

6 Stochastic integrals

Introduction

Unlike traditional calculus, where integration involves deterministic functions, stochastic integration deals with integrals where one or both of the functions involved are stochastic processes. The most commonly used form of stochastic integration is the Itô integral where we integrate with respect to a standard Brownian motion.

In this chapter, we extend the concept of integration to the stochastic setting. This leads to the development of stochastic integrals, particularly the Itô integral, which provides a framework for integrating with respect to Brownian motion. These integrals form the foundation of stochastic calculus, enabling us to define and solve stochastic differential equations (SDEs) that describe a wide range of stochastic processes.

6.1 Stochastic integral

We start with a reminder of what we mean by integration.

In Section 1.12, integration was considered as an approximation to the area under a curve.

Given the curve $y = f(x)$, the area under the curve between the vertical lines $x = a$ and $x = b$ and the x-axis can be approximated by dividing it into strips. This approximation should improve as the number of strips increases. As the number of strips becomes infinite, the shaded area closes in on a limit for all continuous $f(x)$. This limit is the integral.

$$\int_a^b f(x)dx := \lim_{n\to\infty} \sum_{i=0}^{n-1} f(x_i)\Delta x_i$$

We can do something similar when integrating with respect to stochastic processes. Let's first establish a motivation for doing this.

In a finance application, we could let the price of an asset follow standard Brownian motion and the amount of the asset held by a trader be X_t. The value

DOI: 10.4324/9781003477624-8

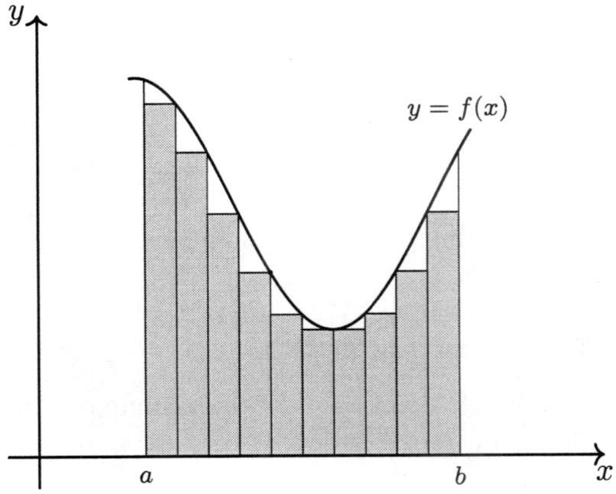

Figure 6.1 Integral in ordinary calculus

of this position at t is then $X_t W_t$. If trades occur at $0 = t_0, \ldots, t_i, t_{i+1}, \ldots, t_n = T$ then letting $\Delta W_{t_i} = W_{t_{i+1}} - W_{t_i}$

$$X_{t_i} \Delta W_{t_i}$$

is the change in value of the position from t_i to t_{i+1} ignoring interest rates and

$$\sum_{i=0}^{n-1} X_{t_i} \Delta W_{t_i}$$

represents the gain in trading over time from 0 to T. If trading is occurring continuously, then in the limit this becomes

$$\lim_{n \to \infty} \sum_{i=0}^{n-1} X_{t_i} \Delta W_{t_i} := \int_0^T X_t \, dW \tag{6.1}$$

This is called the *Itô integral* of the process X. The first thing to note is that this integral which is expressed as the limit of a sum containing increments of standard Brownian motion is itself a random variable. We will see later that the expectation of the *Itô integral* is zero meaning that the expected result of trading in an asset is zero if the asset price follows standard Brownian motion.

If X_t satisfies a number of technical conditions, then this integral is well-defined.[1] One of those is the same as in Section 4.4 where we assumed that we can determine the value of X_t exactly using the information available at t (adaptivity). Another is that

$$\int_0^T X_t^2\,dt < \infty$$

You can assume that all integrands considered in this book satisfy these conditions.[2]

6.2 Basic rules for stochastic integration

With the Itô integral defined as above, we can evaluate our first integral.

Let $X_t = 1$ for all t in Equation (6.1).

$$\begin{aligned}\int_0^T dW &= \lim_{n\to\infty}\Big((W_{t_1} - W_0) + (W_{t_2} - W_{t_1}) + \dots \\ &\qquad + (W_{t_{n-1}} - W_{t_{n-2}}) + (W_T - W_{t_{n-1}})\Big) \\ &= W_T - W_0 \\ &= W_T \end{aligned} \tag{6.2}$$

Similarly, it can easily be shown that the following properties follow from the definition:

Adjacent intervals $$\int_a^c X_t dW = \int_a^b X_t dW + \int_b^c X_t dW \tag{6.3}$$

Linearity $$\int_0^T (aX_t + bY_t)dW = a\int_0^T X_t dW + b\int_0^T Y_t dW \tag{6.4}$$

6.3 Variation

The concept of variation was introduced in Section 1.11, where the total variation was defined as the sum of the absolute value of the increments of the function. This can be extended to sums of squares (*quadratic variation*), cubes of increments (*cubic variation*), and so on.

The quadratic variation of standard Brownian motion underpins the Itô formula. Given $[0, t]$ partitioned into n intervals, the expression

$$\lim_{n\to\infty}\sum_0^{n-1}(W_{t_{i+1}} - W_{t_i})^2$$

is known as the *quadratic variation* of W.

Theorem 6.1. *Given standard Brownian motion W then*

$$\lim_{n\to\infty}\sum_{0}^{n-1}(W_{t_{i+1}} - W_{t_i})^2 = t$$

Proof. We first partition $[0, t]$ into n intervals of equal length $\frac{t}{n}$.[3]

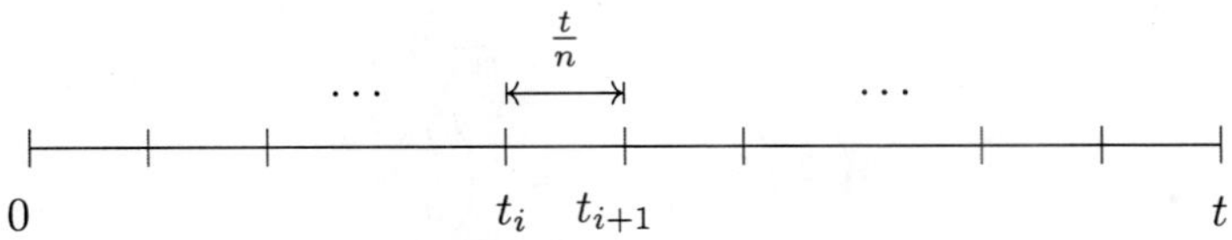

We will then show that as the interval size decreases

$$\sum_{0}^{n-1}(W_{t_{i+1}} - W_{t_i})^2 \to t$$

From Theorem 2.9, a sequence of random variables X_n converges in the mean square sense to k if $\mathbb{E}X_n \to k$ and $var\, X_n \to 0$. So the task is to show

$$\mathbb{E}\sum_{0}^{n-1}(W_{t_{i+1}} - W_{t_i})^2 \to t \text{ and } Var\sum_{0}^{n-1}(W_{t_{i+1}} - W_{t_i})^2 \to 0$$

From part (iii) of the definition of standard Brownian motion (Definition 4.1), we have

$$W_{t_{i+1}} - W_{t_i} \sim N\left(0, \frac{t}{n}\right) = \sqrt{\frac{t}{n}}Z$$

where Z is the standard normal distribution.

Therefore,

$$\mathbb{E}(W_{t_{i+1}} - W_{t_i})^2 = \mathbb{E}\left(\sqrt{\frac{t}{n}}Z\right)^2 = \frac{t}{n}\mathbb{E}Z^2 = \frac{t}{n}$$

$$\implies \mathbb{E}\sum_{0}^{n-1}(W_{t_{i+1}} - W_{t_i})^2 = n \times \frac{t}{n} = t$$

since $\mathbb{E}Z^2 = 1$ (Section 2.7).

To show the variance becomes vanishingly small as the interval size decreases (i.e., letting $n \to \infty$ so $\frac{t}{n} \to 0$), we first note

$$\mathbb{E}(W_{t_{i+1}} - W_{t_i})^4 = \mathbb{E}\left(\sqrt{\frac{t}{n}}Z\right)^4 = \left(\frac{t}{n}\right)^2 \mathbb{E}Z^4 = 3\left(\frac{t}{n}\right)^2 \qquad \text{from Section 2.7.}$$

Then,

$$\begin{aligned} var(W_{t_{i+1}} - W_{t_i})^2 &= \mathbb{E}(W_{t_{i+1}} - W_{t_i})^4 - (\mathbb{E}(W_{t_{i+1}} - W_{t_i})^2)^2 \\ &= 3\left(\frac{t}{n}\right)^2 - \left(\frac{t}{n}\right)^2 = 2\left(\frac{t}{n}\right)^2 \end{aligned}$$

Summing these independent terms,

$$Var\sum_0^{n-1}(W_{t_{i+1}} - W_{t_i})^2 = 2n\left(\frac{t}{n}\right)^2 = \frac{2t^2}{n} \to 0 \text{ as } n \to \infty$$

So we have the result

$$\sum_0^{n-1}(W_{t_{i+1}} - W_{t_i})^2 \to t$$

Therefore, over the time interval $[0, t]$ the limit of the sum of the squared increments in standard Brownian motion equals the length of the time interval.

$$\lim_{n\to\infty}\sum_0^{n-1}\left(W_{t_{i+1}} - W_{t_i}\right)^2 = t$$

$$\text{or } \lim_{n\to\infty}\sum_0^{n-1}(\Delta W_{t_i})^2 = t$$

We write this informally in the differential form seen in Section 5.2

$$(dW)^2 = dt$$

□

Using a recursive argument, it can be shown that cubic and higher-order variation of standard Brownian motion is zero. For example:

$$\sum_0^{n-1}\left|W_{t_{i+1}} - W_{t_i}\right|^3 \le \delta\sum_0^{n-1}(W_{t_{i+1}} - W_{t_i})^2$$

where δ is the biggest absolute increment $|W_{t_{i+1}} - W_{t_i}|$ over all n intervals. In the limit, δ becomes infinitesimally small while the sum of the squared increments becomes t from the theorem above. Hence, the product goes to zero. Now repeat the argument for the fourth, fifth powers, etc.

In general then, for $N > 2$

$$\lim_{n\to\infty} \sum_{0}^{n-1} \left| W_{t_{i+1}} - W_{t_i} \right|^N = 0 \tag{6.5}$$

Theorem 6.2. *Partition* $[0, t]$ *into n intervals then*

$$(i)\quad \lim_{n\to\infty} \sum_{0}^{n-1} (t_{i+1} - t_i)^2 = 0$$

$$(ii)\quad \lim_{n\to\infty} \sum_{0}^{n-1} (t_{i+1} - t_i)(W_{t_{i+1}} - W_{t_i}) = 0$$

Proof. This time we will make the proof more general by relaxing the restriction that the n intervals are of equal length.

(i) $$\sum_{0}^{n-1} (t_{i+1} - t_i)^2 \leq \gamma \sum_{0}^{n-1} (t_{i+1} - t_i)$$

where γ is the biggest increment $(t_{i+1} - t_i)$ over all n intervals. The sum of the intervals $\sum_0^{n-1}(t_{i+1} - t_i)$ is always t. In the limit, γ becomes infinitesimally small and the product $\gamma\, t$ then goes to zero.

(ii) $$\left| \sum_{0}^{n-1} (t_{i+1} - t_i)(W_{t_{i+1}} - W_{t_i}) \right| \leq \sum_{0}^{n-1} |(t_{i+1} - t_i)(W_{t_{i+1}} - W_{t_i})| \leq \delta \sum_{0}^{n-1} (t_{i+1} - t_i)$$

where δ is the biggest absolute increment $|W_{t_{i+1}} - W_{t_i}|$ over all n intervals. The sum of the intervals $\sum_0^{n-1}(t_{i+1} - t_i)$ is always t. In the limit, δ becomes infinitesimally small and the product $\delta\, t$ goes to zero.[4] □

In differential shorthand, we then have the other two results seen in Section 5.2

$$(dt)^2 = 0 \text{ and } dt\, dW = 0$$

6.4 Itô formula in integral form

Instead of evaluating stochastic integrals from first principles every time we can use many of the results from the previous chapter.

Taking the Taylor expansion from Section 1.10 for each time interval $0 = t_0, .., t_i, t_{i+1}, ., t_n = T$:

$$\begin{aligned}
f(W_{t_1}) - f(W_0) &= f'(W_0)(W_{t_1} - W_0) + \frac{f''(W_0)}{2!}(W_{t_1} - W_0)^2 + \ldots \\
f(W_{t_2}) - f(W_{t_1}) &= f'(W_0)(W_{t_2} - W_{t_1}) + \frac{f''(W_{t_1})}{2!}(W_{t_2} - W_{t_1})^2 + \ldots \\
\ldots & \quad \ldots \\
f(W_{t_{n-1}}) - f(W_{t_{n-2}}) &= f'(W_{t_{n-2}})(W_{t_{n-1}} - W_{t_{n-2}}) + \frac{f''(W_{t_{n-2}})}{2!}(W_{t_{n-1}} - W_{t_{n-2}})^2 + \ldots \\
f(W_T) - f(W_{t_{n-1}}) &= f'(W_{t_{n-1}})(W_T - W_{t_{n-1}}) + \frac{f''(W_{t_{n-1}})}{2!}(W_T - W_{t_{n-1}})^2 + \ldots
\end{aligned}$$

It is important to understand that the derivatives of $f(x)$ are determined with respect to x then evaluated at the values W takes at each point. They are not derivatives with respect to W.

Summing both sides:

$$f(W_T) - f(W_0) = \sum_{i=0}^{n-1} f'(W_{t_j})(W_{t_{j+1}} - W_{t_j}) + \sum_{i=0}^{n-1} \frac{f''(W_{t_j})}{2!}(W_{t_{j+1}} - W_{t_j})^2 + \ldots$$

In the limit, the first term on the right-hand side converges to the Itô integral Equation (6.1) while the second term can be shown[5] to converge to

$$\int_0^T \frac{f''(W_t)}{2} dt$$

using Theorem 6.1. And the higher-order terms can be shown to disappear using result 6.5.

We then have the Itô formula in integral form:

$$f(W_T) - f(W_0) = \int_0^T f'(W_t) dW + \int_0^T \frac{f''(W_t)}{2} dt \tag{6.6}$$

The second term on the right is what makes stochastic calculus different to ordinary calculus. Without this term, we would simply have the fundamental theorem of calculus – integration and differentiation as reverse processes.

The Itô formula can be written in the differential form as the expression we used in the previous chapter

$$df(W_t) = f'(W_t) dW + \frac{f''(W_t)}{2} dt$$

This has no clearly defined mathematical meaning. Whereas in ordinary calculus, the differential of a function was the change in the function's value due to changes in its independent variables (Section 1.8), the stochastic differential $df(W_t)$ is not well-defined since the derivative of W with respect to t is not well-defined. The Itô formula in differential form is simply a shorthand for the Itô formula in integral form which we find convenient to work with in calculations.

Exercise 6.1. *Find* $\int_0^T W_t \, dW$

Proof. From Example 5.1, we have

$$d(W_t^2) = 2W_t dW + dt$$

Using the integral form

$$\begin{aligned} W_T^2 - W_0^2 &= \int_0^T 2W_t dW + \int_0^T dt \\ \implies W_T^2 &= 2\int_0^T W_t dW + T \end{aligned}$$

so that

$$\int_0^T W_t dW = \frac{1}{2}W_T^2 - \frac{1}{2}T.$$

□

Using the rules of ordinary calculus, we would have expected to get the first term on the right-hand side. The second term is the one that is peculiar to stochastic calculus.

In some situations where we have not previously found a result using the differential form, we can rearrange the integral form to evaluate the Itô integral.

Exercise 6.2. *Find*

(i) $\int_0^T W_t^2 \, dW$

(ii) $\int_0^T e^{W_t} dW$

Proof. Rewrite Equation (6.6) as

$$\int_0^T f'(W_t)\,dW = f(W_T) - f(W_0) - \int_0^T \frac{f''(W_t)}{2}dt$$

(i) Let $f'(W_t)$ in the Itô formula be W_t^2 then

$$\int_0^T W_t^2 dW = \frac{1}{3}(W_T^3 - W_0^3) - \int_0^T \frac{2W_t}{2}dt = \frac{1}{3}W_T^3 - \int_0^T W_t dt$$

(ii) Let $f'(W_t)$ be e^{W_t} then

$$\int_0^T e^{W_t} dW = (e^{W_T} - e^{W_0}) - \int_0^T \frac{e^{W_t}}{2}dt = e^{W_T} - 1 - \frac{1}{2}\int_0^T e^{W_t}dt$$

□

6.5 Integral form of Itô processes

By similar reasoning to above, whenever we have an Itô process X_t in differential form:

$$dX = \mu(X_t, t)dt + \sigma(X_t, t)dW$$

where $\mu(X_t, t)$ and $\sigma(X_t, t)$ then we have the equivalent integral form:

$$X_T - X_0 = \int_0^T \mu(X_t, t)dt + \int_0^T \sigma(X_t, t)dW$$

We will generally change the subscripts in order to have an expression for X_t, that is

$$X_t = X_0 + \int_0^t \mu(X_s, s)ds + \int_0^t \sigma(X_s, s)dW$$

Exercise 6.3. *In the cash management problem of Constantinides and Richard (1978) a firm's cash reserve X_t follows the process*

$$dX = \mu dt + \sigma dW$$

where μ and σ are constants.

Assume X_t is expressed in millions of dollars and the units of time are quarters. If $\mu = 0.3$, $\sigma = 1.9$ and $X_0 = 2.8$ what is the probability that the firm will run out of reserves and resort to a line of credit after 3 years?

Proof. First express the reserve process in integral form

$$X_t - X_0 = \int_0^t \mu ds + \int_0^t \sigma dW$$
$$\Longrightarrow X_t - 2.8 = \int_0^t 0.3\, ds + \int_0^t 1.9\, dW$$
$$\Longrightarrow X_t = 2.8 + 0.3t + 1.9W_t \qquad \text{from Equation (6.2)}$$

Remember from Chapter 4 that W_t is normally distributed with mean zero and variance t.

Then, given $t = 4 \times 3 = 12$, Z is normally distributed with mean $2.8 + 0.3 \times 12 = 6.4$ and variance $1.9^2 \times 12 = 43.3$.

The probability the firm will run out of reserves after 3 years is

$$P(X_t < 0) = P\left(Z < \frac{0 - 6.4}{\sqrt{43.3}}\right) = 0.1654 \quad \text{or} \quad 16.54\%$$

The process

$$dX = \mu dt + \sigma dW$$

where μ and σ are constants is known as arithmetic Brownian motion.

□

The Itô process in differential form is a simple type of *stochastic differential equation* (SDE). These are differential equations in which one or more of the terms is a stochastic process. As a result, the solution X_t is also a stochastic process. We would like our solution to be explicit, that is, X_t written as a function of t and W_t only. Finding this solution is not normally as straightforward as in the previous example since the integral form may also have expressions involving X_t on the right-hand side. We will use a substitution in each of the following SDEs to get round this obstacle.

Exercise 6.4. *Consider the SDE*

$$dX = \mu X_t dt + \sigma X_t dW$$

where μ and σ are constants. Find the solution.

Proof. From Example 5.4, we have

$$d(\log X_t) = (\mu - \frac{1}{2}\sigma^2)dt + \sigma dW$$

This SDE has no expressions involving X_t on the right-hand side. The integral form is then:

$$\begin{aligned}
\log X_t - \log X_0 &= \int_0^t (\mu - \frac{1}{2}\sigma^2)ds + \sigma \int_0^t dW \\
&= (\mu - \frac{1}{2}\sigma^2)t + \sigma W_t \\
\Longrightarrow \log X_t &= \log X_0 + (\mu - \frac{1}{2}\sigma^2)t + \sigma W_t \\
\Longrightarrow X_t &= X_0 e^{(\mu - \frac{1}{2}\sigma^2)t + \sigma W_t}
\end{aligned}$$

$\{X_t\}_{t \geq 0}$ is called *geometric Brownian motion*. An important feature of the process is that it always takes positive values (assuming $X_0 > 0$). Geometric Brownian motion is the most popular model of continuous-time finance especially for modeling stock prices.

□

Exercise 6.5. *The Ornstein-Uhlenbeck process is described by the following SDE*

$$dr = \alpha(\mu - r_t)dt + \sigma dW$$

with initial value r_0 *and constants* $\alpha, \mu, \sigma > 0$. *Find the solution.*

Proof. Using the time-dependent Itô formula

$$\begin{aligned}
d(e^{\alpha t} r_t) &= \alpha e^{\alpha t} r_t dt + e^{\alpha t} dr \qquad \text{since } \frac{\partial^2 f}{\partial r^2} = 0 \\
&= \alpha e^{\alpha t} r_t dt + e^{\alpha t}\Big(\alpha(\mu - r_t)dt + \sigma dW\Big) \\
&= \Big(\alpha e^{\alpha t} r_t + e^{\alpha t}\alpha(\mu - r_t)\Big)dt + e^{\alpha t}\sigma dW \\
&= e^{\alpha t}\alpha\mu dt + \sigma e^{\alpha t} dW
\end{aligned}$$

The SDE no longer has expressions involving r_t on the right-hand side. The integral form is:

$$e^{\alpha t} r_t - r_0 = \int_0^t e^{\alpha s} \alpha \mu ds + \int_0^t \sigma e^{\alpha s} dW$$
$$\implies e^{\alpha t} r_t = r_0 + e^{\alpha t} \mu - \mu + \int_0^t \sigma e^{\alpha s} dW$$

and

$$r_t = \mu + (r_0 - \mu) e^{-\alpha t} + \sigma \int_0^t e^{\alpha(s-t)} dW$$

The Ornstein-Uhlenbeck process is used to model the short rate in the Vasicek interest rate model. The short rate is the interest rate that applies over an infinitesimally short period of time.

□

What we do if we do not know the correct substitution to make? There are a number of general approaches to take that apply depending on the form of the SDE.[6]

For example, whenever the process has the linear form

$$dX = \left(\alpha(t) X_t + \beta(t) \right) dt + \sigma(W_t, t) dW$$

then use the time-dependent Itô formula to evaluate

$$d(e^{-\int_0^t \alpha(s) ds} X_t)$$

This has the effect of removing the X_t term from the right-hand side.

In Exercise 6.5, $-\alpha$ corresponds to the $\alpha(t)$ term so we used the time-dependent Itô formula to evaluate

$$d(e^{\int_0^t \alpha ds} r_t) = d(e^{\alpha t} r_t)$$

The solution to the Ornstein-Uhlenbeck process has an Itô integral on the right-hand side. It is not obvious how we can find the statistical properties of this solution. This is the focus of the next section.

6.6 Statistical properties of the solution

The Itô integral is itself a stochastic process with statistical properties. The following theorem shows it has mean zero and a variance that is easily computed.

6.6.1 Statistical properties of Itô integral

Theorem 6.3. *If the Itô integral $\int_0^T X_t dW$ also satisfies the condition $\int_0^T \mathbb{E}X_t^2 dt < \infty$ then:*

(i) (Zero mean)

$$\mathbb{E}\left(\int_0^T X_t dW\right) = 0$$

(ii) (Isometry)

$$\mathbb{E}\left(\int_0^T X_t dW\right)^2 = \int_0^T \mathbb{E}X_t^2 dt$$

Proof. The extra condition $\int_0^T \mathbb{E}X_t^2 dt < \infty$ known as *square integrability* ensures the first and second moments of the Itô integral exist.

Recall the partition $0 = t_0, .., t_i, t_{i+1}, ., t_n = T$ and the definition of the Itô integral:

$$\lim_{n\to\infty} \sum_{i=0}^{n-1} X_{t_i}(W_{t_{i+1}} - W_{t_i}) = \int_0^T X_t dW \tag{6.7}$$

The proof proceeds by showing the zero-mean and isometry properties hold for the sum on the left-hand side and then proving that these properties hold in the limit giving the result for the integral on the right-hand side. Proving that these properties hold in the limit is not a trivial step and takes us beyond the scope of this text.[7]

(i) We can pass the expectation operator through the summation in (6.7) and use the fact that the terms are independent to obtain the zero-mean result.

$$\begin{aligned}\mathbb{E}\sum_{i=0}^{n-1} X_{t_i}(W_{t_{i+1}} - W_{t_i}) &= \sum_{i=0}^{n-1} \mathbb{E}\left(X_{t_i}(W_{t_{i+1}} - W_{t_i})\right) \\ &= \sum_{i=0}^{n-1} \mathbb{E}X_{t_i}\mathbb{E}(W_{t_{i+1}} - W_{t_i}) \\ &= 0 \qquad \text{from Definition 4.1(iii)}\end{aligned}$$

The result holds for any n. With extra work, it can be shown to hold also in the limit

$$\lim_{n\to\infty} \mathbb{E}\left(\sum_{i=0}^{n-1} X_{t_i}(W_{t_{i+1}} - W_{t_i})\right) = \mathbb{E}\left(\int_0^T X_t dW\right) = 0$$

(ii) For the isometry property, we first square out the sum on the left-hand side in Equation (6.7) gathering together the square terms and the cross terms in a double sum then pass the expectation operator through:

$$\mathbb{E}\left(\sum_{i=0}^{n-1} X_{t_i}(W_{t_{i+1}} - W_{t_i})\right)^2 = \sum_{i=0}^{n-1} \mathbb{E}\left(X_{t_i}^2(W_{t_{i+1}} - W_{t_i})^2\right)$$
$$+2\sum_{i<j} \mathbb{E}\left(X_{t_i}X_{t_j}(W_{t_{i+1}} - W_{t_i})(W_{t_{j+1}} - W_{t_j})\right)$$

The second sum disappears since it can be split into independent terms:

$$\mathbb{E}\left(X_{t_i}X_{t_j}(W_{t_{i+1}} - W_{t_i})(W_{t_{j+1}} - W_{t_j})\right) = \mathbb{E}\left(X_{t_i}X_{t_j}(W_{t_{i+1}} - W_{t_i})\right)\mathbb{E}(W_{t_{j+1}} - W_{t_j}) = 0$$

The terms are independent in the first sum also and then we use a result established before in Exercise 4.3 (ii).

$$\sum_{i=0}^{n-1} \mathbb{E}\left(X_{t_i}^2(W_{t_{i+1}} - W_{t_i})^2\right) = \sum_{i=0}^{n-1} \mathbb{E}X_{t_i}^2\,\mathbb{E}(W_{t_{i+1}} - W_{t_i})^2$$
$$= \sum_{i=0}^{n-1} \mathbb{E}X_{t_i}^2\,(t_{i+1} - t_i)$$

The result holds for any n. Extra steps are required to show it also holds in the limit. These steps are left to more advanced texts. The last term above becomes $\int_0^T \mathbb{E}X_t^2 dt$ in the limit.

$$\lim_{n\to\infty} \mathbb{E}\left(\sum_{i=0}^{n-1} X_{t_i}(W_{t_{i+1}} - W_{t_i})\right)^2 = \mathbb{E}\left(\int_0^T X_t dW\right)^2 = \int_0^T \mathbb{E}X_t^2 dt$$

□

Using the two properties from the previous theorem, we then have the expectation as well as the variance of the Itô integral

$$var\left(\int_0^T X_t dW\right) = \mathbb{E}\left(\int_0^T X_t dW\right)^2 - \left(\mathbb{E}\int_0^T X_t dW\right)^2$$
$$= \mathbb{E}\left(\int_0^T X_t dW\right)^2 = \int_0^T \mathbb{E}X_t^2\, dt$$

> **Exercise 6.6.** *Find the expectation and variance of* $\int_0^T W_t\, dW$.

Proof. We check first that square integrability applies using the result from Exercise 4.3 (i):

$$\int_0^T \mathbb{E}W_t^2 dt = \int_0^T t\, dt = \frac{1}{2}T^2 < \infty$$

The expectation is zero from the previous theorem. The variance is

$$var\left(\int_0^T W_t\, dW\right) = \int_0^T \mathbb{E}W_t^2 dt = \frac{1}{2}T^2$$

Note that by computing the variance we find that $\int_0^T \mathbb{E}W_t^2 dt$ is finite demonstrating square integrability. The check at the start was therefore unnecessary.

□

Consider the special case when X_t is deterministic, that is, non-random. The expectation of the Itô integral is zero as before but we can now drop the expectation operator in the variance expression:

$$var\left(\int_0^T X_t dW\right) = \int_0^T X_t^2\, dt$$

Furthermore, the X_{t_i} term below in the discrete version of the Itô integral is now non-random so the product $X_{t_i}(W_{t_{i+1}} - W_{t_i})$ is normally distributed. The sum

$$\sum_{i=0}^{n-1} X_{t_i}(W_{t_{i+1}} - W_{t_i})$$

is then a sum of normal variables and as a result is itself normal. Klebaner (2012) shows that in the limit the Itô integral is then normal too.

6.6.2 *Statistical properties of solutions to Itô processes*

Now that the statistical properties of the Itô integral have been established we can determine the statistical properties of solutions to Itô processes.

Exercise 6.7. *Given the Ornstein-Uhlenbeck process in Exercise 6.5, find the probability distribution of the solution.*

Proof. From before we have

$$\begin{aligned}
r_t &= \mu + (r_0 - \mu)e^{-\alpha t} + \sigma \int_0^t e^{\alpha(s-t)}\, dW \\
\implies \mathbb{E}\, r_t &= \mu + (r_0 - \mu)e^{-\alpha t} + \sigma \mathbb{E}\left(\int_0^t e^{\alpha(s-t)}\, dW \right) \\
&= \mu + (r_0 - \mu)e^{-\alpha t} && \text{from Theorem 6.3(i)} \\
\text{and } var\, r_t &= var\left(\sigma \int_0^t e^{\alpha(s-t)}\, dW \right) \\
&= \sigma^2 \int_0^t e^{2\alpha(s-t)}\, ds && \text{from Theorem 6.3(ii)} \\
&= \frac{\sigma^2}{2\alpha} e^{-2\alpha t}(e^{2\alpha t} - 1) \\
&= \frac{\sigma^2}{2\alpha}(1 - e^{-2\alpha t})
\end{aligned}$$

From the discussion above, r_t is normally distributed since the integrand $e^{\alpha(s-t)}$ in the Itô integral is deterministic.

In summary then

$$r_t \sim N\left(\mu + (r_0 - \mu)e^{-\alpha t}, \frac{\sigma^2}{2\alpha}(1 - e^{-2\alpha t}) \right)$$

□

Exercise 6.8. *Find the statistical properties of the solution to the following SDE*

$$dX = -\frac{X_t}{1-t}dt + dW$$

where $0 \leq t < 1$.

Proof. As the SDE is of the type discussed in Section 6.5, we can use the substitution

$$e^{\int_0^t \frac{1}{1-s}ds} = e^{-\log(1-t)} = \frac{1}{1-t}$$

Using the time-dependent Itô formula and finding the integral form

$$\begin{aligned} d\left(\frac{1}{1-t}X_t\right) &= \frac{X_t}{(1-t)^2}dt + \frac{1}{1-t}dX \qquad \text{from time-dependent formula} \\ &= \frac{1}{1-t}dW \\ \Longrightarrow \frac{X_t}{1-t} - X_0 &= \int_0^t \frac{1}{1-s}dW \\ \Longrightarrow X_t &= (1-t)X_0 + (1-t)\int_0^t \frac{1}{1-s}dW \end{aligned}$$

Then using Theorem 6.3

$$\begin{aligned} \mathbb{E}X_t &= (1-t)X_0 \\ \text{and } var\, X_t &= (1-t)^2 \int_0^t \frac{1}{(1-s)^2}ds \\ &= (1-t)^2 \times \left(\frac{1}{1-t} - 1\right) = t(1-t) \qquad \text{from Example 1.15} \end{aligned}$$

And X_t is normally distributed since the integrand is deterministic.

Note that at $t = 1, X_t = 0$. The process can be redefined on any interval $(0, T)$ and fixed to reach a particular value X_T by replacing $-\frac{X_t}{1-t}$ with $\frac{X_T - X_t}{T-t}$ in the SDE. This process is known as a *Brownian bridge*. Its main application in finance is in Monte Carlo simulations.

□

6.6.3 *Statistical properties without finding solution*

In the examples in the previous subsection, the statistical properties of the solution were found by first finding the solution to the SDE. It is possible to find the mean and variance of the solution without solving the SDE explicitly. The technique is demonstrated in the next example.

First, an important result known as Fubini's theorem is given without proof. This states that under quite general conditions, it is possible to pass the expectation operator through an integral of a stochastic process with respect to time.

Theorem 6.4. *If X_t satisfies the conditions given in Section 6.1, then*

$$\mathbb{E}\left(\int_0^T X_t dt\right) = \int_0^T \mathbb{E}X_t dt$$

Given an Itô process X_t in the linear form:

$$dX = \Big(\alpha(t)X_t + \beta(t)\Big)dt + \sigma(X_t, t)dW$$

then, temporarily dropping the dependency of α, β, on t and on X_t and t σ, the integral form of the process is:

$$X_t = X_0 + \int_0^t (\alpha X_s + \beta)ds + \int_0^t \sigma dW$$

Then using theorems 6.3 and 6.4

$$\begin{aligned}\mathbb{E}\,X_t &= X_0 + \mathbb{E}\int_0^t (\alpha X_s + \beta)ds \\ &= X_0 + \int_0^t (\alpha \mathbb{E}X_s + \beta)ds\end{aligned}$$

The expectation of X_t is a deterministic function of time. Letting $m_1(t) := \mathbb{E}\,X_t$:

$$m_1 = X_0 + \int_0^t (\alpha m_1 + \beta)ds$$

Differentiating both sides with respect to time and applying the first fundamental theorem of calculus (Theorem 1.1):

$$\begin{aligned}&\frac{dm_1}{dt} = \alpha m_1 + \beta \\ &\implies \frac{dm_1}{dt} - \alpha m_1 = \beta\end{aligned}$$

This is a first-order differential equation in m_1 with integrating factor $I = e^{-\int \alpha(t)dt}$ (Section 1.16)

$$\implies Im_1 = \int I\beta(t)\,dt \tag{6.8}$$

Then proceed by solving for m_1.

The approach is demonstrated using an example.

> **Exercise 6.9.** *Find the expectation of the solution to the following SDE*
>
> $$dX_t = a(b - X_t)dt + cX_t dW_t$$
>
> *where a, b, and c are constants.*

Proof. Here

$$\alpha(t) = -a \text{ and } \beta(t) = ab$$

The integrating factor is $I = e^{\int a dt} = e^{at}$.

$$\implies e^{at} m_1 = \int e^{at} ab\, dt = be^{at} + C$$

When $t = 0$, we have $m_1(0) = b + C \implies C = m_1(0) - b = X_0 - b$.

$$\implies m_1 = b + (X_0 - b)e^{-at}$$

□

We can then repeat the process to find the second moment $m_2(t) = \mathbb{E}X_t^2$ and hence find $varX_t = m_2 - m_1^2$.

Given an Itô process X_t in linear form:

$$dX = \Big(\alpha(t)X_t + \beta(t)\Big)dt + \sigma(X_t, t)dW$$

Then, temporarily dropping the dependency of α, β, and σ for clarity

$$\begin{aligned} d(X_s^2) &= 2X_s dX + (dX)^2 \\ &= 2X_s\Big((\alpha X_s + \beta)ds + \sigma dW\Big) + \sigma^2 ds \\ &= \Big(2X_s(\alpha X_s + \beta) + \sigma^2\Big)ds + 2X_s\sigma dW \end{aligned}$$

$$\implies X_t^2 - X_0^2 = \int_0^t \Big(2X_s(\alpha X_s + \beta) + \sigma^2\Big)dt + \int_0^t 2X_t\sigma dW$$

Letting $m_2(t) := \mathbb{E}X_t^2$

$$m_2 = X_0^2 + \mathbb{E}\int_0^t \left(2X_s(\alpha X_s + \beta) + \sigma^2\right)dt$$
$$= X_0^2 + \int_0^t \left(2\alpha m_2 + 2\beta m_1 + \mathbb{E}\sigma^2\right)dt$$

Differentiating both sides with respect to time

$$\frac{dm_2}{dt} = 2\alpha(t)m_2 + 2\beta(t)m_1 + \mathbb{E}\sigma^2(X_t, t) \tag{6.9}$$

This is a first-order differential equation. Now proceed by solving for m_2.

6.6.4 *Differential form and the martingale property*

The Itô integral has the important property that it is a martingale. When the Itô process is in integral form, the Itô integral part only contributes randomness to the solution. This will become relevant when considering risk-neutral pricing of derivatives later.

Theorem 6.5. *The Itô integral $\int_0^T X_t dW$ satisfying the condition $\int_0^T \mathbb{E}X_t^2 dt < \infty$ is a martingale.*

Proof. The proof is similar to Theorem 6.3. Once again the result is proven for the sum in the Itô integral and then shown to hold in the limit.

Use the partition $0 = t_0, .., t_i, t_{i+1}, ., t_n = T$ and the definition of the Itô integral:

$$\lim_{n\to\infty}\sum_{i=0}^{n-1} X_{t_i}(W_{t_{i+1}} - W_{t_i}) = \int_0^T X_t dW$$

First, we will show that for any t_i[8]

$$\mathbb{E}_{t_i}\sum_{j=0}^{n-1} X_{t_j}(W_{t_{j+1}} - W_{t_j}) = \sum_{j=0}^{j=i-1} X_{t_j}(W_{t_{j+1}} - W_{t_j})$$

which in the limit becomes

$$\mathbb{E}_{t_i}\int_0^T X_t dW = \int_0^{t_i} X_t dW$$

For any t_i

$$\mathbb{E}_{t_i}\sum_{j=0}^{n-1} X_{t_j}(W_{t_{j+1}} - W_{t_j}) = \mathbb{E}_{t_i}\left(\sum_{j=0}^{j=i-1} X_{t_j}(W_{t_{j+1}} - W_{t_j}) + \sum_{j=i}^{n-1} X_{t_j}(W_{t_{j+1}} - W_{t_j})\right)$$
$$= \sum_{j=0}^{j=i-1} X_{t_j}(W_{t_{j+1}} - W_{t_j}) + \mathbb{E}_{t_i}\left(\sum_{j=i}^{n-1} X_{t_j}(W_{t_{j+1}} - W_{t_j})\right)$$

We have dropped the expectation operator in the first term because X at timepoints before t_i is known.

The second term is zero since

$$\mathbb{E}_{t_i}\sum_{j=i}^{n-1} X_{t_j}(W_{t_{j+1}} - W_{t_j}) = \sum_{j=i}^{n-1} \mathbb{E}_{t_i}\left(X_{t_j}(W_{t_{j+1}} - W_{t_j})\right)$$
$$= \sum_{j=i}^{n-1} \mathbb{E}_{t_i}\mathbb{E}_{t_j}\left(X_{t_j}(W_{t_{j+1}} - W_{t_j})\right) \qquad \text{from Equation (4.2)}$$
$$= \sum_{j=i}^{n-1} \mathbb{E}_{t_i} X_{t_j}\mathbb{E}_{t_j}(W_{t_{j+1}} - W_{t_j}) \qquad \text{since } X_{t_j} \text{ known at } t_j$$
$$= 0 \qquad \text{from Definition 4.1(iii)}$$

So

$$\mathbb{E}_{t_i}\sum_{j=0}^{n-1} X_{t_j}(W_{t_{j+1}} - W_{t_j}) = \sum_{j=0}^{j=i-1} X_{t_j}(W_{t_{j+1}} - W_{t_j})$$

It can be shown that the martingale property continues to hold in the limit. Once again this is not a trivial step and the reader is referred to more advanced texts such as Oksendal (2013) or Shreve (2004) for details.

The finiteness condition for the martingale property follows from square integrability. □

Given an Itô process X_t in differential form:

$$dX = \mu(X_t, t)dt + \sigma(X_t, t)dW$$

we can show that if there is no drift term in the differential form i.e. $\mu(X_t, t) = 0$ then the process is a martingale.

The Itô process X in differential form is now:

$$dX = \sigma(X_t, t)dW$$

In integral form, we have:

$$\begin{aligned} X_T &= X_0 + \int_0^T \sigma(X_t, t)dW \\ \implies \mathbb{E}_t X_T &= X_0 + \int_0^t \sigma(X_t, t)dW \qquad \text{from Theorem 6.5} \\ &= X_t \end{aligned}$$

The converse proposition – if the process is a martingale, then it has no drift – will be shown later in Section 10.1.

We will assume from now on that X is a martingale is equivalent to the process having no drift.

6.7 Simulation

6.7.1 Arithmetic Brownian motion

The arithmetic Brownian motion process $dX = \mu dt + \sigma dW$ was seen in Example 6.3.

We can simulate arithmetic Brownian motion in R. As before, break the time interval from $t = 0$ to $t = 4$ into small steps of $dt = 0.01$.

$$\begin{aligned} \text{T} &< -4 \\ \text{n} &< -400 \\ \text{dt} &< -T/n \\ \text{t} &< -c\,(1:n) \\ \text{t} &< -t * dt \end{aligned}$$

We then model dX. First generate a column vector of normally distributed random numbers $dW = N(0, \sqrt{dt})$. Then construct $dX = \mu dt + \sigma dW$.

```
dW    <- rnorm(n, 0, sqrt(dt))
mu     <- 0.15
sigma <- 0.1
dX    <- mu*dt + sigma* dW
```

Now plot the results.

```
Xt      <- cumsum(dX)
plot(t,Xt,type="l",xlab="Time",ylab="X_t")
```

Simulation of arithmetic Brownian motion

```
T      <- 4
n      <- 400
dt     <- T/n
t      <- c(1:n)
t      <- t*dt
dW     <- rnorm(n, 0, sqrt(dt))
mu     <- 0.15
sigma  <- 0.1
dX     <- mu*dt + sigma* dW
Xt     <- cumsum(dX)
plot(t,Xt,type= "l",xlab= "Time",ylab= "X_t")
```

Arithmetic Brownian motion in integral form is

$$X_t = X_0 + \int_0^t \mu ds + \int_0^t \sigma dW = X_0 + \mu t + \sigma W_t$$

Since $W_t \sim N(0,t)$, then X_t has distribution $N(\mu t, \sigma^2 t)$ from the statistical properties of random variables given in Table 2.1. We notice that the mean and variance of Y_t will change with t given any particular μ and σ.

Some realizations of arithmetic Brownian motion are given in Figure 6.2. They can be seen to follow a linear trend due to the μt part with varying degrees of superimposed additional randomness due to the σW_t part.

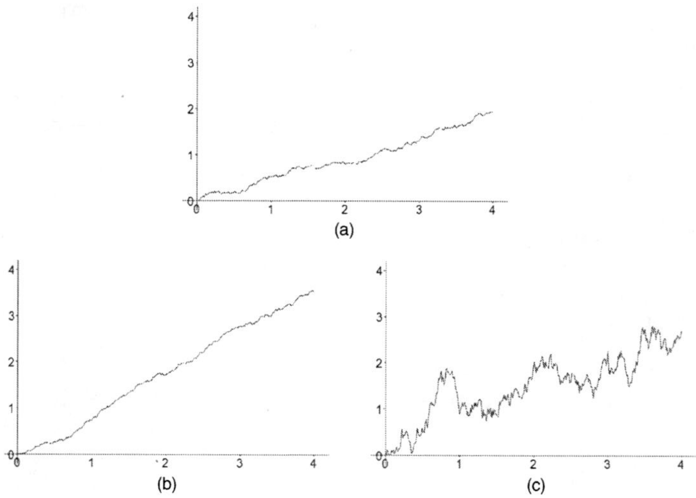

Figure 6.2 Some realizations of arithmetic Brownian motion. (a) $\mu = 0.15, \sigma = 0.1$ (b) $\mu = 0.4, \sigma = 0.1$ (c) $\mu = 0.4, \sigma = 0.6$

Note how they change as the parameters μ and σ change. The coefficient on the t term increases from $\mu = 0.15$ in Figure 6.2(a) to $\mu = 0.4$ in Figure 6.2(b) resulting in the mean of Y_t increasing from $0.15t$ to $0.4t$ at every time t. So the graph gets steeper.

Moving from Figure (b) to (c), the coefficient on the W_t term increases from $\sigma = 0.1$ to $\sigma = 0.6$ resulting in the variance of Y_t increasing from 0.1^2t to 0.6^2t at every time t. So the graph gets noisier.

As already explained, standard Brownian motion in practice does not tend to usefully describe stock behavior due to the possibility of negative values. But arithmetic Brownian motion also has the limitation that the process can become negative where μ and σ have been chosen to reflect real-world data.

6.7.2 *Geometric Brownian motion*

The geometric Brownian motion process $dX = \mu X_t dt + \sigma X_t dW$ was seen in Exercise 6.4. The simulation is coded in R similar to before.

The only complication is that the difference term dX depends on the level X_t. This is achieved by using a loop in R.

Simulation of geometric Brownian motion

```
T      <- 4
n      <- 400
dt     <- T/n
t      <- c(1:n)
t      <- t*dt
dW     <- rnorm(n, 0, sqrt(dt))
mu     <- 0.05
sigma  <- 0.20
X0     <- 0.5
Xt     <- c(X0)
for (i in 2:n) {
          Xt[i] <- Xt[i-1] + mu*Xt[i-1]*dt + sigma*Xt[i-1]*
                dW[i-1]
}
plot(t,Xt,type="l",xlab="Time",ylab="X_t")
```

The solution to geometric Brownian motion was found to be

$$X_t = X_0 e^{(\mu - \frac{1}{2}\sigma^2)t + \sigma W_t}$$

We see from this expression that positive X_0 implies X_t is always positive since the exponential function is always positive. In Chapter 7, the expected growth rate is shown to be $e^{\mu t}$ and so we will have exponential growth as long as $\mu > 0$ as is typically the case. Once again the addition of a W_t term gives us a degree

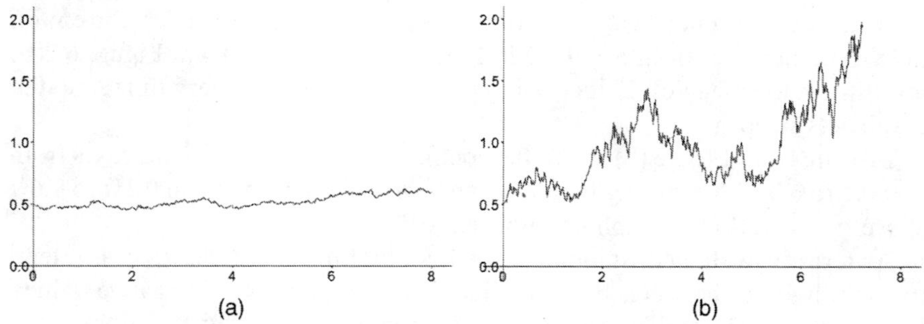

Figure 6.3 Some realizations of geometric Brownian motion. (a) Low growth, low volatility (b) High growth, high volatility

of noisy variation around this growth trend. We will refer to the parameters of the model as *expected return* μ and *volatility* σ when discussing the geometric Brownian process.

Stocks are presumed to grow exponentially over time and this feature cannot be captured by arithmetic Brownian motion due to its linearity. The graph in Figure 6.4 shows how the S&P 500 has moved over time. We notice that there has been long-term exponential growth in this composite of US stocks. Graphs of many individual stocks also behave similarly at least in periods of economic stability.

One simple explanation for exponential growth is that companies reinvest earnings from their assets into purchasing more assets which generate further earnings which are then reinvested and so on. That process is by nature exponential.

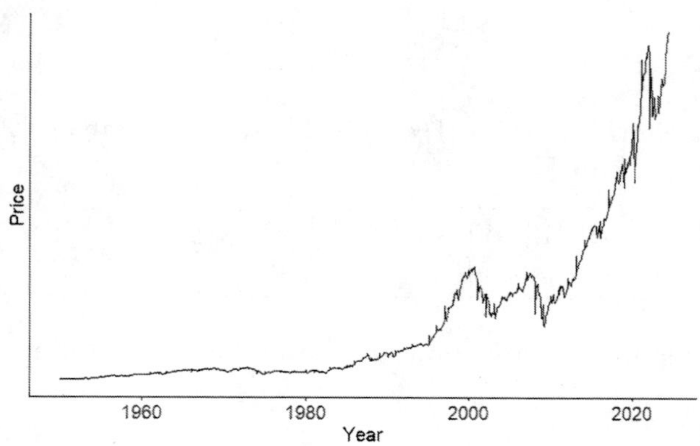

Figure 6.4 S&P 500 since 1950. *Note*: Real total return price. Updated from Shiller (2015)

Geometric Brownian motion is the most common model for stock prices and forms the basis of the famous Black–Scholes model for pricing options. We will see it again and again throughout this book.

6.7.3 *Ornstein-Uhlenbeck process*

The Ornstein-Uhlenbeck process

$$dr = \alpha(\mu - r_t)dt + \sigma dW$$

was seen in Exercise 6.5. The simulation is coded below.

Simulation of the Ornstein-Uhlenbeck process

```
T       <- 4
n       <- 400
dt      <- T/n
t       <- c(1:n)
t       <- t*dt
dW      <- rnorm(n, 0, sqrt(dt))
alpha   <- 5
mu      <- 0.05
sigma   <- 0.02
r0      <- 0.10
rt      <- c(r0)
for (i in 2:n) {
        rt[i] <- rt[i-1] + alpha*(mu-rt[i-1])*dt +
                            sigma*dW[i-1]
}
plot(t,rt,type="l",xlab="Time",ylab= "r_t")
```

This model is used commonly to model interest rates, currency exchange rates and commodity prices as well as price-earning ratios and inflation.

The solution for r_t was found to have the distribution:

$$r_t \sim N\left(\mu + (r_0 - \mu)e^{-\alpha t}, \frac{\sigma^2}{2\alpha}(1 - e^{-2\alpha t})\right)$$

The term $(r_0 - \mu)e^{-\alpha t}$ will decline to zero as t gets large. So, in the long-run, the process r_t will move to a mean level given by μ. This feature is known as *mean reversion*. The parameter α determines the speed of adjustment to the long-run mean level with large α corresponding to a rapid move to μ. The variance term superimposes a degree of noise on this process with value $\frac{\sigma^2}{2\alpha}$ in the long-run.

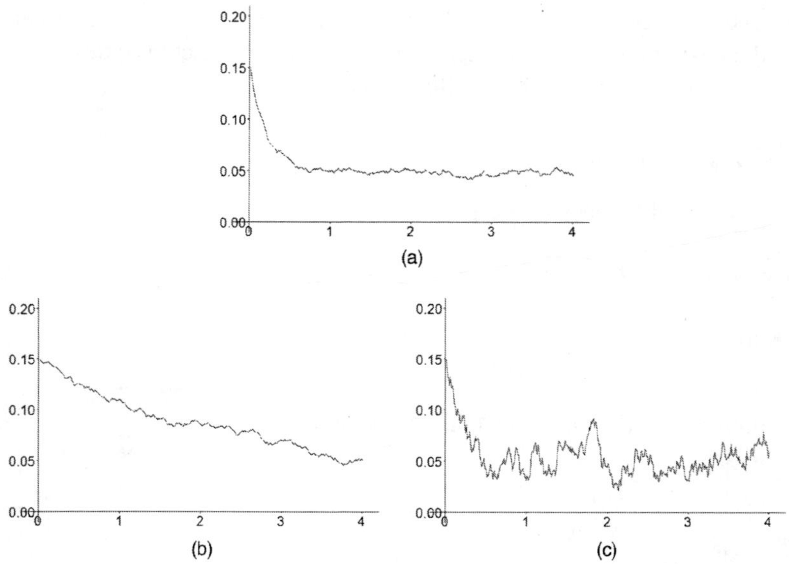

Figure 6.5 Some realizations of the Ornstein-Uhlenbeck process ($r_0 = 10\%, \mu = 5\%$). (a) Fast adjustment (b) Slow adjustment (c) Fast adjustment with high σ

This model is attractive for interest rates as many models in economics and finance make the assumption of constant interest rates over time not least the Black–Scholes model.[9] However, empirical evidence for mean reversion has been mixed with, for example, Chan et al. (1992) finding only weak evidence in the case of US short-term rates whereas Dahlquist (1996) report mean reversion playing an important role in European interest-rate dynamics.

6.8 Summary

This chapter extends ordinary calculus to the stochastic setting, where functions depend on random processes such as Brownian motion. The key development is the Itô integral, which allows integration with respect to a stochastic process and forms the foundation of stochastic calculus and SDEs used throughout quantitative finance.

The Itô integral of X with respect to W is the limit of a sum of the random increments $X_t \Delta W$. It is itself a random variable with zero expected value and a variance term expressed as an integral with respect to time. Armed with this knowledge, the statistical properties of solutions to SDEs can be determined.

The martingale property of the Itô integral was established. This implies that Itô processes with no drift term are martingales and the converse is also true – if an Itô process is a martingale then it has no drift term. This will prove to be a very useful result.

The chapter concluded with simulations of three important processes and their graphical properties.

Key things to remember

The following results will be useful throughout the course

- Given X_t a function of time t only then from Section 6.6.1

$$\int_0^T X_t dW \sim N(0, \sigma^2) \text{ where } \sigma^2 = \int_0^T X_t^2 dt$$

- The solution to Geometric Brownian Motion process: Exercise 6.4.
- The solution to the Ornstein-Uhlenbeck process (Exercise 6.5) comes up again in Chapter 14 on interest rate derivatives.
- An Itô process with no drift is a martingale: Section 6.6.4.

Further reading

Klebaner (2012) offers an accessible and example-driven introduction, well-suited for readers building intuition and practical understanding. For a more rigorous and finance-oriented treatment, Shreve (2004) presents a detailed account of stochastic integration and its role in continuous-time asset pricing. Oksendal (2013) remains a classic reference, combining mathematical precision with clear exposition and numerous applied examples. Finally, Calin (2015) provides a more intuitive narrative with a wealth of worked examples.

Notes

1. That is, the *Itô integral* defined in this way always provides a unique result for a given stochastic process X.
2. The other condition (progressive) explained in Klebaner (2012).
3. We do not require the partition points to be equally spaced but it is assumed here for convenience. The more general proof follows the same logic. See Shreve (2004) pp.100–103.
4. Given $[0, t]$ partitioned into n intervals, the expression

$$\lim_{n\to\infty} \sum_0^{n-1} (t_{i+1} - t_i)(W_{t_{i+1}} - W_{t_i})$$

   is known as the *quadratic covariation* of t and W_t.
5. See Klebaner (2012) theorem 4.14.
6. Chapter 8 of Calin (2015) provides details with plenty of clear worked examples.
7. See Klebaner (2012) or Oksendal (2013) for further details
8. Shreve (2004) shows the more general result $\mathbb{E}_t \int_0^T X_t dW = \int_0^t X_t dW$ where $t \in [t_i, t_{i+1})$
9. Kim and Ji (2011) also identify the Solow growth model; consumption-based capital asset pricing models and the Fisher equation.

Questions

1 Find $\int_0^T W_t^3 dW$.

Solution

$$\int_0^T W_t^3 dW = \frac{1}{4}(W_T^4 - W_0^4) - \int_0^T \frac{3W_t^2}{2} dt = \frac{1}{4}W_T^4 - \frac{3}{2}\int_0^T W_t^2 dt$$

2 Apply the time-dependent Itô formula to find $d(tW_t)$. Use your result to evaluate

$$\int_0^T (W_t^2 - t)dW$$

Solution

$$d(tW_t) = W_t dt + t dW \implies TW_T = \int_0^T W_t dt + \int_0^T t dW$$

Rearranging and using the result from Exercise 6.2(i)

$$\begin{aligned}\int_0^T (W_t^2 - t)dW &= \int_0^T W_t^2 dW - \int_0^T t\,dW \\ &= \left(\frac{1}{3}W_T^3 - \int_0^T W_t dt\right) - \left(TW_T - \int_0^T W_t dt\right) \\ &= \frac{1}{3}W_T^3 - TW_T\end{aligned}$$

3 A firm's cash reserve X_t follows the process

$$dX = \mu dt + \sigma dW$$

where $\mu = 0.3$, $\sigma = 1.1$ and $X_0 = 1.7$. Assume X_t is expressed in millions of dollars and the units of time are years.

(a) What is the probability distribution of the cash reserve after 3 years?

(b) At the beginning of year 4, revenues stabilize at a higher level and the parameters of the process become $\mu = 0.5$ and $\sigma = 0.7$. What is the probability distribution of the cash reserve after 5 years?

Solution

(a)

$$\begin{aligned} X_t &= X_0 + \int_0^t \mu ds + \int_0^t \sigma dW \\ &= 1.7 + 0.3t + 1.1W_t \\ \implies X_3 &\sim N(2.6, 3.63) \end{aligned}$$

(b)

$$\begin{aligned} X_5 - X_3 &= 0.5t + 0.7W_t \\ \implies X_5 - X_3 &\sim N(1, 0.98) \end{aligned}$$

So $X_5 \sim N(2.6 + 1\,,\, 3.63 + 0.98) = N(3.6, 4.61)$ since the two stages are independent.

4 The Itô process, X_t, is defined by the SDE:

$$dX = \frac{1}{2}X_t(1 - X_t)(1 - 2X_t)dt - X_t(1 - X_t)dW$$

where $X_0 = 0.5$. By applying Itô's formula to

$$\log\left(\frac{1}{X_t} - 1\right)$$

solve for X_t.

Solution

$$\begin{aligned} d\left(\log(\frac{1}{X_t} - 1)\right) &= \left(\frac{1}{\frac{1}{X_t} - 1}\right)\frac{-1}{X_t^2}dX + \frac{1}{2}\frac{1 - 2X_t}{(X_t - X_t^2)^2}(dX)^2 \\ &= \frac{-1}{X_t - X_t^2}\left(\frac{1}{2}X_t(1 - X_t)(1 - 2X_t)dt - X_t(1 - X_t)dW\right) + \\ &\quad \frac{1}{2}\frac{1 - 2X_t}{X_t^2(1 - X_t)^2}X_t^2(1 - X_t)^2 dt \\ &= \frac{-1}{X_t(1 - X_t)}\left(\frac{1}{2}X_t(1 - X_t)(1 - 2X_t)dt - X_t(1 - X_t)dW\right) + \frac{1}{2}(1 - 2X_t)dt \\ &= -\frac{1}{2}(1 - 2X_t)dt + dW + \frac{1}{2}(1 - 2X_t)dt \end{aligned}$$

$$= dW$$

Therefore

$$\log\left(\frac{1}{X_t} - 1\right) - \log\left(\frac{1}{0.5} - 1\right) = W_t$$

$$\implies X_t = \frac{1}{1 + e^{W_t}}$$

5 Solve the linear SDE defined by

$$dX = (2X_t + 1)dt + e^{2t}dW$$

by applying the time-dependent Itô formula to $e^{-2t}X_t$.

Solution

$$\begin{aligned} d(e^{-2t}X_t) &= -2e^{-2t}X_t dt + e^{-2t}dX \\ &= -2e^{-2t}X_t dt + e^{-2t}\Big((2X_t + 1)dt + e^{2t}dW\Big) \\ &= e^{-2t}dt + dW \\ \implies e^{-2t}X_t - X_0 &= \int_0^t e^{-2s}ds + \int_0^t dW \\ &= -\frac{1}{2}(e^{-2t} - 1) + W_t \\ \implies X_t &= e^{2t}X_0 + \frac{1}{2}(e^{2t} - 1) + e^{2t}W_t \end{aligned}$$

6 Bolder (2001) estimated the following short-rate model using ten years of data on Canadian zero-coupon rates

$$dr = 0.06(0.056 - r_t)dt + 0.02dW$$

Find the probability distribution of r_t at time 10 given $r_0 = 0.05$.

Solution This is an Ornstein-Uhlenbeck process with $\alpha = 0.06, \mu = 0.056, \sigma = 0.02$.

$$r_t \sim N\left(\mu + (r_0 - \mu)e^{-\alpha t}, \frac{\sigma^2}{2\alpha}(1 - e^{-2\alpha t})\right) = N(0.0527, 0.0023)$$

7 Find the probability distribution of

(a) $\int_0^1 t dW$

(b) $\int_0^T \sqrt{t} dW$

Solution

(a) $\mathbb{E}(\int_0^1 t dW) = 0$ and $var(\int_0^1 t dW) = \int_0^1 t^2 dt = \frac{1}{3}$. So $\int_0^1 t dW \sim N(0, \frac{1}{3})$ since the integrand is deterministic.

(b) $\mathbb{E}(\int_0^T \sqrt{t} dW) = 0$ and $var(\int_0^T \sqrt{t} dW) = \int_0^T t dt = \frac{1}{2}T^2$. So $\int_0^T \sqrt{t} dW \sim N(0, \frac{1}{2}T^2)$ since the integrand is deterministic.

8 Given the SDE

$$dX = e^{-\frac{t}{2}} dt + e^{-\frac{t}{2} + W_t} dW, \quad X_0 = 0$$

Solve for X_t and find its mean.

Solution

$$X_t = \int_0^t e^{-\frac{s}{2}} ds + \int_0^t e^{-\frac{s}{2} + W_s} dW$$

$$= 2(1 - e^{-\frac{t}{2}}) + \int_0^t e^{-\frac{s}{2} + W_s} dW$$

$$\Longrightarrow \mathbb{E}X_t = 2(1 - e^{-\frac{t}{2}})$$

9 You are given the following SDE:

$$dX = 0.06X_t dt + 0.25X_t dW \text{ with } X_0 = 1$$

Calculate $\mathbb{E}(\log X_t)$ using Itô's formula.

[QFIQF Spring 2019 Q1(c)]

Solution This is geometric Brownian motion (Exercise 6.4)

$$\log X_t = \log X_0 + (\mu - \frac{1}{2}\sigma^2)t + \sigma W_t$$
$$\implies \mathbb{E}(\log X_t) = \log X_0 + (\mu - \frac{1}{2}\sigma^2)t = (0.06 - \frac{1}{2}0.25^2)t = 0.02875t$$

10 Show that $Z_t = \int_0^t W_s\, ds$ is a normally distributed random variable with mean 0 and variance $t^3/3$.

Solution from Question 6.8, we have

$$tW_t = \int_0^t W_s ds + \int_0^t sdW$$
$$\implies \int_0^t W_s ds = tW_t - \int_0^t sdW$$
$$= \int_0^t tdW - \int_0^t sdW = \int_0^t (t-s)dW$$

As the integrand is deterministic, this is normal with mean 0 and variance $= \int_0^t (t-s)^2 ds = -1/3(t-s)^3\Big|_0^t = t^3/3$.

11 Use Itô's formula to find an expression for $\mathbb{E}W_t^4$.

Solution

Starting from Exercise 5.2, we have

$$d(W_t^4) = 4W_t^3 dW + 6W_t^2 dt$$
$$\implies W_t^4 = 4\int_0^t W_s^3 dW + 6\int_0^t W_s^2 ds$$

$$\begin{aligned}\implies \mathbb{E}W_t^4 &= 6\mathbb{E}\int_0^t W_s^2 ds\\ &= 6\int_0^t \mathbb{E}(W_s^2)\, ds\\ &= 6\int_0^t s\, ds = 6\left(\frac{1}{2}t^2\right) = 3t^2\end{aligned}$$

12 Apply Itô's formula to find $d(W_t e^{W_t})$ and hence find $\mathbb{E}(W_t e^{W_t})$.

Solution

$$\begin{aligned}d(W_t e^{W_t}) &= (e^{W_t} + W_t e^{W_t})dW + \frac{1}{2}(W_t e^{W_t} + 2e^{W_t})dW_t^2\\ &= (\frac{1}{2}W_t e^{W_t} + e^{W_t})dt + (e^{W_t} + W_t e^{W_t})dW\end{aligned}$$

Since $\mathbb{E}(e^{W_t}) = e^{\frac{t}{2}}$ (Equation 2.9) we have

$$\begin{aligned}W_t e^{W_t} &= \int_0^t (\frac{1}{2}W_s e^{W_s} + e^{W_s})ds + \int_0^t (e^{W_s} + W_s e^{W_s})dW\\ \implies \mathbb{E}(W_t e^{W_t}) &= \int_0^t \frac{1}{2}\mathbb{E}(W_s e^{W_s}) + \mathbb{E}e^{W_s}\, ds\\ \implies m &= \int_0^t \frac{1}{2}m + e^{\frac{s}{2}}\, ds\\ \implies \frac{dm}{dt} - \frac{1}{2}m &= e^{\frac{t}{2}}\\ \implies e^{\frac{-t}{2}}m &= \int e^{\frac{-s}{2}} e^{\frac{s}{2}}\, dt = t + C\\ \implies m &= e^{\frac{t}{2}}t\end{aligned}$$

13 Let X_t be the process such that

$$dX = \mu X_t dt + \sigma X_t dW$$

Apply Itô's formula to find $d(X_t^n)$ and hence find $\mathbb{E}(X_t^n)$.

Solution

$$d(X_t^n) = (n\mu + \frac{1}{2}n(n-1)\sigma^2)X_t^n dt + n\sigma X_t^n dW$$

Letting $Y_t = X_t^n$ this process is of the form discussed in Section 6.6.3 with
$\alpha_t = (n\mu + \frac{1}{2}n(n-1)\sigma^2)$ and $\beta_t = 0$.

$$dY = (n\mu + \frac{1}{2}n(n-1)\sigma^2)Y_t dt + n\sigma Y_t dW$$

$$\implies e^{-(n\mu+\frac{1}{2}n(n-1)\sigma^2)t}m = C$$

When $t = 0$, we have $m(0) = C \implies C = X_0^n$

$$\implies m = \mathbb{E}(X_t^n) = X_0^n e^{n(\mu+\frac{1}{2}(n-1)\sigma^2)t}$$

14 Find the variance of the solution to the SDE given in Exercise 6.9 by first writing an expression for $d(X_t^2)$ and then solving a differential equation in $m_2(t) = \mathbb{E}X_t^2$.

Solution
Starting from Equation (6.9) with

$$\alpha(t) = -a\,,\ \beta(t) = ab \text{ and } \sigma(X_t, t) = cX_t$$

$$\frac{dm_2}{dt} = -2am_2 + 2abm_1 + c^2m_2$$

$$\implies \frac{dm_2}{dt} + (2a - c^2)m_2 = 2abm_1$$

This is a first-order differential equation in m_2 with integrating factor $I = e^{(2a-c^2)t}$

$$\implies Im_2 = \int I(2\alpha\mu m_1 + \sigma^2)ds$$

$$\implies e^{(2a-c^2)t}m_2 = \int e^{(2a-c^2)s}\,2ab\left(b + (X_0 - b)e^{-as}\right)ds$$

$$\implies e^{(2a-c^2)t}m_2 = \frac{2ab^2e^{(2a-c^2)t}}{2a-c^2} + \frac{2ab(X_0-b)e^{(a-c^2)t}}{a-c^2} + C$$

$$\implies e^{(2a-c^2)t}m_2 = \frac{2ab^2e^{(2a-c^2)t}}{2a-c^2} + \frac{2ab(X_0-b)e^{(a-c^2)t}}{a-c^2} + X_0^2 - \frac{2ab^2}{2a-c^2}$$
$$-\frac{2ab(X_0-b)}{a-c^2}$$

$$\implies m_2 = e^{-(2a-c^2)t}\left(X_0^2 + 2ab\left(b\left(\frac{e^{(2a-c^2)t}-1}{2a-c^2}\right) + (X_0-b)\left(\frac{e^{(a-c^2)t}-1}{a-c^2}\right)\right)\right)$$

The variance is then $var X_t = m_2 - m_1^2$, where m_1 is given in Exercise 6.9.

15 The process defined by

$$dr = \alpha(\mu - r_t)dt + \sigma\sqrt{r_t}dW$$

where $\sigma, \mu > 0$ is known as the Feller square root process. It is used to model the short rate in the Cox–Ingersoll–Ross interest rate model. Find $\mathbb{E}r_t$ and $var\, r_t$.

Solution

Finding the solution first. This has the general linear form in Section 6.5 so

$$d(e^{\alpha t}r_t) = \alpha e^{\alpha t}\mu dt + \sigma e^{\alpha t}\sqrt{r_t}dW$$

$$\implies e^{\alpha t}r_t - r_0 = \int_0^t \alpha e^{\alpha s}\mu ds + \int_0^t \sigma e^{\alpha s}\sqrt{r_s}dW$$

$$\implies r_t = e^{-\alpha t}r_0 + \mu(1 - e^{-\alpha t}) + e^{-\alpha t}\int_0^t \sigma e^{\alpha s}\sqrt{r_t}dW$$

Hence

$$\mathbb{E}r_t = \mu + e^{-\alpha t}(r_0 - \mu)$$

and

$$\begin{aligned}
var\, r_t &= e^{-2\alpha t} var\left(\int_0^t \sigma e^{\alpha s}\sqrt{r_s}dW\right) \\
&= \sigma^2 e^{-2\alpha t}\int_0^t e^{2\alpha s}\mathbb{E}r_s ds \qquad \text{from 6.3(ii)} \\
&= \sigma^2 e^{-2\alpha t}\int_0^t \left((r_0 - \mu)e^{\alpha s} + \mu e^{2\alpha s}\right)ds \\
&= \sigma^2 e^{-2\alpha t}\left(\frac{r_0 - \mu}{\alpha}(e^{\alpha t} - 1) + \frac{\mu}{2\alpha}(e^{2\alpha t} - 1)\right)
\end{aligned}$$

Alternatively using the approach of Section 6.6.3. Starting from Equation (6.8) with

$$\alpha(t) = -\alpha\,,\ \beta(t) = \alpha\mu \text{ and } \sigma(r_t, t) = \sigma\sqrt{r_t}$$

Integrating factor is $I = e^{\alpha t}$ and

$$e^{at} m_1 = \int e^{\alpha t} \alpha \mu \, dt = \mu e^{\alpha t} + C$$

$$\implies \mathbb{E} r_t = m_1 = \mu + (r_0 - \mu) e^{-\alpha t}$$

Starting from Equation (6.9)

$$\frac{dm_2}{dt} = -2\alpha m_2 + 2\alpha \mu m_1 + \sigma^2 m_1$$

$$\implies \frac{dm_2}{dt} + 2\alpha m_2 = (2\alpha\mu + \sigma^2) m_1$$

This is a first-order differential equation in m_2 with integrating factor $I = e^{2\alpha t}$

$$\implies I m_2 = \int I(2\alpha\mu + \sigma^2) m_1 \, ds$$

$$\implies e^{2\alpha t} m_2 = (2\alpha\mu + \sigma^2) \int e^{2\alpha s} \Big(\mu + (r_0 - \mu) e^{-\alpha s} \Big) ds$$

$$= (2\alpha\mu + \sigma^2) \left(\frac{\mu(e^{2\alpha t} - 1)}{2\alpha} + \frac{(r_0 - \mu)(e^{\alpha t} - 1)}{\alpha} \right) + r_0^2$$

$$\implies m_2 = \frac{(2\alpha\mu + \sigma^2)}{\alpha} \left(\frac{\mu(1 - e^{-2\alpha t})}{2} + (r_0 - \mu)(e^{-\alpha t} - e^{-2\alpha t}) \right) + e^{-2\alpha t} r_0^2$$

$$= \sigma^2 e^{-2\alpha t} \left(\frac{\mu}{2\alpha}(e^{2\alpha t} - 1) + \frac{r_0 - \mu}{\alpha}(e^{\alpha t} - 1) \right) + \Big(\mu + (r_0 - \mu) e^{-\alpha t} \Big)^2$$

Then

$$var\, r_t = m_2 - m_1^2 = \sigma^2 e^{-2\alpha t} \left(\frac{r_0 - \mu}{\alpha}(e^{\alpha t} - 1) + \frac{\mu}{2\alpha}(e^{2\alpha t} - 1) \right)$$

16 Let

$$X_t = t^2 W_t - 2 \int_0^t s W_s ds$$

Show that X_t is a martingale.

[QFIQF Fall 2018 Q2]

Solution

$$d(t^2 W_t) = 2tW_t dt + t^2 dW$$
$$\Longrightarrow t^2 W_t = 2\int_0^t sW_s ds + \int_0^t s^2 dW$$

$X_t = t^2 W_t - 2\int_0^t sW_s ds = \int_0^t s^2 dW$ is a martingale from Theorem 6.5

17 Let X_t be the Itô process given by

$$dX = \mu(X_t, t)dt + \sigma(X_t, t)dW$$

What condition must be satisfied by a function $f(X_t, t)$ in order for it to be a martingale?

Solution From Equation (5.6)

$$df(t, X_t) = (f_t + f_x\mu + \frac{1}{2}f_{xx}\sigma^2)dt + f_x\sigma dW$$

So for $f(X_t, t)$ to be a martingale

$$f_t + f_x\mu(X_t, t) + \frac{1}{2}f_{xx}\sigma^2(X_t, t) = 0$$

18 Find a deterministic function $g(t)$ such that $W_t^3 + g(t)W_t$ is a martingale.

Solution
Apply the time-dependent Itô formula:

$$\begin{aligned} d(W_t^3 + g(t)W_t) &= g'(t)W_t dt + (3W_t^2 + g(t))dW + \frac{1}{2}6W_t dt \\ &= (g'(t)W_t + 3W_t)dt + (3W_t^2 dW + g(t))dW \end{aligned}$$

For $W_t^3 + g(t)W_t$ to be a martingale,

$$\begin{aligned} g'(t)W_t + 3W_t &= 0 \\ \implies g'(t) &= -3 \\ \implies g(t) &= -3t + C \end{aligned}$$

19 Find the expected value of $\int_0^T W_t dW$? What is the expected value at time τ of $\int_0^T W_t dW$ where $0 < \tau < T$?

Solution The Itô integral has zero mean so the expected value of $\int_0^T W_t dW$ is zero.

The Itô integral is a martingale so $\mathbb{E}_\tau\left(\int_0^T W_t dW\right) = \int_0^\tau W_t dW = \frac{1}{2}(W_\tau^2 - \tau)$ from Exercise 6.1.

20 Let $X_t = \int_0^t W_s^2 ds + \alpha(W_t, t)$.

(a) Derive a SDE for X_t using Itô's formula.

Assume from this point on that $\alpha(W_t, t) = -tW_t^2 + \beta(t)$, with $\beta(t)$ a deterministic function of time.

(b) Identify an appropriate $\beta(t)$, if it exists, that makes X_t a martingale.

[QFIQF Fall 2022 Q1]

Solution

(a) Using the time-dependent Itô formula

$$\begin{aligned} dX &= W_t^2 dt + \left(\frac{\partial \alpha}{\partial t} dt + \frac{\partial \alpha}{\partial w} dW + \frac{1}{2}\frac{\partial^2 \alpha}{\partial w^2}(dW)^2\right) \\ &= \left(W_t^2 + \frac{\partial \alpha}{\partial t} + \frac{1}{2}\frac{\partial^2 \alpha}{\partial w^2}\right) dt + \frac{\partial \alpha}{\partial w} dW \end{aligned}$$

(b) For X_t to be a martingale,

$$\begin{aligned} W_t^2 + \frac{\partial \alpha}{\partial t} + \frac{1}{2}\frac{\partial^2 \alpha}{\partial w^2} &= 0 \\ \implies W_t^2 - W_t^2 + \frac{d\beta}{dt} - t &= 0 \\ \implies \beta(t) &= \frac{1}{2}t^2 + C \end{aligned}$$

21 Demonstrate that the quadratic variation of standard Brownian motion W is equal to t.

$$\lim_{n\to\infty}\sum_{0}^{n-1}(W_{t_{i+1}} - W_{t_i})^2 = t$$

(a) Using the code below, simulate the evolution of W over the interval [0,1].

```
T     <- 1
n     <- 10
dt    <- T/n
t     <- c(1:n)
t     <- t*dt

dW    <- rnorm(n, 0, sqrt(dt))
```

(b) Calculate the sum $\sum_0^{n-1}(W_{t_{i+1}} - W_{t_i})^2$.

(c) Repeat the simulation and calculation a number of times. What is the average result? What is the variance of the result?

(d) Divide the interval into a progressively larger number of timesteps $n = 10, 100, 1000, \ldots$. Check that the sum

$$\sum_{0}^{n-1}(W_{t_{i+1}} - W_{t_i})^2$$

approaches 1 in the limit.

Solution

From Theorem 6.1 for all n

$$\mathbb{E}\sum_{0}^{n-1}(W_{t_{i+1}} - W_{t_i})^2 = t \implies \text{Expectation is 1 on interval [0,1]}$$

and

$$var\sum_{0}^{n-1}(W_{t_{i+1}} - W_{t_i})^2 = \frac{2t^2}{n} \implies \implies \text{Variance is } \frac{2}{10}$$

$$= 0.2 \text{ when } n = 10$$

```
T    <- 1
n    <- 10
```

```
dt      <- T/n
t       <- c(1:n)
t       <- t*dt

dW      <- rnorm(n, 0, sqrt(dt))
sq_dW <- dW^2
QV      <- sum(sq_dW)
print(paste("Quadratic variation is ", QV))
```

22 In Exercise 6.1, it was shown that

$$\int_0^T W_t\,dW = \frac{1}{2}W_T^2 - \frac{1}{2}T$$

Demonstrate this is the case using the code from the previous question on the interval [0,1].

Solution

Approximate the integral as a finite sum for n a large number.

$$\sum_{i=0}^{n-1} W_{t_i}\Delta W_{t_i} \approx \int_0^T W_t\,dW$$

where $\Delta W_{t_i} = W_{t_{i+1}} - W_{t_i}$

```
T       <- 1
n       <- 1000000
dt      <- T/n
t       <- c(1:n)
t       <- t*dt

dW      <- rnorm(n, 0, sqrt(dt))
Wt      <- cumsum(dW)

# integral as a finite sum
prod  <- rep(0,n)
for (i in 2:n) {
        prod[i] <- Wt[i-1]* dW[i]
}
```

```
int    <- sum(prod)

# integral value
WT     <- Wt[length(Wt)] # final value
rhs    <- 0.5*WT^2 - 0.5

print(paste("Stochastic integral of W wrt W is ", int))
print(paste("1/2 W^2 - 1/2 T is ", rhs))
```

Part III

Pricing derivatives

7 Geometric Brownian motion

Introduction

In Chapter 4, standard Brownian motion was rejected as a model of stock behavior. There were a number of issues:

- Brownian motion is certain to become negative eventually. Shares always have non-negative value.
- Changes in security prices are not independent of the current security price. Large changes are more common when the current price is high.

In Chapter 6, geometric Brownian motion was suggested as an improvement since:

- Share prices do not become negative in this model.
- The exponential growth implied by the model appears plausible in theory and is observed in practice at least for some stocks and for some time periods.

As geometric Brownian motion is widely used to model stock prices, this process will be examined in more detail in this chapter. The first sections outline results which will prove useful when tackling option pricing later in this book. Then parameter estimation based on real-world data is examined. There are a number of limitations to this model that analysts must bear in mind in applications. These are highlighted in a final section.

7.1 Working with the lognormal model

Geometric Brownian motion process for security price S_t in differential form is

$$dS = \mu S_t dt + \sigma S_t dW$$

where μ, σ are constants.

DOI: 10.4324/9781003477624-10

Solving for S_t allows the statistical properties of the security price to be determined. The solution to geometric Brownian motion at time t was found in Section 6.4 to be

$$S_t = S_0 e^{(\mu - \frac{1}{2}\sigma^2)t + \sigma W_t}$$

Rearranging and taking the logarithm of both sides

$$\log \frac{S_t}{S_0} = (\mu - \frac{1}{2}\sigma^2)t + \sigma W_t$$

Since $W_t \sim N(0, t)$, then $\sigma W_t \sim N(0, \sigma^2 t)$ and

$$\log \frac{S_t}{S_0} = \log S_t - \log S_0 \sim N((\mu - \frac{\sigma^2}{2})t, \sigma^2 t) \tag{7.1}$$

which implies

$$\log S_t \sim N(\log S_0 + (\mu - \frac{\sigma^2}{2})t, \sigma^2 t) \tag{7.2}$$

From the discussion of the lognormal distribution in Section 2.6.4, $\log S_t$ is normally distributed and so S_t has a lognormal distribution.

The approach outlined in the next example will help in the derivation of the European call and put option pricing formulas later.

Exercise 7.1. *A stock currently has price \$10. Assume the price follows geometric Brownian motion with expected return and volatility 15% and 20% respectively. What is the probability that a European put option on the stock with strike \$9 and maturity 1 month will be exercised?*

Proof. A European put option is exercised when the stock price at maturity S_T is lower than the strike price K. The task is to find the probability that

$$S_T = S_0 e^{(\mu - \frac{1}{2}\sigma^2)T + \sigma W_T} < K$$

Or rearranging, we seek the probability that

$$W_T < \frac{\log(K/S_0) - (\mu - \frac{1}{2}\sigma^2)T}{\sigma}$$

Remember that if a random variable $X \sim N(m, s^2)$ then $P(X < c) = P(Z < \frac{c-m}{s})$, where Z has the standard normal distribution.

Since $W_T \sim N(0, T)$

$$
\begin{aligned}
P\left(W_T < \frac{\log(K/S_0) - (\mu - \frac{1}{2}\sigma^2)T}{\sigma}\right) \\
&= P\left(Z < \frac{\log(K/S_0) - (\mu - \frac{1}{2}\sigma^2)T}{\sigma\sqrt{T}}\right) \\
&= P\left(Z < \frac{1}{0.2\sqrt{1/12}}\left(\log\frac{9}{10} - (0.15 - \frac{1}{2}0.2^2)1/12\right)\right) \\
&= P(Z < -2.013) \\
&= 0.022
\end{aligned}
$$

□

Exercise 7.2. *A stock currently has price \$30. Assume the price follows geometric Brownian motion with expected return and volatility 18% and 30% respectively. Calculate a 95% confidence interval for the stock price in three months, time.*

Proof.

$$
\begin{aligned}
\log S_t &\sim N(\log S_0 + \left(\mu - \frac{\sigma^2}{2}\right)t, \sigma^2 t) \\
&\sim N(\log 30 + (0.18 - \frac{1}{2}0.3^2)0.25, 0.3^2 \times 0.25) \\
&\sim N(3.435, 0.0225)
\end{aligned}
$$

95% confidence interval for log stock price

$$(3.435 \pm 1.96 \times 0.15) = (3.141, 3.729)$$

95% confidence interval for stock price

$$(23.13, 41.64)$$

□

7.2 Statistical properties

Since the solution to geometric Brownian motion at time t is in exponential form, the security price S_t at any point of time will be positive as long as S_0 is positive

$$S_t = S_0 e^{(\mu - \frac{1}{2}\sigma^2)t + \sigma W_t}$$

Share prices do not become negative in reality therefore geometric Brownian motion appears a suitable model for share price processes.

The expectation and variance of S_t are easily found.

In Section 2.7, it is shown that for a normally distributed variable $Y \sim N(m, s^2)$

$$\mathbb{E}e^{rY} = e^{rm + \frac{r^2 s^2}{2}}$$

So $\mathbb{E}e^Y = e^{m + \frac{1}{2}s^2}$ and

$$\begin{aligned} var\, e^Y &= \mathbb{E}(e^Y)^2 - (\mathbb{E}e^Y)^2 = \mathbb{E}e^{2Y} - (e^{m+\frac{1}{2}s^2})^2 = e^{2m+2s^2} - e^{2m+s^2} \\ &= (e^{s^2} - 1)e^{2m+s^2} \end{aligned}$$

Let $Y_t = \log S_t \sim N(\log S_0 + (\mu - \frac{1}{2}\sigma^2)t, \sigma^2 t)$ then substituting for $m = \log S_0 + (\mu - \frac{1}{2}\sigma^2)t$ and $s^2 = \sigma^2 t$

$$\begin{aligned} \mathbb{E}S_t = \mathbb{E}e^{Y_t} &= e^{\log S_0 + (\mu - \frac{1}{2}\sigma^2)t + \frac{\sigma^2 t}{2}} \\ &= S_0 e^{\mu t} \end{aligned} \tag{7.3}$$

The expected value of the security price is then growing exponentially over time.

Using the result above for $var(e^Y)$, the variance of S_t is

$$\begin{aligned} var(S_t) &= (e^{\sigma^2 t} - 1)e^{2\log S_0 + 2(\mu - \frac{1}{2}\sigma^2)t + \sigma^2 t} \\ &= (e^{\sigma^2 t} - 1)S_0^2 e^{2\mu t} \end{aligned} \tag{7.4}$$

The variance of S_t also grows over time as the evolution of the process becomes more uncertain.

From Equation (7.3), the expected return per unit time is μ using continuous compounding:

$$\mathbb{E}\left(\frac{S_t}{S_0}\right) = e^{\mu t}$$

Hence, our use of the label 'expected return' for the μ parameter in geometric Brownian motion model.[1]

Geometric Brownian motion has the property that returns over non-overlapping intervals are independent. For any particular realization of W, the continuously-compounded return (or *log return*) over the interval $[0, t]$ is

$$\log \frac{S_t}{S_0} = \left(\mu - \frac{1}{2}\sigma^2\right) t + \sigma W_t$$

and the log return over the interval $[t, t + \Delta t]$ is

$$\log \frac{S_{t+\Delta t}}{S_t} = \left(\mu - \frac{1}{2}\sigma^2\right) \Delta t + \sigma(W_{+\Delta t} - W_t)$$

The return over $[t, t + \Delta t]$ is independent of the return over $[0, t]$ since $W_{+\Delta t} - W_t$ is independent of W_t. Independent returns are an attractive feature as it is consistent with the Efficient Market Hypothesis which implies that past returns do not predict future returns. All known information is already priced in and future returns should be uncorrelated with the past. The market then cannot be beaten because all important information determining prices is incorporated into current share prices.

Exercise 7.3. *You are given the following information on a stock:*

μ	*10%*
r	*3%*
σ	*20%*
Stock price	*100*
Dividend rate	*0%*

Assume that the stock price S follows geometric Brownian motion:

$$dS = \mu S_t dt + \sigma S_t dW$$

where W is a standard Brownian motion.

Calculate the amount Y you need to invest in the stock at time zero so that the expected value is \$250,000 in 15 years.

[FETE - Fall 2012 Q10(a)]

Proof.

$$\mathbb{E}(S_t) = S_0 e^{\mu t} = 100 \times e^{0.1 \times 15} = \$448.17$$

To have \$250,000 in 15 years you need to purchase $\frac{250000}{448.17} = 557.8254$ shares.

The amount Y you need to invest in the stock at time zero is therefore $100 \times 557.8254 = \$55782.54$. □

7.3 Parameter estimation and simulation

(Equation 7.1) can be rewritten as

$$\log \frac{S_{t+\Delta t}}{S_t} \sim N\left(\left(\mu - \frac{1}{2}\sigma^2\right)\Delta t, \sigma^2 \Delta t\right)$$

Then given data on stock price S_t, the parameters μ and σ can be determined from the realized behavior of $\log \frac{S_{t+\Delta t}}{S_t}$. The example below shows how this is done.

Exercise 7.4. *The table below contains weekly prices for Apple Inc. (AAPL) in dollars. Assuming prices follow geometric Brownian motion estimate the expected return and volatility.*

Week	S_t	$u_t = \log \frac{S_{t+\Delta t}}{S_t}$
17/02/2025	*245.55*	*–0.0152*
24/02/2025	*241.84*	*–0.0115*
03/03/2025	*239.07*	*–0.1132*
10/03/2025	*213.49*	*0.0221*
17/03/2025	*218.27*	*–0.0017*
24/03/2025	*217.90*	*–0.1456*
31/03/2025	*188.38*	*0.0506*
07/04/2025	*198.15*	*–0.0059*
14/04/2025	*196.98*	*0.0606*
21/04/2025	*209.28*	*–0.0190*
28/04/2025	*205.35*	*–0.0338*
05/05/2025	*198.53*	*0.0621*
12/05/2025	*211.26*	—

Proof. The values of $u_t = \log \frac{S_{t+\Delta t}}{S_t}$ have already been calculated for the sample. The next step is to estimate the population parameters $(\mu - \frac{\sigma^2}{2})\Delta t$ and $\sigma^2 \Delta t$.

The best estimator for the mean of the population from which the sample is drawn is the sample mean:

$$\bar{u} = \frac{-0.0152 + \ldots + 0.0621}{12} = -0.0125$$

The best estimator for the population standard deviation is

$$\sqrt{\tfrac{1}{n-1}\sum_{i=1}^{n}(u_t-\bar{u})^2}$$

$$=\sqrt{\tfrac{1}{12}\left((-0.0152+0.0125)^2+\ldots+(0.0621+0.0125)^2\right)}=0.0637$$

The gap between data points is one week therefore $\Delta t = \frac{1}{52}$. Setting the estimators equal to the population parameters

$$\sigma\sqrt{\frac{1}{52}}=0.0637 \implies \sigma=0.459$$

and

$$\left(\mu-\frac{0.459^2}{2}\right)\left(\frac{1}{52}\right)=-0.0125 \implies \mu=-0.546$$

□

There are a number of limitations to this approach. The first limitation is that the sample mean is essentially based on two values – the starting price and end price. This can be seen by using the properties of logs:

$$\begin{aligned}\bar{u}&=\frac{\log(S_2/S_1)+\ldots+\log(S_n/S_{n_1})}{n}\\&=\frac{\log(S_2/S_1\times S_3/S_2\times\ldots\times S_{n-1}/S_{n_2}\times S_n/S_{n_1})}{n}\\&=\frac{\log(S_n/S_1)}{n}\end{aligned}$$

No matter how many data points are used the expected return estimate will be very imprecise. Brigo et al. (2009) state that the imprecision of the estimate of μ is one of the reasons for the success of risk neutral valuation. This term is replaced by the risk-free rate when modeling stock behavior under the risk-neutral probability measure.

The second limitation is that estimates are based on historic data. Whether the model is being applied to asset pricing or risk management the evolution of price over coming weeks or months is the focus. There is no guarantee that geometric Brownian motion parameters are constant over time therefore estimates using historic data may not be appropriate for modeling future prices.

An alternative approach is to calculate *implied volatilities* based on option prices. This estimate will be forward-looking as it relies on the value of stock price volatility expected over the duration of the option. An example is given below.

Exercise 7.5. *Suppose the price of a 3-month European call option on a non-dividend paying stock is \$3.16 when the current stock price is \$30, the strike price is \$28 and the risk-free rate is 10%. What volatility is implied by the option price?*

Proof. The European call option price on a non-dividend paying stock is

$$c_t = S_t\Phi(d_1) - e^{-r(T-t)}K\Phi(d_2)$$

where

$$d_1 = \frac{\log(S_t/K) + (r + \frac{1}{2}\sigma^2)(T-t)}{\sigma\sqrt{T-t}}, d_2 = d_1 - \sigma\sqrt{T-t}$$

With $S_t = 30, K = 28, r = 0.1$ and $T - t = 0.25$ we could try various values for σ to get close to the answer

- For $\sigma = 0.20$, $c_t = 30 \times \Phi(0.99) - e^{-0.1(0.25)} \times 28 \times \Phi(0.89) = 2.958$. This is too low.
- For $\sigma = 0.30$, $c_t = 3.384$. This is too high.

And keep on trying until we get to the answer $\sigma = 0.25$.

This can be done quickly in Excel. Set up the call option pricing formula in excel with the input values in separate cells. This has been done for you in an excel sheet *call.xlsx* provided online.

- Click on the *Goal Seek* feature in Excel found under *Data/What-if Analysis*.
- Then set the price cell to the value of 3.16 by changing the cell where the σ value is stored.
- *Set cell* becomes the cell where the call price value is stored [highlighted orange].
- *To value* becomes 3.16.
- *By changing cell* becomes the cell where the σ value is stored [highlighted yellow].
- Click *OK* to get the answer of $\sigma = 0.25$. □

It should not matter which option we use to calculate the implied volatility. But in practice it does. At any point in time the implied volatility varies by both time to maturity and the strike price. Implied volatilities are plotted for call options on Amazon stock in Figure 7.1 clearly demonstrating a skew with high volatilities where out-of-the-money tapering to low volatilities where in-the-money.[2] Other relationships sometimes occur – a u-shaped graph or 'smile' is also possible. Also, these implied volatilities do not tend to agree with

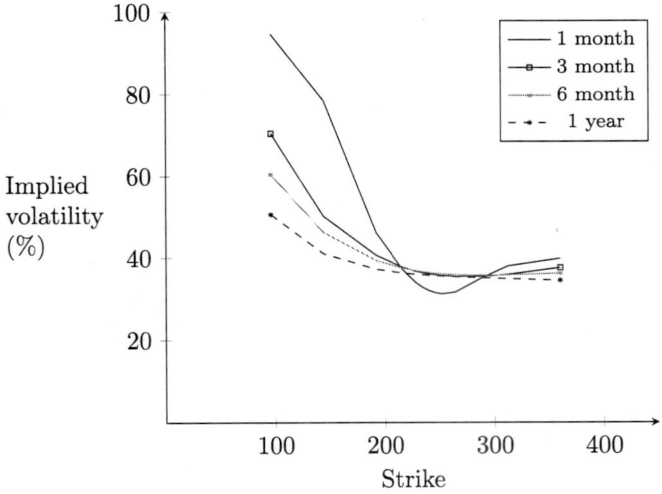

Figure 7.1 Volatility skew: implied volatility for call options on Amazon stock 7 Nov 2025. *Source*: Bloomberg

historical volatilities. This is a major challenge to geometric Brownian Motion assumption of constant volatilities. Jarrow and Chatterjea (2024) advises the use of Black–Scholes model purely as a statistical tool to best fit prices of available options and then used in turn to price other options on the same stock. Hedging with a model calibrated to historical data although common practice is theoretically unsound.

The relationship of implied volatilities with maturities is also seen in Figure 7.1. The degree of skewness is less pronounced for longer maturities. Variation of implied volatility with maturity is less problematic than variation with strike price as expectations of volatility over different time horizons should be different.

7.4 Limitations of geometric Brownian motion

Geometric Brownian Motion is widely used for modeling stock prices. The model assumes time-invariant parameters as well as returns that are normally distributed and independent over time. These assumptions may make it unrealistic as a model of actual stock price behavior.

7.4.1 Normal returns

From the previous section, the continuously compounded return over the interval $[t, t + \Delta t]$ is

$$(\mu - \frac{1}{2}\sigma^2)\Delta t + \sigma(W_{t+\Delta t} - W_t)$$

which has a normal distribution.

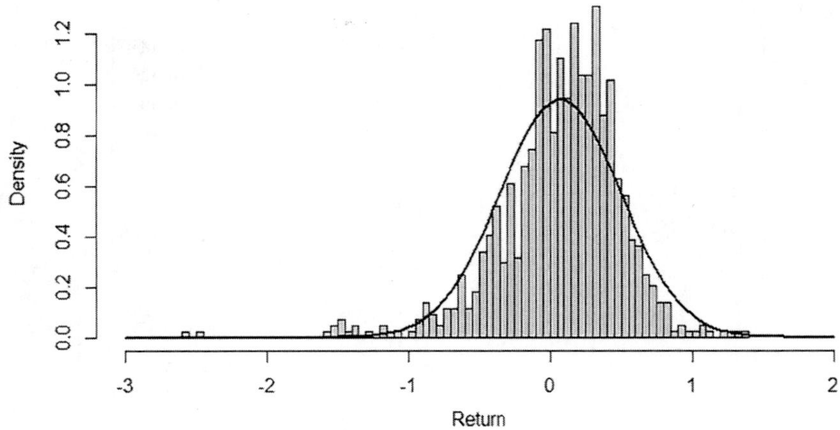

Figure 7.2 Monthly returns of S&P 500 since 1950. *Note*: Continuously-compounded and annualized. Data from Shiller (2015)

Are stock returns normal in practice? Figure 7.2 shows the distribution of monthly returns for the Standard and Poor's (S&P) Composite Stock Price Index since 1950. As an index of stock market prices based on 500 major firms S&P returns will be less susceptible to the impacts of firm- or industry-specific events. Overlaid on top of this distribution is a normal distribution with the same mean and standard deviation. The returns distribution is seen to deviate from normality. Monthly returns close to zero where prices are changing very little occur more often than we would expect from normally distributed returns. Also rare events particularly crashes occur more often than predicted by the normal distribution. This is a serious limitation as investors, creditors, and analysts using this model will underestimate risk metrics when determining potential losses. In conclusion, the actual distribution of returns is then more peaked with fatter tails or *leptokurtic* than an equivalent normal distribution.

The central limit theorem would indicate that returns will conform to normality as the number of different factors influencing returns increases. The return distribution over shorter terms such as daily or weekly will therefore exhibit greater deviations from normality. And geometric Brownian model will also provide a worse fit than indicated in the figure if applied at the firm-level rather than to a composite index.

7.4.2 Independent returns

If returns are independent, then there should be no correlation between the return in one period and the return in the previous period. Early studies (e.g. Fama, 1965; Kendall and Hill, 1953) found this to be the case with no significant correlation between successive daily or weekly stock returns indicating independence in short period returns.

More recent studies have focused on price movements over longer periods and find very different results. Jegadeesh and Titman (1993) present evidence of a momentum effect with positive (negative) stock returns over the last six months followed by positive (negative) returns over the next six months. This is perhaps due to investors buying strong performing stocks pushing up their prices and selling off non-performers pushing down their prices. Studies indicate the opposite is true over longer periods. Fama and French (1988) find a negative autocorrelation in returns over several years suggesting reversion to a mean. The effect is stronger for 3-5-year returns than one-year returns and for smaller stocks compared to larger stocks. Poterba and Summers (1988) attribute mean reversion to *noise trading* – that is investor irrationality – causing transient price distortions which disappear in the longer term as the stock reverts to a price reflecting the firm's financial fundamentals.

7.4.3 Time-invariant parameters

The parameters of geometric Brownian model, expected return μ and volatility σ, are presumed to be time invariant.

The empirical evidence shows that volatility in fact changes over time often in complex and persistent ways.

Figure 7.3 shows volatility over the previous three months calculated every week during the first half of 2025 for the companies Tesla and Amazon. Volatility for these two companies fluctuated markedly during this period due to a combination of uncertainty over new ventures, changing US trade policy and intense scrutiny from financial markets.

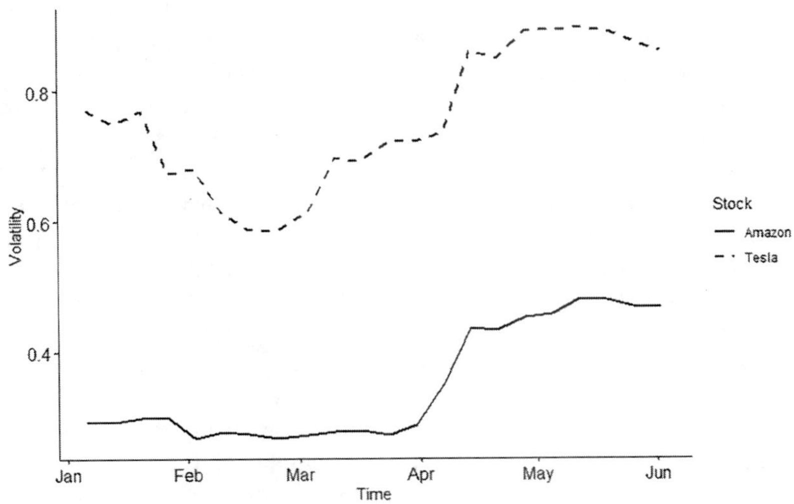

Figure 7.3 Three-month historic volatility Jan–June 2025 based on daily data. *Source*: S&P Capital IQ.

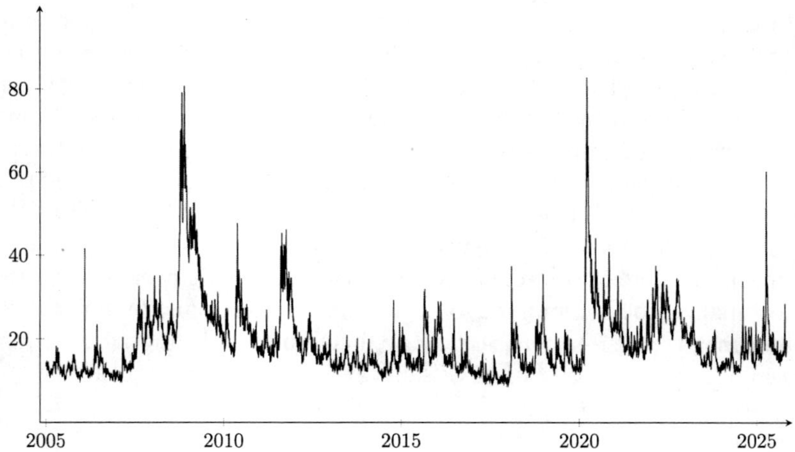

Figure 7.4 Chicago Board Options Exchange's Volatility Index (VIX)
Source: CBOE.

In addition to idiosyncratic events affecting volatility at firm-level, volatility will also increase due to macroeconomic events. Volatility responds to business cycles increasing during recessions or crises and low during periods of economic expansion. The Chicago Board Options Exchange (CBOE) reports a volatility measure for the S&P 500 index called the 'VIX'. It calculates implied volatility based on option prices though the methodology is different to that described in Section 7.3.[3] The VIX is graphed in Figure 7.4 for the past 20 years. Volatility peaked during the 2007–2008 Global Financial Crisis and again during the COVID pandemic while often persistently low for long periods in between. Occasional spikes were due to other economy-wide factors such as Treasury announcements of interest rate changes or geopolitical crises.

The expected return parameter, μ, in the lognormal model is also assumed constant. There are good financial reasons why μ should vary.

Investors in assets such as stocks and commodities demand a risk premium when taking on additional risk compared to a risk-free investment such as government bonds. Because investors are risk-averse, they require compensation for bearing uncertainty. The riskier the asset, the higher the premium required to entice investors to hold it.

The expected return in the lognormal model should therefore reflect underlying prevailing interest rates as well as the compensation required for risk relative to the risk-free rate. The risk premium may vary over time as the perceived level of the firm's risk changes. Premiums will rise during recessions and will also change as the nature of the business changes or the firm's debt levels change. Investors taste for a particular asset or industry can also fluctuate independently of business fundamentals.

Despite these limitations, geometric Brownian motion is still a popular model for stock price evolution. The model could be improved upon by incorporating stochastic volatility especially volatility clustering as in GARCH models. The assumption that returns follow a normal distribution can be replaced by using a statistical distribution that accommodates fat tails such as Student's t distribution. Price discontinuities to reflect how markets react suddenly to news or earnings reports are added in jump-diffusion models. However, the downside of making the model more realistic is the loss in mathematical tractability. Although not perfect, geometric Brownian motion is a reasonable approximation for short intervals and its relative simplicity helps build intuition before using more complex models. For applications such as pricing derivatives sensitive to volatility behavior or jumps or for pricing long-dated derivatives or options on illiquid assets other models are more appropriate. We will also see later that interest-rate models use stochastic processes allowing for mean reversion which is not a feature of geometric Brownian motion.

7.5 Summary

The lognormal distribution is a standard assumption in many financial models – including the Black–Scholes model for equity option prices – as well as for modeling stock and other asset prices in portfolio management. Its strengths are that prices are always positive and grow exponentially. Normally distributed returns are also a reasonable approximation to reality. And it is easy to work with mathematically requiring few parameters.

There are a number of important limitations to the model that an analyst should take into account – absence of crashes, jumps and volatility clustering. If these features are important to the empirical application, then more complex models should be preferred.

Key things to remember

- Why standard Brownian motion is not a good model for asset prices: Chapter 4.
- Why GBM is an improvement: Chapter 6.
- Statistical properties of GBM: Section 7.2
- Issues with GBM as a model for asset prices: Section 7.4.

Further reading

McDonald (2013) provides accessible, well-explained examples on lognormal parameter estimation with financial interpretation. Brigo et al. (2009) go deeper into modeling real-world deviations from the lognormal distribution. Hull (2021) is an excellent resource on the limitations of this model (e.g., volatility smiles, fat tails) and more advanced approaches.

Notes

1. Given the price of an asset X_0 at time zero and X_t at time t the continuously compounded return r (or *log return*) over the interval $[0, t]$ is found by solving $\frac{X_t}{X_0} = e^r$, that is, $r = \log \frac{X_t}{X_0}$.
2. Amazon stock price was \$240.14 on 7 Nov 2025.
3. Details in McDonald (2013).

Questions

1 A stock currently has price \$60 and follows geometric Brownian motion with expected return and volatility 10% and 20% respectively. What is the probability that the stock price will be greater than \$70 in 2 years?

Solution Since $S_T = S_0 e^{(\mu - \frac{1}{2}\sigma^2)T + \sigma W_T} > 70$ we seek the probability that

$$Z > \frac{\log(70/S_0) - (\mu - \frac{1}{2}\sigma^2)T}{\sigma\sqrt{T}}$$
$$= \frac{1}{0.2\sqrt{2}}\left(\log\frac{70}{60} - (0.1 - \frac{1}{2}0.2^2) \times 2\right)\Bigg) = -0.021$$

This is
$P(Z > -0.021) = 0.508$

2 You are given the following information for the price of a stock that follows geometric Brownian motion.

a. The current stock price is 0.25.
b. The stock's volatility is 0.35.
c. The continuously compounded expected rate of stock-price appreciation is 15%.

Calculate the upper limit of the 90% lognormal confidence interval for the price of the stock in 6 months.

[MFE - Sample Q50]

Solution

$$\log S_t \sim N(\log 0.25 + (0.15 - \frac{1}{2}0.35^2)0.5, 0.35^2 \times 0.5)$$
$$\sim N(-1.342, 0.06125)$$

Upper limit of 90% confidence interval for log stock price is $-1.342 + 1.64 \times \sqrt{0.06125} = -0.935$.

Upper limit of 90% confidence interval for stock price is 0.393.

3 The current price of a stock is \$10, the expected return is 12% per annum, and the volatility is 30% per annum. What is the expectation and variance of the stock price in 2 years?

Solution

$$\mathbb{E}(S_T) = S_0 e^{\mu T} = 10e^{0.12\times 2} = 12.71$$

$$var(S_T) = (e^{\sigma^2 T} - 1)S_0^2 e^{2\mu T} = (e^{0.3^2\times 2} - 1)10^2 e^{2\times 0.12\times 2} = 31.87$$

4 You are given the following diffusion process for a stock price:

$dS = \mu S_t dt + \sigma S_t dW$ where W is standard Brownian motion

You are given the following information:

S_0	\$30
μ	10%
σ	30%

(a) Calculate the expected stock price and the standard deviation of the stock price 9 months from now.
(b) Calculate the 95% confidence interval for the stock price 9 months from now.
(c) Calculate the probability that 9 months from now the stock price will have fallen by 50%.

You are given that 9 months later, the stock price has actually fallen by 50%.

(d) Assess your model in light of this experience.

[FETE - Fall 2009 Q11]

Solution

(a)
$$\mathbb{E}(S_T) = 30e^{0.1\times 0.75} = 32.34 \text{ and } s.d.(S_T)$$
$$= \sqrt{(e^{0.3^2\times 0.75} - 1)30^2 e^{2\times 0.1\times 0.75}} = 8.55$$

(b)
$$\log S_t \sim N(\log 30 + (0.1 - \frac{1}{2}0.3^2)0.75, 0.3^2 \times 0.75)$$
$$\sim N(3.442, 0.0675)$$

95% c.i. for log stock price is $3.442 \pm 1.96 \times \sqrt{0.0675} = (2.933, 3.952)$.

95% c.i. for stock price is (18.79,52.02).

(c) We seek the probability that

$$Z < \frac{\log(0.5) - (0.1 - \frac{1}{2}0.3^2)0.75}{0.3\sqrt{0.75}} = -2.827$$

This is $P(Z < -2.827) = 0.002$ i.e. highly unlikely.

(d) Price crashes occur in reality more often than predicted by geometric Brownian motion. The model works well in short-time intervals but does not generate reliable results in the longer run such as the 9 months considered in this example.

5 An investor has decided to model a firm's shares using the continuous-time lognormal model. She estimates the expected return parameter as 6% and volatility as 15%. The share price is currently \$1.50 and no dividends will be paid over the next year.

Calculate the probability that:

(a) Shares fall in value over the next year.

(b) Shares yield a return of greater than 25% over the next year.

Solution

(a)
$$S_T = S_0 e^{(\mu - \frac{1}{2}\sigma^2)T + \sigma W_T} < S_0$$

$$\Longrightarrow W_T < \frac{\log(1) - (\mu - \frac{1}{2}\sigma^2)T}{\sigma}$$

We seek the probability that

$$Z < \frac{\log(1) - (0.06 - \frac{1}{2}0.15^2)1}{0.15\sqrt{1}} = -0.325$$

i.e. $P(Z < -0.325) = 0.373$

(b)
$$S_T = S_0 e^{(\mu - \frac{1}{2}\sigma^2)T + \sigma W_T} > 1.25 S_0$$

$$\Longrightarrow W_T > \frac{\log(1.25) - (\mu - \frac{1}{2}\sigma^2)T}{\sigma}$$

We seek the probability that

$$Z > \frac{\log(1.25) - (0.06 - \frac{1}{2}0.15^2)1}{0.15\sqrt{1}} = 1.163$$

i.e. $P(Z > 1.163) = 0.122$

6 Assume the price of a non dividend-paying stock follows geometric Brownian motion.

The price of the stock in seven consecutive months is:

Month	*Price*
1	54
2	56
3	48
4	55
5	60
6	58
7	62

Estimate the continuously compounded expected rate of return on the stock.

[MFE - Sample Q51]

Solution

Month	*Price*	$log(S_t/S_{t-1})$
1	54	—
2	56	0.0364
3	48	–0.1542
4	55	0.1361
5	60	0.0870
6	58	–0.0339
7	62	0.0667

The estimator for the population mean is $\bar{u} = \frac{0.0364+....+0.0667}{6} = 0.0230.$

The estimator for the population standard deviation is

$$\sqrt{\frac{1}{5}\left((0.0364 - 0.0230)^2 + ... + (0.0667 - 0.0230)^2\right)} = 0.1035$$

$\Delta t = \frac{1}{12}$. Setting the estimators equal to the population parameters

$$\sigma\sqrt{\frac{1}{12}} = 0.1035 \implies \sigma = 0.359$$

and

$$\left(\mu - \frac{0.359^2}{2}\right)\left(\frac{1}{12}\right) = 0.0230 \implies \mu = 0.341$$

The continuously compounded expected rate of return on the stock is 0.341.

7 The Multnomah Options Company has just written European call option contracts on a non-dividend paying stock. You are given:

Exercise Price = \$150
Current Stock Price = \$145
Risk free rate = 2%
Time to maturity = 18 months
Call option price = \$25

You are given the following information:

Volatility	$\Phi(d_1)$	$\Phi(d_2)$
32%	0.574	0.418
34%	0.579	0.414
36%	0.584	0.409
38%	0.589	0.405

Estimate the implied volatility given the above option price.

[FETE - Fall 2012 Q4]

Solution The call price is

$$c_t = S_t\Phi(d_1) - e^{-r(T-t)}K\Phi(d_2).$$

$S_t = 145, r = 0.02, T - t = 1.5, K = 150$ so calculate c_t for each volatility.

Volatility	$\Phi(d_1)$	$\Phi(d_2)$	c_t
32%	0.574	0.418	22.4
34%	0.579	0.414	23.7
36%	0.584	0.409	25.1
38%	0.589	0.405	26.5

The implied volatility is then 36% as this call price is closest to \$25.

8 A call option on a non-dividend-paying stock has a price of \$0.24. The current stock price is \$10, the strike price is \$12, the time to maturity is two months, and the risk-free interest rate is 6% per annum. What is the implied volatility?

Solution The call price is

$$c_t = S_t\Phi(d_1) - e^{-r(T-t)}K\Phi(d_2).$$

where $d_1 = \frac{\log(S_t/K)+(r+\frac{1}{2}\sigma^2)(T-t)}{\sigma\sqrt{T-t}}, d_2 = d_1 - \sigma\sqrt{T-t}$

Here $r = 0.065, T - t = 1/6, S_t = 10, K = 12, c_t = 0.24 \implies \sigma = 0.495$ by goal-seek in excel or trial-and-error.

9 Consider the model for geometric Brownian motion. You are given the following three statements on variances, conditional on knowing S_t, the stock price at time t. Which of them are true?

(i) $\text{var}(\log S_{t+h}|S_t) = \sigma^2 h, \qquad h > 0$

(ii) $\text{var}(\frac{dS}{S_t}|S_t) = \sigma^2 dt$

(iii) $\text{var}(S_{t+dt}|S_t) = S_t^2\sigma^2 dt$

[MFE - Sample Q11]

Solution All are true.

(i) From (Equation 7.2), $\log S_{t+h} \sim N(\log S_t + (\mu - \frac{\sigma^2}{2})h, \sigma^2 h)$

(ii) $\frac{dS}{S_t} = \mu dt + \sigma dW$ and $\text{var}(\sigma dW) = \sigma^2 dt$ from the definition of standard Brownian motion.

(iii) Using (Equation 7.4),

$$var(S_{t+\Delta t}) = (e^{\sigma^2\Delta t} - 1)S_t^2 e^{2\mu\Delta t}$$

Letting $\Delta t \to 0$ and using the Taylor expansion (Equation 1.4)

$$var(S_{t+dt}) = (\sigma^2 dt)S_t^2 \times 1 = S_t^2\sigma^2 dt$$

The same result can be found by letting $S_{t+dt} = S_t + dS$ then

$$\begin{aligned}\text{var}(S_{t+dt}|S_t) &= \text{var}(dS) = \text{var}(\mu S_t dt + \sigma S_t dW)\\ &= \text{var}(\sigma S_t dW) = S_t^2\sigma^2 dt\end{aligned}$$

10 The price of a stock is governed by the stochastic differential equation:

$$\frac{dS}{S_t} = 0.03dt + 0.2dW$$

where W is a standard Brownian motion. Consider the geometric average

$$G = (S_1 \times S_2 \times S_3)^{1/3}$$

Find the variance of $\log G$.

[MFE - Sample Q37]

Solution

From (Equation 7.1)

$$\begin{aligned}\log G &= \frac{1}{3}\Big(\log S_1 + \log S_2 + \log S_3\Big)\\ &= \frac{1}{3}\Big(\log S_0 + (\mu - \frac{\sigma^2}{2}) + \sigma W_1 + \log S_0 + 2(\mu - \frac{\sigma^2}{2})\end{aligned}$$

$$+ \sigma W_2 + \log S_0 + 3(\mu - \frac{\sigma^2}{2}) + \sigma W_3\Big)$$

$$= \frac{1}{3}\Big(3 \log S_0 + 6(\mu - \frac{1}{2}\sigma^2) + \sigma(W_1 + W_2 + W_3)\Big)$$

$$\Longrightarrow \text{var}(\log G) = \frac{\sigma^2}{9}\text{var}(W_1 + W_2 + W_3)$$

Since W_1, W_2, W_3 are all correlated, we proceed in increments as in Question 5(iii) in Chapter 4.

$$\text{var}(\log G) = \frac{\sigma^2}{9}\text{var}\Big(3W_1 + 2(W_2 - W_1) + (W_3 - W_2)\Big)$$

$$= \frac{\sigma^2}{9}(9 + 4 + 1) = \frac{14}{225}$$

as from the question we are given $\sigma = 0.2$.

11 (a) A stock has price \$20, expected return 5% and volatility 20%. Simulate the stock price path over 5 years using weekly time steps in R. Plot your results. Rerun the code and observe how the simulated stock price path changes.

(b) What is the probability that a European call option on this stock with maturity 5 years and strike price \$30 will be in-the-money?

Solution

(a) Using the code given in Section 6.7.2, set T = 5, n = T*52, mu = 0.05, sigma = 0.20, X0 = 20.

(b) Create a loop to simulate 10,000 times. Initialize parameter values and add the following code. The proportion of simulations where the stock price is above the strike price will be approximately 0.28. The same answer can be found numerically by following the method of Question 1.

```
T      <- 5
n      <- T*52
dt     <- T/n
t      <- c(1:n)
t      <- t*dt
mu     <- 0.05
sigma  <- 0.20
X0     <- 20
Xt     <- c(X0)
ITM <- vector(mode = ``numeric'', length = 10000)
```

```
K       <- 30
for (j in 1:10000) {
for (i in 2:n) {
  dWt    <- rnorm(n, 0, sqrt(dt))
  Xt[i]  <-  Xt[i-1] + mu*Xt[i-1]*dt
  + sigma*Xt[i-1]*dWt[i-1]
}
if (Xt[n] > K) {
  ITM[j] <- 1  # store 1 if it's greater than K
} else {
  ITM[j] <- 0  # otherwise 0
}
}

pITM = sum(ITM)/10000
print(paste(''Prob that option ITM is '', pITM))
```

8 The binomial model

Introduction

In this chapter, we will use a simple binomial model to price options. Although the exposition is quite basic it introduces us to a number of key concepts going forward: *change of probability measure*, *risk-neutral probability measure*, the *Black-Scholes formula* and our derivative valuation formula

$$V_0 = e^{-r}\mathbb{E}^{\mathbb{Q}} C_1$$

We will meet all these ideas again when we move from discrete changes in stock price to continuous changes in stock price.

The other reason why the binomial model is of interest is that some complex derivatives do not have closed-form solutions and can only be priced using discrete models. Variants on models presented here provide reasonably good approximate solutions to many option pricing problems.

8.1 Basic model

In this section, we consider a very simple binomial model with only one period. We will then extend this model to multiple periods in the next section. In practice, multi-period models provide prices close to models based on continuous time processes.

The share price S_t can move to only two values over any discrete time interval. It is therefore a simple discrete stochastic process. Initially, the share has price S_0 where $S_0 > 0$. After one period, the share price moves to either $S_0 u$ with probability p or $S_0 d$ with probability $1 - p$ where we assume $u > d$. Usually u is taken to be greater than one and d less than one but it is not necessary for the derivations that follow. Share price movements are obviously much more complicated than this but the basic model demonstrates the logic of the binomial asset-pricing model.

DOI: 10.4324/9781003477624-11

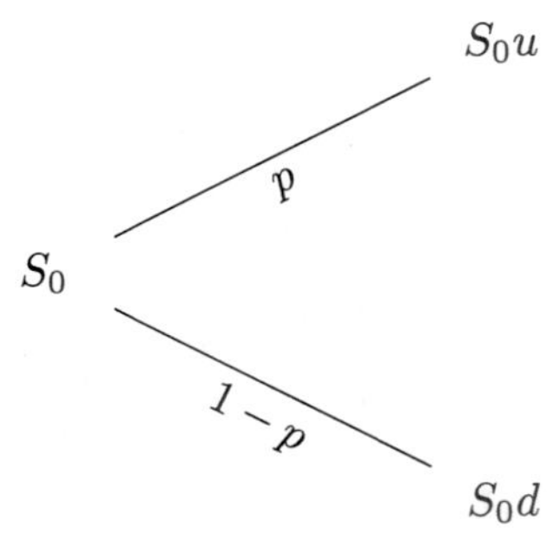

Figure 8.1 Basic one-period model

Our one-period model is given in Figure 8.1.

A market participant can borrow or invest at the same interest rate r. After one period, one dollar becomes $1 + R$ dollars with discrete compounding or e^r with continuous compounding. Although discrete compounding is more intuitive in a discrete model, continuous compounding will be used throughout this chapter in order to maintain the parallel with continuous option pricing models discussed later in the book. In practice, r is generally positive but our arguments still work as long as $r > -1$.

By the principle of no arbitrage, the inequality $d < e^r < u$ must hold since otherwise we have a zero probability of losing money and a positive probability of making money.

If $e^r \leq d$, then the investor can borrow S_0 in cash in order to buy the stock. In the next period, she then earns at least $S_0(d - e^r)$. So she has a zero probability of losing money. Also she has a positive probability of making money since it is possible that the share price goes up. She will earn $S_0(u - e^r)$ after repaying the loan in this case.

Likewise, if $u \leq e^r$, then the investor can short the stock and invest the proceeds S_0 at rate r. In the next period, she then earns at least $S_0(e^r - u)$. So she has a zero probability of losing money. Also she has a positive probability of making money if the share price goes down. She will earn $S_0(e^r - d)$ after buying back the stock in this case.

The converse is also true. If the inequality $d < e^r < u$ holds then there are no risk-free arbitrage opportunities. We must show that there is no possible portfolio of cash and shares with zero initial value that has a zero probability of losing money and a positive probability of making money. Create a portfolio of ϕ units of stock by borrowing ϕS_0 units of cash. In this way, the initial portfolio value is

$$V_0 = \phi S_0(\text{stock}) - \phi S_0(\text{borrowing}) = 0$$

At time one, the portfolio has value

$$V_1 = \begin{cases} V_1(1) = \phi u S_0 - e^r \phi S_0 = \phi S_0(u - e^r) & \text{if price moves up} \\ V_1(2) = \phi d S_0 - e^r \phi S_0 = \phi S_0(d - e^r) & \text{if price moves down} \end{cases}$$

If the portfolio has a positive probability of making money, then the portfolio value in the event of an up-movement $V_1(1)$ is greater than zero and the inequality $e^r < u$ implies that the number of shares contained in the portfolio ϕ must be greater than zero. Then this result together with the inequality $d < e^r$ implies the portfolio value in the event of a down-movement $V_1(2)$ is less than zero. So the portfolio has a non-zero probability of losing money for any number of shares ϕ we choose to create the portfolio. In other words, there are no risk-free arbitrage opportunities.

Our task is to find the value of the derivative (C_0) which pays c_u if the price of the underlying stock goes up and c_d if the price of the underlying stock goes down. Let us denote the payoff after one period by random variable C_1. The arbitrage pricing theory approach involves replicating the option by creating a portfolio of cash and shares with the same value at time one.

Suppose we hold ϕ units of stock and ψ units of cash initially. The value of this portfolio is V_0. So

$$V_0 = \phi S_0 + \psi$$

We will choose ϕ and ψ in order that the portfolio value at time one, V_1, replicates the pay-off C_1 of the derivative. There are no restrictions on the value of the number of units of stock. The parameter ϕ can be fractional or negative and is unlimited with either positive or negative sign. Without this flexibility, we would not be able to replicate all derivatives at all times. Choosing (ϕ, ψ) is called a *replicating strategy* because it replicates the derivative payoff without risk. More generally, a *hedging strategy* is an investment strategy that reduces the risk carried by the issuer of the contract.

Setting the value of the replicating portfolio at time one to equal the derivative pay-off C_1:

$$\left.\begin{aligned} \phi S_0 u + \psi e^r &= c_u \\ \phi S_0 d + \psi e^r &= c_d \end{aligned}\right\} \Longrightarrow \phi = \frac{c_u - c_d}{S_0(u - d)}, \psi = e^{-r}\frac{c_d u - c_u d}{u - d} \tag{8.1}$$

Then the value of the replicating portfolio at time zero is:

$$V_0 = \frac{c_u - c_d}{u - d} + e^{-r}\frac{c_d u - c_u d}{u - d} = e^{-r}\Bigg(\underbrace{\frac{e^r - d}{u - d}}_{q} c_u + \underbrace{\frac{u - e^r}{u - d}}_{1-q} c_d\Bigg) \tag{8.2}$$

The value of the replicating portfolio at time zero must be the same as the value of the derivative (i.e. $V_0 = C_0$). Otherwise, there would be an arbitrage opportunity. For example, if the value of the derivative was initially higher than the value of this portfolio, then the investor could sell the derivative, set up the replicating portfolio and invest the difference risk-free. The replicating portfolio would then cover the obligations of the derivative after one period at no cost.

The symbol q is used to represent the first quotient in the parentheses and the second quotient is then $1 - q$ algebraically. We note that the probability of a stock price increase (p) did not enter into our analysis. The fair price is unaffected by this probability since the hedging strategy is the same no matter how likely the stock price is to increase. The price of the option only depends on the spread in the two possible outcomes of the stock price at exercise time (S_0u and S_0d).

We met probability measures in Section 2.1. Our q behaves like a probability since q and $1-q$ sum to one and are positive. From our no-arbitrage inequality:

$$d < e^r < u \implies q = \frac{e^r - d}{u - d} > 0 \text{ and } 1 - q = \frac{u - e^r}{u - d} > 0$$

Then we can write Equation (8.2) more simply as,

$$V_0 = C_0 = e^{-r}(qc_u + (1 - q)c_d) = e^{-r}\mathbb{E}^{\mathbb{Q}}C_1 \tag{8.3}$$

where $\mathbb{E}^{\mathbb{Q}}$ is the expectation using the risk-neutral $\mathbb{Q}$ probability measure. Note that q is unrelated to p or C_1.

Exercise 8.1. *A stock price can move from its current value of \$100 up to \$120 with probability $\frac{1}{2}$ or down to \$70 with probability $\frac{1}{2}$ over one period.*

(i) Find the value of a one-period European call option that has an exercise price of \$110 given the continuously compounded risk-free rate of return is 3% per time period.

(ii) What are the stock and cash holdings ϕ and ψ for the replicating portfolio ?

Proof. The probability of the stock price movements does not enter into our analysis.

(i) $c_u = 120 - 110 = 10; c_d = 0$ and $u = \frac{120}{100} = 1.2; d = \frac{70}{100} = 0.7$

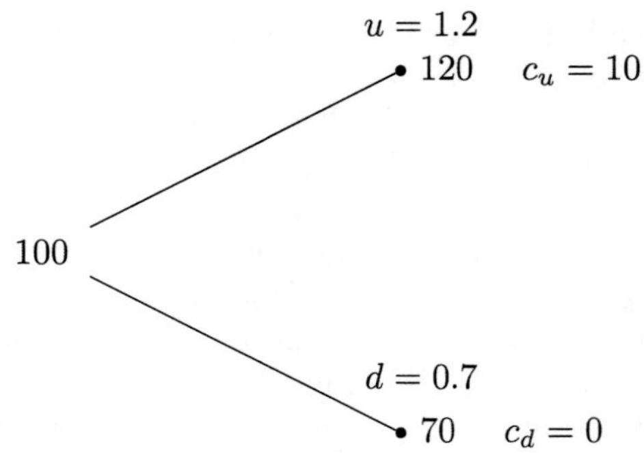

Figure 8.2 One-period model for Exercise 8.4

The risk-neutral probability is

$$q = \frac{e^r - d}{u - d} = \frac{e^{0.03} - 0.7}{1.2 - 0.7} = 0.661$$

Then from Equation (8.2), the value of the derivative is

$$\begin{aligned} C_0 &= e^{-r}\Big(qc_u + (1-q)c_d\Big) \\ &= e^{-0.03}\Big(0.661 \times 10 + 0.339 \times 0\Big) = 6.414 \text{ or } \$6.41 \end{aligned}$$

(ii) From Equation (8.1), the stock and cash holdings for the replicating portfolio are
$\phi = \frac{c_u - c_d}{S_0(u-d)} = \frac{10-0}{100(1.2-0.7)} = 0.2$, that is, hold one-fifth of a share

and

$\psi = e^{-r}\frac{c_d u - c_u d}{u-d} = e^{-0.03}\frac{0 \times 1.2 - 10 \times 0.7}{1.2 - 0.7} = -13.586$, that is, borrow \$13.59 cash

□

It would have seemed intuitive to price a derivative by discounting the value of the expected payoff. But this does not work with real-world probabilities. It works only when we use the $\mathbb{Q}$ probability measure.

$\mathbb{Q}$ is known as the *risk-neutral probability measure* because

$$\mathbb{E}^{\mathbb{Q}} S_1 = S_0(qu + (1-q)d)) = S_0\left(\frac{u(e^r - d)}{u - d} + \frac{d(u - e^r)}{u - d}\right) = S_0 e^r$$

and under $\mathbb{Q}$ the expected return on the risky stock is the same as that on a risk-free investment in cash.

$$\mathbb{E}^{\mathbb{Q}}\frac{S_1}{S_0} = e^r \tag{8.4}$$

Investors are therefore indifferent between risky and risk-free investments or, in other words, they are *risk-neutral.*

We can also say that the discounted stock price is a martingale when we use the $\mathbb{Q}$ probability measure. Let B_t be the process of investing B_0 at the risk-free rate, that is, $B_t = B_0 e^{rt}$. Then rearranging Equation (8.4)

$$S_0 = e^{-r}\mathbb{E}^{\mathbb{Q}}S_1 \implies \frac{S_0}{B_0} = \mathbb{E}^{\mathbb{Q}}\frac{S_1}{B_1}$$

Dividends are easily accommodated within the binomial model. Following the same logic as above, it can be shown that if dividends are received continuously at rate δ per period, the risk-neutral probability is modified to $q = \frac{e^{r-\delta}-d}{u-d}$.

The assumptions made in the binomial model were

1. There are no risk-free arbitrage opportunities.
2. No market frictions – no taxes or transaction costs.
3. Borrowing and investing occurs at the same rate r.
4. Competitive and well-functioning markets
   - No maximum or minimum units of trading.
   - No limits to short selling (e.g., due to credit limits and collaterals).
5. The share price can move to only two values over any discrete time interval.

All these same assumptions except for the last one on price movements will underpin option pricing in the continuous case considered later.

8.2 Two-period binomial model

The basic one-period model is obviously a very crude representation of reality. But we can make things more realistic by increasing the number of time-steps. Let's start by using a two-period model for a European-style derivative. The share price S_0 moves to S_0u_0 or S_0d_0 over the first time interval and from there it moves to one of two values in the second time interval.

The model is displayed in Figure 8.3. In this general model, the u and d values in the second period can be different sizes on each branch (e.g. $u_2(1)$ in upper branch and $u_2(2)$ in lower branch). In practice, they will often be the same. Note

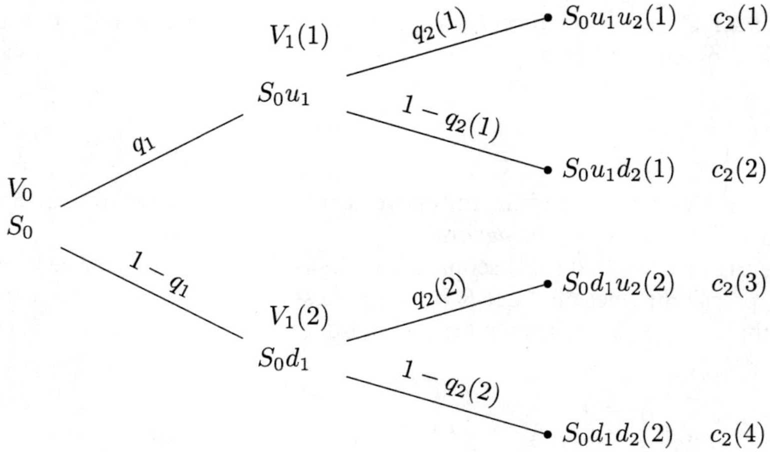

Figure 8.3 Two-period binomial tree

that we have not put in the real-world probability of stock price movements in the figure because we now know they are not needed.

The value once again is determined by setting up a replicating portfolio of shares and cash. The initial portfolio value is then the same as the derivative price by arbitrage arguments. We will find the value of the derivative which pays $c_2(j)$ if the price of the stock after two periods is in state j.

Working backwards, the value of the contract after one period from the logic of the previous section is:

$$V_1(1) = e^{-r}\Big(q_2(1)c_2(1) + (1 - q_2(1))c_2(2)\Big)$$

$$V_1(2) = e^{-r}\Big(q_2(2)c_2(3) + (1 - q_2(2))c_2(4)\Big)$$

where $q_2(1) = \frac{e^r - d_2(1)}{u_2(1) - d_2(1)}$ and $q_2(2) = \frac{e^r - d_2(2)}{u_2(2) - d_2(2)}$

Repeating the logic again, the current derivative value is then

$$\begin{aligned}
V_0 = C_0 &= e^{-r}\Big(q_1 V_1(1) + (1 - q_1)V_1(2)\Big) \\
&= e^{-2r}\Big(q_1\Big(q_2(1)c_2(1) + (1 - q_2(1))c_2(2)\Big) \\
&\quad + (1 - q_1)\Big(q_2(2)c_2(3) + (1 - q_2(2))c_2(4)\Big)\Big) \\
&= e^{-2r}\mathbb{E}^{\mathbb{Q}} C_2
\end{aligned}$$

Exercise 8.2. *A stock price can move from its current value of \$100 up to \$120 or down to \$72 over the first period. Following an up-movement, the stock price can move up to \$160 or down to \$100 over the second period. Following a down-movement, the stock price can move up to \$96 or down to \$60 over the second period. Find the value of a two-period European call option that has an exercise price of \$150 given the continuously compounded risk-free rate of return is 3% per time period.*

Proof.

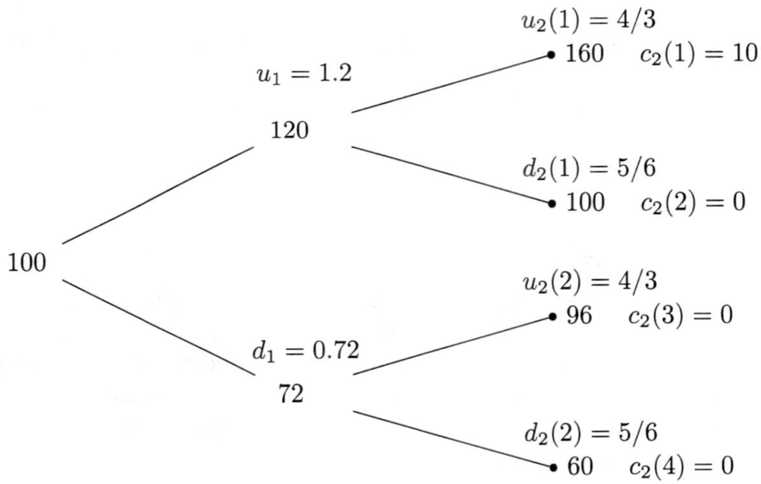

Figure 8.4 Two-period binomial tree for Exercise 8.2

First $q_2(1) = \frac{e^r - d_2(1)}{u_2(1) - d_2(1)} = \frac{e^{0.03} - 5/6}{4/3 - 5/6} = 0.3942 = q_2(2)$. The q values in the upper and lower branches are equal here because of the stock price values given but they do not have to be.

The value of the contract after one period is:

$$V_1(1) = e^{-r}\Big(q_2(1)c_2(1) + (1 - q_2(1))c_2(2)\Big)$$
$$= e^{-0.03}\Big(0.3942 \times 10 + (1 - 0.3942) \times 0\Big) = 3.826$$
$$V_1(2) = e^{-r}\Big(q_2(2)c_2(3) + (1 - q_2(2))c_2(4)\Big)$$
$$= e^{-0.03}\Big(0.3942 \times 0 + (1 - 0.3942) \times 0\Big) = 0$$

The current derivative value is:

$$q_1 = \frac{e^r - d_1}{u_1 - d_1} = \frac{e^{0.03} - 0.72}{1.2 - 0.72} = 0.6468$$

$$V_0 = C_0 = e^{-r}\Big(q_1 V_1(1) + (1 - q_1) V_1(2)\Big)$$

$$= e^{-0.03}\Big(0.6468 \times 3.826 + (1 - 0.4017) \times 0\Big) = 2.401 \text{ or } \$2.40$$

Or in one step:

$$V_0 = C_0 = e^{-2r}\mathbb{E}^{\mathbb{Q}} C_2$$

$$= e^{-2\times 0.03}\Big(0.6468 \times 0.3942 \times 10 + 0.6468 \times (1 - 0.3942) \times 0$$

$$+ (1 - 0.6468) \times 0.3942 \times 0 + (1 - 0.6468) \times (1 - 0.3942) \times 0\Big)$$

$$= 2.401$$

□

If the derivative was American-style, the no-arbitrage argument could still be applied working backwards from the second period. The only modification is that we have to select the higher of the early exercise price and the value of the replicating portfolio at each node. Examples of pricing American-style options with a binomial model are given in the end of the chapter exercises.

8.3 *n*-period binomial model

Generalizing to n-periods, we can show that for European-style derivatives

$$C_0 = e^{-nr}\,\mathbb{E}^{\mathbb{Q}} C_n \tag{8.5}$$

As before, we can work backwards to construct a replicating strategy $\{\phi_i(j), \psi_i(j)\}$ after any number of time periods i at any node j by solving two simple simultaneous equations involving terms from the next period. The replicating portfolio is therefore being rebalanced dynamically at every time interval and for each node. This is always possible under the binomial model assumptions. The derivatives market is then said to be *complete* meaning that the payoff from any derivative can be replicated by a portfolio of the underlying stock and cash.

The binomial model is very flexible and can accommodate complications such as path-dependent derivatives: American options and derivatives on shares paying dividends. See the questions at the end of the chapter for examples.

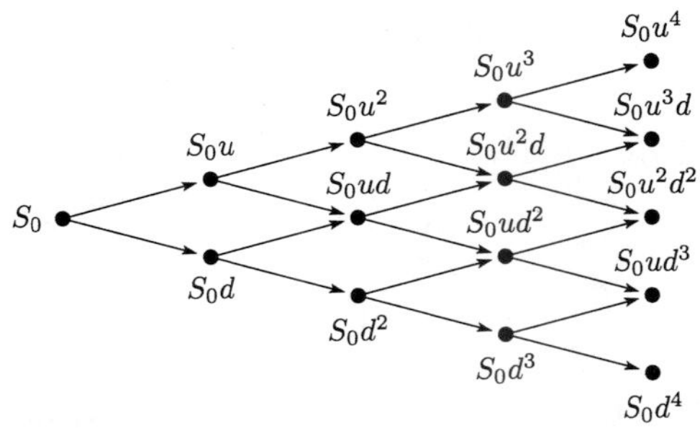

Figure 8.5 Recombining binomial trees

8.4 Recombining binomial trees

In order to price accurately, a binomial model may have a very large number of periods. Since the number of states after n periods is 2^n the computations may become unmanageable.

We can simplify our tree when we have a large number of periods with some loss of precision by using a *recombining binomial tree* or *binomial lattice*. This way we have less states at each point in time, that is, $n + 1$ instead of 2^n.

We do this by letting u and d be constant for all states and for all time periods. We have $u_i(j) = u$ and $d_i(j) = d$ for all periods i and nodes j making the risk-neutral probability also constant, $q_i(j) = q$. The tree now becomes much simpler (Figure 8.5) and calculations are also greatly simplified.

The binomial theorem describes the values that the stock price can take after n-periods. After k up-movements, the price will take the form $S_0u^kd^{n-k}$. The derivative value at payoff will be a function of this price. Using risk-neutral probabilities, the probability the stock reaches this price is $\frac{n!}{k!(n-k)!}q^k(1-q)^{n-k}$. The initial value of a European-style derivative is

$$\begin{aligned}
V_0 &= e^{-nr}\mathbb{E}^{\mathbb{Q}}C_n \\
&= e^{-nr}\sum_{k=0}^{n} f(S_n)P_{\mathbb{Q}}S_n \\
&= e^{-nr}\sum_{k=0}^{n} f(S_0u^kd^{n-k})\frac{n!}{k!(n-k)!}q^k(1-q)^{n-k}
\end{aligned}$$

> **Exercise 8.3.** *A stock is currently priced \$18 and can move up by 3.6% or down by 2.9% each month. Use a binomial lattice model to find the value of an 8-month European call option on this stock with an exercise price of \$20. The risk-free continuously compounded rate is 0.2% per month.*

Proof. $n = 8, r = 0.002, u = 1.036, d = 0.971, S_0 = 18$

$$q = \frac{e^r - d}{u - d} = \frac{e^{0.002} - 0.971}{1.036 - 0.971} = 0.477$$

$$V_0 = e^{-rn} \sum_{k=0}^{n} f(S_0 u^k d^{n-k}) \frac{n!}{k!(n-k)!} q^k (1-q)^{n-k}$$

Since the payoff from the option is zero unless $k \geq 6$, we do not have to write out all the terms:

$$\begin{aligned} V_0 &= e^{-8\times 0.002}\Big((18 \times 1.036^6 \times 0.971^2 - 20)\frac{8!}{6!2!} \times 0.477^6 \times 0.523^2 \\ &+ (18 \times 1.036^7 \times 0.971^1 - 20)\frac{8!}{7!1!} \times 0.477^7 \times 0.523^1 \\ &+ (18 \times 1.036^8 - 20)0.477^8 \Big) \\ &= e^{-8\times 0.002}(0.98 \times 0.0902 + 2.39 \times 0.0235 + 3.89 \times 0.0027) \\ &= 0.153 \text{ or } \$0.15 \end{aligned}$$

□

8.5 Calibrating binomial models

The binomial lattice can be calibrated to approximate any stochastic process we believe our underlying asset follows.

For example, stocks are usually modeled using geometric Brownian motion. If a stock price process S_t follows geometric Brownian motion, then from before (Equation 7.1):

$$\log \frac{S_{t+\Delta t}}{S_t} \sim N\Big((\mu - \frac{\sigma^2}{2})\Delta t, \sigma^2 \Delta t \Big) \tag{8.6}$$

So the variance of $\log \frac{S_{t+\Delta t}}{S_t}$ is $\sigma^2 \Delta t$, where S_t follows geometric Brownian motion.

Now find an equivalent expression from the binomial lattice and select the parameter values u and d so that the expressions for $\log \frac{S_{t+\Delta t}}{S_t}$ equate. In the binomial lattice

$$\log \frac{S_{t+\Delta t}}{S_t} = \begin{cases} \log \frac{S_t u}{S_t} = \log u & \text{if price moves up} \\ \log \frac{S_t d}{S_t} = \log d & \text{if price moves down} \end{cases}$$

Then assume $u = \frac{1}{d}$ and use the relationship $var(X) = \mathbb{E}X^2 - (\mathbb{E}X)^2$:

$$var\left(\log \frac{S_{t+\Delta t}}{S_t}\right) = p \times (\log u)^2 + (1-p) \times (-\log u)^2 - \left(\mathbb{E}\log \frac{S_{t+\Delta t}}{S_t}\right)^2$$

$$= (\log u)^2 - \left(\mu - \frac{\sigma^2}{2}\right)^2 (\Delta t)^2 \quad \text{from the expectation in 8.6}$$

$$\approx (\log u)^2 \quad \text{for } \Delta t \text{ small}$$

Equating this to the variance term in the continuous time expression given in Equation (8.6):

$$(\log u)^2 \approx \sigma^2 \Delta t \implies u = e^{\sigma\sqrt{\Delta t}} \text{ and } d = \frac{1}{u} = e^{-\sigma\sqrt{\Delta t}}$$

With a little bit of work we can show that the binomial model calibrated in this way leads to the Black-Scholes equation – see the appendix at the end of the chapter for details.

Exercise 8.4. *A non-dividend paying share has volatility $\sigma = 20\%$ p.a. Find the value of a 2-day put option on this stock with an exercise price of \$10. The risk-free continuously compounded rate is 3% p.a. The current share price is \$10.05. You can assume there are 250 trading days in a year.*

Proof. Calibrate the binomial lattice.

$$u = e^{\sigma\sqrt{\Delta t}}$$

$$= e^{0.2\sqrt{\frac{1}{250}}} = 1.013$$

$$d = \frac{1}{u} = 0.987$$

Adjust the risk-free rate from annual to daily.

$$r = \frac{0.03}{250} = 0.00012$$

Find the risk-neutral probability:

$$q = \frac{e^{0.00012} - 0.987}{1.013 - 0.987} = 0.502$$

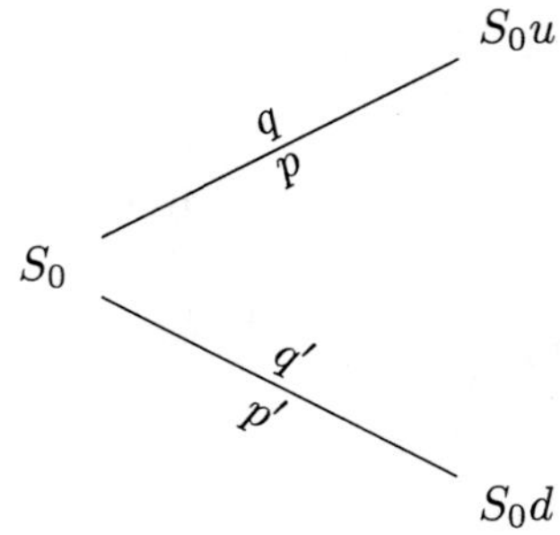

Figure 8.6 Basic one-period model

Calculating the derivative value in one step:

$$C_0 = e^{-2\times 0.00012}\left(0.502^2 \times 0 + 2 \times 0.502 \times (1 - 0.502) \times 0 + (1 - 0.502)^2 \times 0.201\right)$$
$$= 0.0499 \text{ or } 5\,¢$$

□

8.6 Radon-Nikodým derivative

In the exposition of the binomial model, there were two probabilities associated with each price path – an actual probability and a risk-neutral probability. We can then define the **Radon-Nikodým derivative** for each price path by the quotient of the two probabilities associated with that price path.

In the basic one-period binomial model, we have a probability p of an up-movement and a probability $p' = 1 - p$ of a down-movement. Let us denote the real-world probabilities by measure $\mathbb{P}$. Associated with these movements are risk-neutral probabilities q and $q' = 1 - q$ given by

$$q = \frac{e^r - d}{u - d}$$

Let us denote the risk-neutral probabilities by measure $\mathbb{Q}$.

Then the **Radon-Nikodým derivative** of $\mathbb{Q}$ with respect to $\mathbb{P}$ is the stochastic variable Θ defined by

$$\Theta = \begin{cases} \frac{q}{p} & \text{if the price moves up} \\ \frac{q'}{p'} & \text{if the price moves down} \end{cases}$$

When we extend the binomial tree to more than one period, the Radon-Nikodým derivative becomes a stochastic process.

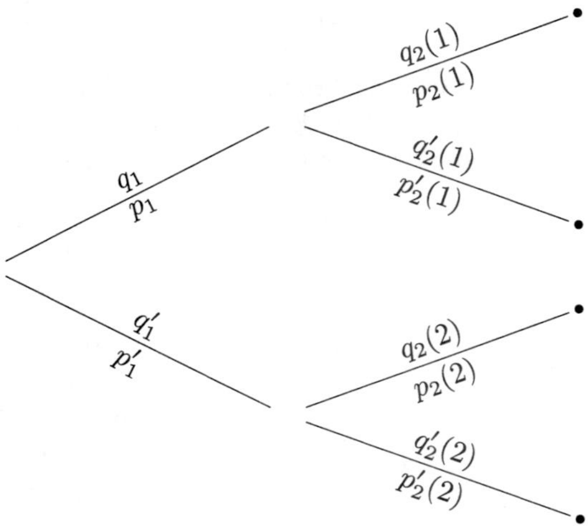

Figure 8.7 Two-period binomial tree

The process for two periods is given as an example. Initially, the share price is given with certainty under both probability measures and $\Theta_0 = 1$. In period one, Θ_1 can take either of two values as above in the one-period model depending on whether the share price moves up or down. In period two, there are four price paths with associated $\mathbb{P}$ and $\mathbb{Q}$ probabilities and hence Θ_2 can take any of four values.

$$\Theta_0 = 1 \quad \Theta_1 = \begin{cases} \frac{q_1}{p_1} \\ \\ \frac{q_1'}{p_1'} \end{cases} \quad \Theta_2 = \begin{cases} \frac{q_1 q_2(1)}{p_1 p_2(1)} \\ \\ \frac{q_1 q_2'(1)}{p_1 p_2'(1)} \\ \\ \frac{q_1' q_2(2)}{p_1' p_2(2)} \\ \\ \frac{q_1' q_2'(2)}{p_1' p_2'(2)} \end{cases}$$

Whenever an event has a non-zero probability of actually happening in the real world, the risk-neutral probability of that event is also non-zero and vice versa. The real-world probability measure and the risk-neutral probability measure are then **equivalent**. Two probability measures $\mathbb{P}$ and $\mathbb{Q}$ on a finite sample space Ω are said to be equivalent if each event that has positive probability under one measure must also have positive probability under the other one.

The Radon-Nikodým derivative of $\mathbb{Q}$ with respect to $\mathbb{P}$ is sometimes written $\frac{d\mathbb{Q}}{d\mathbb{P}}$. In order for $\frac{d\mathbb{Q}}{d\mathbb{P}}$ and $\frac{d\mathbb{P}}{d\mathbb{Q}}$ to exist, the $\mathbb{P}$ and $\mathbb{Q}$ probability measures must be equivalent. Otherwise we get a zero probability in the denominator of the quotient for some event.

The Radon-Nikodým derivative can be used to define the change from one probability measure to another. The risk-neutral expectation of payoffs C_n after n periods can be written in terms of real-world probabilities as:

$$\mathbb{E}^{\mathbb{Q}} C_n = \sum_j q(j) c_n(j) = \sum_j p(j) \frac{q(j)}{p(j)} c_n(j) = \sum_j p(j) \theta_n(j) c_n(j) = \mathbb{E}^{\mathbb{P}} \Theta_n C_n$$

where $p(j)$, $q(j)$ and $\theta_n(j)$ are the real-world probability, risk-neutral probability and the Radon-Nikodým derivative value associated with each price path leading to payoff $c_n(j)$ at node j after n periods. Defining the random variable A_n as

$$A_n(j) :- e^{-nr} \theta_n(j)$$

the risk-neutral pricing formula (8.5) can then be simplified as

$$C_0 = \mathbb{E}^{\mathbb{P}} A_n C_n$$

The advantage of this formulation is that the risk-neutral probability measure does not feature. The derivative value is calculated from a weighted average of the payoffs at expiry where the weights depend on the market state in which the payoff is received. $A_n(j)$ is known as the *state price density* or *state price deflator* or *stochastic discount factor*. These can be modeled in some applications without reference to the corresponding risk-neutral probabilities.

8.7 Summary

The binomial model breaks the evolution of the asset price into small steps where the stock price can either go up or down in each step. The method relies on constructing a portfolio from the underlying asset and cash that replicates the derivative payoff. Then, using the no-arbitrage principle, the value of this portfolio will match the derivative price at all times before maturity. The problem of finding the derivative price reduces to finding the portfolio value. In the one-period model, the solution to two simple simultaneous equations provides this value.

This chapter serves two purposes. On the one hand, the binomial model is an easily understood and versatile approach to option pricing that gives accurate results. European, American and path-dependent options can be handled as well as dividends. Other important features not discussed here such as barriers, variable interest rates and time-dependent volatility can also be accommodated. There are more sophisticated numerical methods such as trinomial trees, finite difference methods and Monte Carlo simulation that are more suitable for complex, high-dimensional derivatives where the binomial model becomes inefficient. These topics are covered in the references.

The other purpose is to prepare the ground for the discussion of option pricing in continuous time. The binomial approach has introduced some important concepts such as *replicating strategies*, *change of probability measure*, *Radon-Nikodým derivatives* and the *risk-neutral probability measure* all of which appear again later. The continuous case will also involve pricing a derivative by following a replicating strategy and applying no-arbitrage arguments. The resulting option pricing formula will have the same formulation in terms of a risk-neutral expectation as presented here. The reader is encouraged to refer back to this chapter to see the intuition behind these ideas as the continuous model develops.

Key things to remember

- Replicating strategies: Section 8.1.
- Change of probability measure: Section 8.1.
- Risk-neutral probability measure: Section 8.1.
- Derivative price is $V_0 = e^{-r}\mathbb{E}^{\mathbb{Q}}C_1$ (one-period model) or $V_0 = e^{-rn}.\mathbb{E}^{\mathbb{Q}}C_n$ (n-period model) where $\mathbb{Q}$ is risk-neutral probability measure
- Evaluating derivative price using two-period binomial model: Exercise 8.2.

Further reading

A detailed theoretical account of the binomial model is given in Shreve (2005). Coleman and Li (2010) also provide theoretical insights. Extra details accessible to students from a finance background are given in Mazzoni (2018) and Wilmott (2013). The intuition behind Radon-Nikodým derivatives is explained in Baxter and Rennie (1996). Various worked examples are elaborated by Lo (2018*a*).

8.8 Coding in R

It is straightforward to code pricing European options using a binomial lattice in R. The code below will price European call or put options for any number of periods given the correct parameters. The choose() function in R is used to calculate $\frac{n!}{i!(n-i)!}$. The parameters chosen replicate Exercise 8.3.

```
European options

S0       <- 18
u        <- 1.036
d        <- 0.971
r        <- 0.002
delta    <- 0          # dividend yield
q        <- (exp(r -delta) - d) / (u - d )
K        <- 20
N        <-  8         # no. of periods

type = 'call'

ST <- replicate((N+1), 0)
VT <- replicate((N+1), 0)
V0 <- replicate((N+1), 0)

for (i in 0:N ) {
        ST[i+1]     <-  S0*u ^i *d ^(N - i)
        if(type ==   'call')
        { VT[i+1] <- max(ST[i+1] - K ,0)   }
        else
        { VT[i+1] <- max(K  - ST[i+1],0)   }
        V0[i+1]    <- exp(-r *N)*VT[i+1]*choose
                         (N ,i)*q ^i *(1-q )^(N - i)
}

Price  <- sum(V0)
Price
```

The code for American-style options is a little more involved as the value of the replicating portfolio at each node must be compared to the early exercise value. We therefore first generate a matrix to capture the share price on the binomial tree. The payoff is calculated at maturity. Then working backwards, the larger of the value of the replicating portfolio and the early exercise value at each node is placed in a second matrix until the initial value is reached.

American options

```
St = matrix(0, nrow=N +1, ncol=N +1)
Vt = matrix(0, nrow=N +1, ncol=N +1)
# share price evolution
for (i in 1:(N +1)) {
        for (j in 1:i ) {
                St[i , j ] = S0 * u ^(j -1)
            * d ^((i -1)-(j -1))
} }
# payoff
type = 'call'

if(type == 'call')
{Vt[N +1,] = pmax(St[N +1,] - K , 0)}
else {Vt[N +1,] = pmax(K - St[N +1,], 0)}

# option valuation

for (i in N :1) {
        for(j in 1:i ) {
                if(type == 'call')
                { Vt[i ,j ] = pmax(((1-q)*Vt[i +1,j]
+ q *Vt[i +1,j +1])/exp(r ), St[i ,j ] - K )}
                else
                { Vt[i ,j ] = pmax(((1-q)*Vt[i +1,j]
+ q *Vt[i +1,j +1])/exp(r), K - St[i ,j ])}
        }
}
Price <- Vt[1 ,1]
Price
```

Questions

1 A stock price is currently \$15. After one month it will be either \$18 or \$14. The continuously compounded risk-free interest rate is 5% p.a. What is the value of a one-month European call option with a strike price of \$16 ? No dividends are payable during this time.

Solution $c_u = 18 - 16 = 2; c_d = 0$ and $u = \frac{18}{15} = 1.2; d = \frac{14}{15} = 0.933$

$$q = \frac{e^r - d}{u - d} = \frac{e^{0.05 \times \frac{1}{12}} - 0.933}{1.2 - 0.933} = 0.266$$

$$\begin{aligned} V_0 &= e^{-r}\left(qc_u + (1-q)c_d\right) \\ &= e^{-0.05\times\frac{1}{12}}\left(0.266 \times 2 + (1 - 0.266) \times 0\right) = \$0.53 \end{aligned}$$

2 Show that there are no-arbitrage opportunities in the market described in the previous question of one stock and risk-free investments.

Solution Since the inequality $d < e^r < u$ holds, there are no-arbitrage opportunities $(0.933 < e^{\frac{0.05}{12}} = 1.004 < 1.2)$.

3 A stock price is currently \$20. Each month it is expected to go up by 3% or down by 4%. The continuously compounded risk-free interest rate is 2% p.a. What is the value of a two-month European put option with a strike price of \$19? No dividends are payable during this time.

Solution

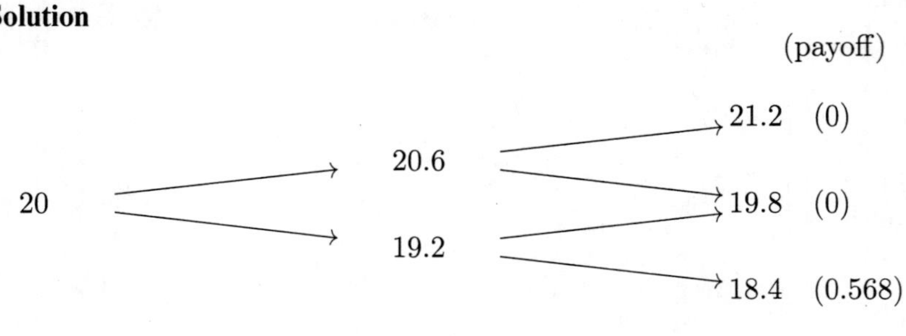

$$r = \frac{0.02}{12}, u = 1.03, d = 0.96, S_0 = 20$$

$$q = \frac{e^r - d}{u - d} = \frac{e^{\frac{0.02}{12}} - 0.96}{1.03 - 0.96} = 0.595$$

$$V_1(1) = e^{-\frac{0.02}{12}}(0.595 \times 0 + (1 - 0.595) \times 0) = 0$$

$$V_1(2) = e^{-\frac{0.02}{12}}(0.595 \times 0 + (1 - 0.595) \times 0.568) = 0.230$$

$$\Longrightarrow V_0 = e^{-\frac{0.02}{12}}(0.595 \times 0 + (1 - 0.595) \times 0.230) = 0.093$$

Or in one step:

$$V_0 = e^{-2\times\frac{0.02}{12}}\Big(0.595^2 \times 0 + 2 \times 0.595 \times (1 - 0.595) \times 0 + (1 - 0.595)^2 \times 0..568\Big) = 0.093$$

Value of the European put is \$0.09.

4 A stock price is currently \$87. Each month it is expected to go up by 2% or down by 5%. The continuously compounded risk-free interest rate is 4% p.a. No dividends are payable during the next two months.
  (i) What is the value of a two-month European put option with a strike price of \$85?
  (ii) What is the value of a two-month American put option with the same strike price ?
  (iii) What is the value of a two-month American call option with the same strike price ?

Solution

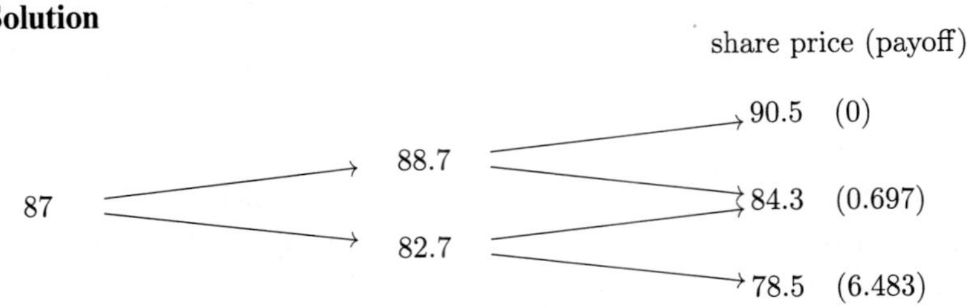

(i) $r = 4\%/12 = 0.0033, u = 1.02, d = 0.95, S_0 = 87$

$$q = \frac{e^{0.0033} - 0.95}{1.02 - 0.95} = 0.762$$

$$V_1(1) = e^{-0.0033}(0.762 \times 0 + (1 - 0.762) \times 0.697) = 0.165$$

$$V_1(2) = e^{-0.0033}(0.762 \times 0.697 + (1 - 0.762) \times 6.483) = 2.067$$

$$\Longrightarrow V_0 = e^{-0.0033}(0.762 \times 0.165 + (1 - 0.762) \times 2.067) = 0.62$$

Or in one step:

$$V_0 = e^{-2\times 0.0033}\Big(0.762^2 \times 0 + 2 \times 0.762 \times (1 - 0.762) \times 0.697 + (1 - 0.762)^2 \times 6.483\Big) = 0.62$$

The value of the European put is \$0.62.

(ii) When the put is American the no-arbitrage approach is modified to select the higher of the early exercise price and the value of the replicating portfolio at each node.

The evolution of the share price is as given in the previous figure.

If the price falls after one month, then exercise the option since the early exercise value is more than the risk-neutral value:

$$K - S_0 d = 85 - 87 \times 0.95 = 2.35 > 2.067$$

$$V_1(1) = 0.165 \text{ as before}$$
$$V_1(2) = 2.35$$
$$\Longrightarrow V_0 = e^{-0.0033}(0.762 \times 0.165 + (1 - 0.762) \times 2.35) = 0.68$$

The value of the American put is \$0.68.

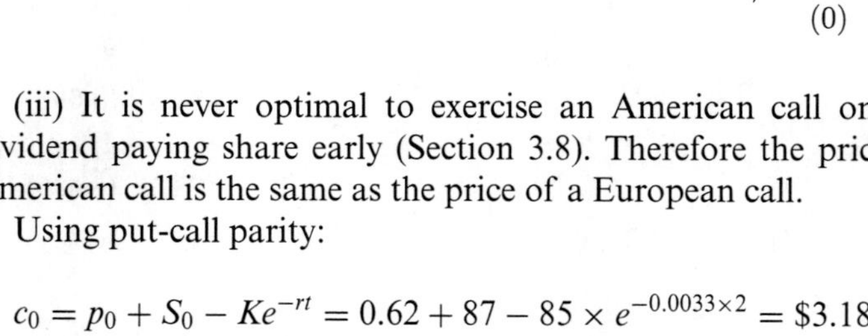

(iii) It is never optimal to exercise an American call on a non-dividend paying share early (Section 3.8). Therefore the price of the American call is the same as the price of a European call.

Using put-call parity:

$$c_0 = p_0 + S_0 - Ke^{-rt} = 0.62 + 87 - 85 \times e^{-0.0033\times 2} = \$3.18$$

or

$$V_0 = e^{-2\times 0.0033}(5.52 \times 0.762^2) = \$3.18$$

5 For the stock in the previous question, estimate the lowest strike price for which it is optimal to exercise an American put option immediately.

Solution

The solution is $K = \$89.61$. There are three approaches.

a. Use trial and error, experimenting with different values of K until the solution is found.
b. Or use the Goal Seek function in Microsoft Excel to find the K for which the difference between the initial value V_0 and the immediate exercise price $K - S_0$ is zero. See the spreadsheet online.
c. It is easily coded in R.

```
S0      <- 87
u       <- 1.02
d       <- 0.95
r       <- 0.04/12
q       <- (exp(r) - d) / (u - d)
N       <- 2

St = matrix(0, nrow=N+1, ncol=N+1)
Vt = matrix(0, nrow=N+1, ncol=N+1)

# share price evolution
V <- function(K) {
        for (i in 1:(N+1)) {
                for (j in 1:i) {
                        St[i, j] = S0*u^(j-1)*d^((i-1)-(j-1))
                }
                }

# payoff
Vt[N+1,] = pmax(K - St[N+1,], 0)
# option valuation
for(j in 1:2) {
        Vt[2,j]= pmax(((1-q)*Vt[3,j]
                        + q*Vt[3,j+1])/exp(r), K - St[2,j])}
        Vt[1,1]= ((1-q)*Vt[2,1] + q*Vt[2,2])/exp(r)
        return(Vt[1,1])
        }

f <- function(K) {
        V(K) - (K - S0)
        }

zero <- uniroot(f,c(0,200))
zero$root
```

6 A stock is currently priced \$20 and can move up by 4% or down by 4% each year. Use a binomial lattice model to find the value of a 10-year European call option on this stock with an exercise price of \$23. The risk-free continuously compounded rate is 2% p.a. The dividend yield is 1% p.a.

Solution

$n = 10, r = 0.02, u = 1.04, d = 0.96, S_0 = 20, \delta = 1\%$

$$q = \frac{e^{r-\delta} - d}{u - d} = \frac{e^{0.02-0.01} - 0.96}{1.04 - 0.96} = 0.626$$

$$V_0 = e^{-10\times 0.02}\Bigg((20 \times 1.04^7 \times 0.96^3 - 23)\frac{10!}{7!3!} \times 0.626^7 \times 0.374^3 +$$
$$(20 \times \times 1.04^8 \times 0.96^2 - 23)\frac{10!}{8!2!} \times 0.626^8 \times 0.374^2 +$$
$$(20 \times 1.04^9 \times 0.96^1 - 23)\frac{10!}{9!1!} \times 0.626^9 \times 0.374^1$$
$$+ (20 \times 1.04^{10} - 23) \times 0.626^{10}\Bigg)$$
$$= 0.569 \text{ or } \$0.60$$

7 A stock is currently priced \$100 and can move up by 3% or down by 4% each week. The risk-free continuously compounded rate is 6% p.a. No dividends are payable during the next five weeks. Asian options are available on this stock where the payoff depends on the arithmetic average of the share price recorded at the end of each of the five weeks. Find the value of an Asian call option when the strike price is (i) \$108 and (ii) \$107. Assume there are 52 weeks in a year.

Solution

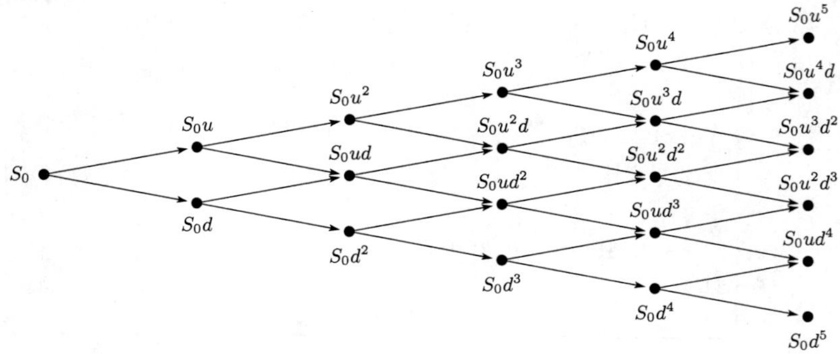

The payoff here is of the form $[A(n) - K]^+$ where

$$A(n) = \frac{1}{n}\sum_{i=1}^{n} S_i$$

and n the number of periods, K the strike price. The payoff is path-dependent since most of the share prices at the end of week five can be reached in multiple ways.

$$q = \frac{e^{\frac{0.06}{52}} - 0.96}{1.03 - 0.96} = 0.588$$

(i) At $K = \$108$, the payoff will be zero unless there are five consecutive up-movements.

Arithmetic average = $(103+106.1+109.3+112.6+115.9)/5 = 109.37$
Payoff = 109.37 – 108 = 1.37

$$V_0 = e^{-5\times\frac{0.06}{52}}(1.37 \times 0.588^5) = \$0.10$$

(ii) At $K = \$107$, the payoff will be zero unless there are five consecutive up-movements or also four consecutive up-movements with a final down-movement.

Average for four ups then one down = $(103+106.1+109.3+112.6+108.0)/5 = 107.79$
Payoff = 107.79 – 107 = 0.79

$$V_0 = e^{-5\times\frac{0.06}{52}}(2.37 \times 0.588^5 + 0.79 \times 0.588^4 \times 0.412) = \$0.20$$

8 A non-dividend paying stock is currently priced \$56 with volatility 20% p.a. The risk-free continuously compounded rate is 6% p.a.

(a) Use a two-step binomial tree to determine the value of a 2-month European call option with a strike price of \$60.

(b) What are the initial stock and cash holdings ϕ and ψ for the replicating portfolio ?

Solution

(a)

$$u = e^{\sigma\sqrt{\delta t}} = e^{0.2\sqrt{\frac{1}{12}}} = 1.059, d = \frac{1}{u} = 0.944$$

$$q = \frac{e^r - d}{u - d} == \frac{e^{\frac{0.06}{12}} - 0.944}{1.059 - 0.944} = 0.529$$

The value of the option $V_0 = e^{-2\times\frac{0.06}{12}}(2.85 \times 0.529^2) = \$0.79$.
The tree is given below:

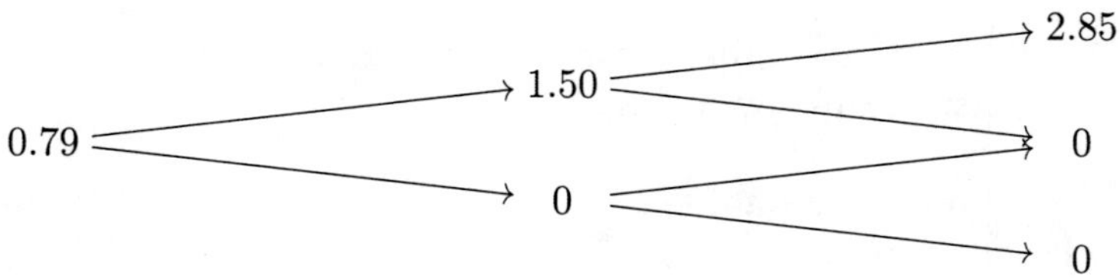

(b) We have $V_1(1) = 1.50$ and $V_1(2) = 0$

$$\phi = \frac{V_1(1) - V_1(2)}{S_0(u - d)} = \frac{1.50}{56(1.059 - 0.944)} = 0.23$$

that is, purchase 0.23 shares and

$$\psi = e^{-r}\frac{V_1(2)u - V_1(1)d}{u - d} = e^{-\frac{0.06}{12}}\frac{0 \times 1.059 - 10 \times 0.944}{1.059 - 0.944} = -12.21$$

that is, borrow \$12.21 cash.

9 A stock price can move up with probability 0.6 or down with probability 0.4 over each period. Associated with these movements are risk-neutral probabilities 0.5 for an up-movement and 0.5 for a down-movement. Determine the Radon-Nikodým process of $\mathbb{Q}$ with respect to $\mathbb{P}$ using a three-period binomial model.

Solution

$$\Theta_0 = 1 \quad \Theta_1 = \begin{cases} \frac{q}{p} = \frac{5}{6} \\ \frac{q'}{p'} = \frac{5}{4} \end{cases} \quad \Theta_2 = \begin{cases} \frac{q^2}{p^2} = \frac{25}{36} \\ \frac{qq'}{pp'} = \frac{25}{24} \\ \frac{q'q}{p'p} = \frac{25}{24} \\ \frac{q'^2}{p'^2} = \frac{25}{16} \end{cases} \quad \Theta_3 = \begin{cases} \frac{q^3}{p^3} = \frac{125}{216} \\ \frac{q^2q'}{p^2p'} = \frac{125}{144} \\ \frac{qq'q}{pp'p} = \frac{125}{144} \\ \frac{q'q^2}{p'p^2} = \frac{125}{144} \\ \frac{qq'^2}{pp'^2} = \frac{125}{96} \\ \frac{q'qq'}{p'pp'} = \frac{125}{96} \\ \frac{q'^2q}{p'^2p} = \frac{125}{96} \\ \frac{q'^3}{p'^3} = \frac{125}{64} \end{cases}$$

Appendix

A.1 Continuous-time limit of binomial model

Start with the pricing Formula 8.3 but allow the calculation to occur over the interval $[t, t+\delta t]$.

$$V_t = e^{-r\delta t}(qc_u + (1-q)c_d)$$

$$\Longrightarrow \frac{e^{r\delta t}-d}{u-d}c_u + \frac{u-e^{r\delta t}}{u-d}c_d - e^{r\delta t}V_t = 0 \qquad \text{from definition of } q$$

$$\Longrightarrow \frac{e^{r\delta t}-e^{-\sigma\sqrt{\delta t}}}{e^{\sigma\sqrt{\delta t}}-e^{-\sigma\sqrt{\delta t}}}c_u + \frac{e^{\sigma\sqrt{\delta t}}-e^{r\delta t}}{e^{\sigma\sqrt{\delta t}}-e^{-\sigma\sqrt{\delta t}}}c_d - e^{r\delta t}V_t = 0 \qquad \text{from Section 8.5}$$

The Taylor expansion of the exponential function for small x implies $e^x \approx 1+x$.

$$\frac{\sigma\sqrt{\delta t}+(r-\frac{1}{2}\sigma^2)\delta t}{2\sigma\sqrt{\delta t}}c_u + \frac{\sigma\sqrt{\delta t}-(r-\frac{1}{2}\sigma^2)\delta t}{2\sigma\sqrt{\delta t}}c_d - (1+r\delta t)V_t = 0$$

$$\Longrightarrow \frac{1}{2}(c_u+c_d) + \frac{(r-\frac{1}{2}\sigma^2)\sqrt{\delta t}}{2\sigma}(c_u-c_d) - (1+r\delta t)V_t = 0$$

Now find expressions for the payoffs c_u and c_d. The value of the replicating portfolio V_t is a function of the share price and time, that is, $V_t = F(S_t, t)$. This portfolio replicates the payoffs c_u and c_d so $c_{u,d} = F(S_{t+\delta t}, t+\delta t)$.

$$c_{u,d} = F(e^{\pm\sigma\sqrt{\delta t}}S_t, t+\delta t)$$

Apply the Taylor expansion in two variables to the function $F(S_t, t)$. To simplify the notation we drop the subscripts. Remember that F and all its derivatives are evaluated at (S_t, t).

$$\begin{aligned}
c_{u,d} &\approx F + (e^{\sigma\sqrt{\delta t}}-1)S\frac{\partial F}{\partial S} + \delta t\frac{\partial F}{\partial t} + \frac{1}{2}(e^{\pm\sigma\sqrt{\delta t}}-1)^2 S^2\frac{\partial^2 F}{\partial S^2} \\
&\approx F + (e^{\pm\sigma\sqrt{\delta t}}-1)S\frac{\partial F}{\partial S} + \delta t\frac{\partial F}{\partial t} + \frac{1}{2}(e^{\pm 2\sigma\sqrt{\delta t}} - 2e^{\pm\sigma\sqrt{\delta t}} + 1)S^2\frac{\partial^2 F}{\partial S^2} \\
&\approx F + (\pm\sigma\sqrt{\delta t} + \frac{1}{2}\sigma^2\delta t)S\frac{\partial F}{\partial S} + \delta t\frac{\partial F}{\partial t} + \frac{1}{2}(1 \pm 2\sigma\sqrt{\delta t} + 2\sigma^2\delta t - 2 \\
&\qquad - \pm 2\sigma\sqrt{\delta t} - \sigma^2\delta t + 1)S^2\frac{\partial^2 F}{\partial S^2} \\
&\approx F + (\pm\sigma\sqrt{\delta t} + \frac{1}{2}\sigma^2\delta t)S\frac{\partial F}{\partial S} + \delta t\frac{\partial F}{\partial t} + \frac{1}{2}\sigma^2(\delta t)S^2\frac{\partial^2 F}{\partial S^2}
\end{aligned}$$

Substituting back into the equation above for c_u and c_d

$$\frac{1}{2}(c_u + c_d) + \frac{(r - \frac{1}{2}\sigma^2)\sqrt{\delta t}}{2\sigma}(c_u - c_d) - (1 + r\delta t)F = 0$$

$$\Longrightarrow \left(F + (\frac{1}{2}\sigma^2\delta t)S\frac{\partial F}{\partial S} + \delta t\frac{\partial F}{\partial t} + \frac{1}{2}\sigma^2(\delta t)S^2\frac{\partial^2 F}{\partial S^2}\right)$$

$$+ \frac{(r - \frac{1}{2}\sigma^2)\sqrt{\delta t}}{2\sigma}\left(2\sigma\sqrt{\delta t}S\frac{\partial F}{\partial S}\right) - (1 + r\delta t)F = 0$$

$$\Longrightarrow \delta t\frac{\partial F}{\partial t} + \frac{1}{2}\frac{\partial^2 F}{\partial S^2}\sigma^2(\delta t)S^2 + r(\delta t)\frac{\partial F}{\partial S}S - r(\delta t)F = 0$$

$$\Longrightarrow \frac{\partial F}{\partial t} + \frac{1}{2}\frac{\partial^2 F}{\partial S^2}\sigma^2 S^2 + r\frac{\partial F}{\partial S}S - rF = 0$$

This is the Black–Scholes equation.

9 Black–Scholes equation

Introduction

The Black–Scholes equation is one of the most influential results in modern financial mathematics providing the theoretical framework for pricing European-style options. Black and Scholes (1973) originated the theory while Merton (1973) provided a rigorous mathematical derivation and extended the model to more general settings. Before this breakthrough, option trading was regarded as speculation or gambling but the Black–Scholes pricing framework introduced an approach grounded in financial theory providing legitimacy to derivative trading and contributing to its explosive growth. Scholes and Merton were awarded the 1997 Nobel Prize in Economics for this work.

The model used to develop the theory rests on the same key assumptions as the binomial model – no risk-free arbitrage opportunities as well as competitive, well-functioning and frictionless markets. In addition, we have that the underlying asset can be traded continuously and its price follows a geometric Brownian motion. Many of the ideas met in the previous chapter – *replicating strategy, change of probability measure, risk-neutral probability measure* – appear here also but the exposition inevitably becomes more complicated as we allow for the underlying asset price to evolve continuously over time with a continuous probability distribution of price changes.

The Black–Scholes equation is obtained by simply considering the functions of time and asset price that can arise as values of self-financing portfolios which by a no-arbitrage argument replicate the derivative price. The elements needed to derive the equation are very similar to the binomial case: changes in replicating portfolio value are driven by changes in the value of the portfolio assets and no-arbitrage arguments are employed to equate the replicating portfolio value with the derivative value as time changes. This led to two simultaneous equations in the binomial model which were solved for the option price. Analogous steps are followed in the continuous case although they look different due to the stochastic calculus. The result is not a set of simultaneous equations but a differential equation. This is the Black–Scholes equation.

DOI: 10.4324/9781003477624-12

The equation is dependent on the underlying assumptions. We will examine how the analysis changes as these assumptions change. Solving the equation to usefully price options is not straightforward and is considered here but largely left to the following chapter.

9.1 Self-financing portfolios

A *self-financing* portfolio has the property that changes in portfolio value are driven by changes in the value of the portfolio assets. There is no possibility of adding to the portfolio through labor income or subtracting from the portfolio by consumption or transaction costs.

We will now derive a very simple expression for the change in value of a self-financing portfolio. It is assumed that the portfolio assets do not pay dividends.

The portfolio value at time t is given by

$$V_t = \sum_{i=1}^{n} \phi_{i,t} S_{i,t}$$

where n is the number of assets, $S_{i,t}$ the price of asset i at time t and $\phi_{i,t}$ the amount of holdings of asset i.

The value of the portfolio changes just as a result of changes in the asset prices. The portfolio may be rebalanced but only in such a way that the overall value does not change. The asset holdings $\phi_{i,t}$ are then a function of time.

The sequence is

(i) time t,

$$V_t = \sum_{i=1}^{n} \phi_{i,t} S_{i,t}$$

(ii) at time $t + \Delta t$ asset values have changed

$$V_{t+\Delta t} = \sum_{i=1}^{n} \phi_{i,t} S_{i,t+\Delta t}$$

(iii) then possibly rebalance the portfolio by choosing new asset holdings $\phi_{i,t+\Delta t}$ in such a way that the overall value does not change

$$V_{t+\Delta t} = \sum_{i=1}^{n} \phi_{i,t+\Delta t} S_{i,t+\Delta t}$$

Just using the first two steps in the sequence, we have:

$$\Delta V_t = V_{t+\Delta t} - V_t = \sum_{i=1}^{n} \phi_{i,t}(S_{i,t+\Delta t} - S_{i,t})$$
$$= \sum_{i=1}^{n} \phi_{i,t} \Delta S_{i,t}$$

And in the limit

$$dV = \sum_{i=1}^{n} \phi_{i,t}\, dS_i \tag{9.1}$$

Portfolio strategies that satisfy this condition are called *self-financing strategies*. For instance, a *buy-and-hold* strategy where holdings are constant over time is self-financing.

9.2 Black–Scholes equation

The model is similar to the binomial case. The assumptions made in the Black–Scholes model are

1. There are no risk-free arbitrage opportunities.
2. No market frictions – no taxes or transaction costs.
3. Borrowing and investing occurs at the same rate r.
4. Competitive and well-functioning markets
   - No maximum or minimum units of trading.
   - No limits to short selling (e.g., due to credit limits and collaterals).
5. The underlying asset can be traded continuously and its price follows a geometric Brownian motion.

These assumptions are necessary for constructing perfect hedges and so that there are no losses through rebalancing.

There are two assets to create the portfolio replicating the derivative. There is a risky asset with price S_t as before. For the moment, we will assume the asset does not pay dividends. We generalize from cash in the binomial model to a riskless asset with price B_t.

The asset price evolutions are:

$$\begin{aligned} dS &= \mu S_t dt + \sigma S_t dW \\ dB &= r B_t dt \end{aligned} \tag{9.2}$$

Merton developed a version of the model with a stochastic interest rate but to keep things simple here we will assume the interest rate r is constant.

The value of the replicating portfolio at time t containing ϕ_t units of the risky asset and ψ_t units of the riskless asset is

$$V_t = \phi_t S_t + \psi_t B_t$$

Now to determine which functions of time and asset price can arise as values of a self-financing replicating portfolio.

From the self-financing property

$$\begin{aligned} dV &= \phi_t dS + \psi_t dB \\ &= \phi_t(\mu S_t dt + \sigma S_t dW) + \psi_t(rB_t dt) \\ &= \underline{(\phi_t \mu S_t + \psi_t r B_t)}dt + \underbrace{\phi_t \sigma S_t}\, dW \end{aligned}$$

Let the derivative value be some function of the risky asset price and time, that is, $C_t = F(S_t, t)$. Applying the time-dependent Itô rule, we get:

$$\begin{aligned} dC &= \frac{\partial F}{\partial t}dt + \frac{\partial F}{\partial S}dS + \frac{1}{2}\frac{\partial^2 F}{\partial S^2}(dS)^2 \\ &= \frac{\partial F}{\partial t}dt + \frac{\partial F}{\partial S}(\mu S_t dt + \sigma S_t dW) + \frac{1}{2}\frac{\partial^2 F}{\partial S^2}\sigma^2 S_t^2 dt \\ &= \underline{(\frac{\partial F}{\partial t} + \frac{\partial F}{\partial S}\mu S_t + \frac{1}{2}\frac{\partial^2 F}{\partial S^2}\sigma^2 S_t^2)}dt + \underbrace{\frac{\partial F}{\partial S}\sigma S_t}\, dW \end{aligned}$$

Using the no-arbitrage property, changes in the replicating portfolio value have to equal changes in the derivative value. Comparing the underlined parts

$$\frac{\partial F}{\partial t} + \frac{\partial F}{\partial S}\mu S_t + \frac{1}{2}\frac{\partial^2 F}{\partial S^2}\sigma^2 S_t^2 = \phi_t \mu S_t + \psi_t r B_t \tag{9.3}$$

$$\frac{\partial F}{\partial S}\sigma S_t = \phi_t \sigma S_t \tag{9.4}$$

And the derivative value must equal the value of the replicating portfolio.

$$F = \phi_t S_t + \psi_t B_t \tag{9.5}$$

Then from Equation (9.4)

$$\phi_t = \frac{\partial F}{\partial S} \tag{9.6}$$

This expression gives the number of units of the risky asset to be held in the replicating portfolio and is known as the *delta* of the derivative.

Substituting this into Equation (9.3)

$$\frac{\partial F}{\partial t} + \frac{1}{2}\frac{\partial^2 F}{\partial S^2}\sigma^2 S_t^2 = \psi r B_t \tag{9.7}$$

From Equation (9.5)

$$\psi_t B_t = F - \phi_t S_t$$

Substituting this into Equation (9.7)

$$\begin{aligned}\frac{\partial F}{\partial t} + \frac{1}{2}\frac{\partial^2 F}{\partial S^2}\sigma^2 S_t^2 &= rF - r\phi_t S_t \\ &= rF - r\frac{\partial F}{\partial S}S_t\end{aligned}$$

Rearranging gives the Black–Scholes partial differential equation

$$\frac{\partial F}{\partial t} + \frac{1}{2}\sigma^2 S_t^2 \frac{\partial^2 F}{\partial S^2} + rS_t\frac{\partial F}{\partial S} - rF = 0 \tag{9.8}$$

Any function F that represents the price of a derivative where S_t is the underlying asset must satisfy this equation. So, given the payoff from the derivative at maturity, the problem of pricing the derivative becomes the problem of finding the solution to a partial differential equation with a boundary condition.

Once we have found the function F that solves the partial differential equation then the self-financing replicating strategy is defined by

$$\phi_t = \frac{\partial F}{\partial S}, \quad \psi_t = \frac{F - \phi_t S_t}{B_t} \tag{9.9}$$

Exercise 9.1. *Show that the following are solutions to the Black–Scholes equation*

(a) $F(S_t) = S_t$
(b) $F(t) = e^{rt}$
(c) $F(S_t, t) = S_t - Ke^{-r(T-t)}$

Proof. The Black–Scholes equation is

$$\frac{\partial F}{\partial t} + \frac{1}{2}\sigma^2 S_t^2 \frac{\partial^2 F}{\partial S^2} + rS_t \frac{\partial F}{\partial S} - rF = 0$$

(a) Given $F(S_t) = S_t$ then

$$\frac{\partial F}{\partial t} = 0, \frac{\partial F}{\partial S} = 1, \text{ and } \frac{\partial^2 F}{\partial S^2} = 0$$

The left-hand side of the Black–Scholes equation becomes

$$0 + \frac{1}{2}\sigma^2 S_t^2 \times 0 + rS_t \times 1 - rS_t = rS_t - rS_t = 0$$

Therefore $F(S_t) = S_t$ satisfies the Black–Scholes equation.

(b) Given $F(t) = e^{rt}$ then

$$\frac{\partial F}{\partial t} = re^{rt}, \frac{\partial F}{\partial S} = 0, \text{ and } \frac{\partial^2 F}{\partial S^2} = 0$$

The left-hand side of the Black–Scholes equation becomes

$$re^{rt} + \frac{1}{2}\sigma^2 S_t^2 \times 0 + rS_t \times 0 - re^{rt} = re^{rt} - re^{rt} = 0$$

Therefore $F(t) = e^{rt}$ satisfies the Black–Scholes equation.

(c) Given $F(S_t, t) = S_t - Ke^{-r(T-t)}$ then

$$\frac{\partial F}{\partial t} = -rKe^{-r(T-t)}, \frac{\partial F}{\partial S} = 1, \text{ and } \frac{\partial^2 F}{\partial S^2} = 0$$

The left-hand side of the Black–Scholes equation becomes

$$\begin{aligned} &- rKe^{-r(T-t)} + \frac{1}{2}\sigma^2 S_t^2 \times 0 + rS_t \times 1 - r(S_t - Ke^{-r(T-t)}) \\ &= -rKe^{-r(T-t)} + rS_t - rS_t + rKe^{-r(T-t)} = 0 \end{aligned}$$

Therefore $F(S_t, t) = S_t - Ke^{-r(T-t)}$ satisfies the Black–Scholes equation.
This expression is the value of a forward contract on a non-dividend paying stock.[1] □

The assumptions underlying the Black–Scholes model are unrealistic. In reality, options are subject to capital gain taxes. Transaction costs for trading derivatives include commissions paid to a broker, fees levied by exchanges and implicit costs associated with bid-ask spreads. Short selling may be suspended temporarily during a crisis or limited on a particular security following a significant price decline. Extensions of the Black–Scholes model have been developed to allow for these shortcomings. For example, Leland (1985) includes transaction costs while Atmaz and Basak (2019) allow costly short-selling. More complex models often provide greater insight but at the cost of greater complexity and perhaps reduced tractability. All models inevitably are approximations to reality. Analyses of derivative prices tend to show that the Black–Scholes model gives a reasonably good approximation to the market.

9.3 Extensions of the Black–Scholes equation

We can extend the analysis of the previous section in a number of ways.

9.3.1 Continuous dividend stream

If we assume the underlying asset is a stock which pays dividends continuously at a rate q per share, then it can be shown that

$$dV = \phi_t(dS + qS_t dt) + \psi_t dB$$

This is given as an exercise at the end of this chapter. The share price process when no dividend is paid is

$$dS = \mu S_t dt + \sigma S_t dW$$

When dividends are issued, the value of the firm is reduced by the sum of all dividends paid. Each share in the firm's stock is also reduced in value proportionately. If the dividend is paid at rate q per share, the share loses $qS_t dt$ in value over the time interval dt and the new share price process becomes

$$dS = (\mu - q)S_t dt + \sigma S_t dW$$

We then have

$$\begin{aligned} dV &= \phi_t(dS + qS_t dt) + \psi_t dB \\ &= \phi_t\Big((\mu - q)S_t dt + \sigma S_t dW + qS_t dt\Big) + \psi_t(rB_t dt) \\ &= \underline{(\phi_t \mu S_t + \psi_t rB)dt} + \underbrace{\phi_t \sigma S_t}\, dW \quad \text{as before} \end{aligned}$$

Applying the time-dependent Itô rule, we get:

$$\begin{aligned}dC &= \frac{\partial F}{\partial t}dt + \frac{\partial F}{\partial S}dS + \frac{1}{2}\frac{\partial^2 F}{\partial S^2}(dS)^2\\ &= \frac{\partial F}{\partial t}dt + \frac{\partial F}{\partial S}\Big((\mu - q)S_t dt + \sigma S_t dW\Big) + \frac{1}{2}\frac{\partial^2 F}{\partial S^2}\sigma^2 S_t^2 dt\\ &= \underline{\left(\frac{\partial F}{\partial t} + \frac{\partial F}{\partial S}(\mu - q)S_t + \frac{1}{2}\frac{\partial^2 F}{\partial S^2}\sigma^2 S_t^2\right)dt} + \underbrace{\frac{\partial F}{\partial S}\sigma S_t}\, dW\end{aligned}$$

Comparing underlying sections as before we get:

$$\frac{\partial F}{\partial t} + \frac{1}{2}\sigma^2 S_t^2 \frac{\partial^2 F}{\partial S^2} + (r - q)\frac{\partial F}{\partial S}S_t - rF = 0 \tag{9.10}$$

9.3.2 *Storage costs*

The underlying may instead be a commodity with associated storage costs. The commodity price process S_t is assumed to be geometric Brownian motion. If we assume storage costs are

(i) proportional to the commodity price[2] at rate c

(ii) proportional to the storage period

(iii) paid continuously

then the incremental storage cost per unit of the commodity is $-cS_t dt$.

If we assume furthermore that costs are paid from the value of the replicating portfolio, we then have

$$\begin{aligned}dV &= \phi_t(dS - cS_t dt) + \psi_t dB\\ &= \phi\Big((\mu - c)S_t dt + \sigma S_t dW\Big) + \psi_t(rB_t dt)\\ &= \underline{(\phi(\mu - c)S_t + \psi r B_t)dt} + \underbrace{\phi\sigma S_t}\, dW\end{aligned}$$

Application of the time-dependent Itô rule is unchanged from Section 9.2. Then proceeding as before we get:

$$\frac{\partial F}{\partial t} + \frac{1}{2}\frac{\partial^2 F}{\partial S^2}\sigma^2 S_t^2 + (r + c)\frac{\partial F}{\partial S}S_t - rF = 0 \tag{9.11}$$

Extensions of the Black–Scholes equation can be developed for derivatives with other underlying securities such as options on foreign currency. These extensions are explained very clearly in Mazzoni (2018).

9.4 Solving the Black–Scholes equation

We say that a portfolio strategy $F(S_t, t)$ meets the liability given by the option payoff function $P(S_t)$ at time T if the function $F(S_t, t)$ satisfies

$$F(S_T, T) = P(S_T) \text{ for all } S_T$$

For example, the payoff function for a European call option is $P(S_T) = [S_T - K]^+$. Finding the option price means looking for a function $F(S_t, t)$ that satisfies both the Black–Scholes partial differential equation and the above terminal condition which expresses that the liability is met exactly.

This is similar to the type of problem encountered in Section 1.16 where first-order differential equations were solved with initial conditions. But it is more challenging as we are dealing here with a second-order partial differential equation. It is not easy to find a solution. Partial differential equations (PDEs) are notoriously difficult to solve especially non linear ones. Fortunately, the Black–Scholes PDE has a number of properties that make finding a solution easier.

A PDE is called *linear* if it is linear in the unknown and its derivatives. For example, for a function F of S_t and t, a second-order linear PDE is of the form

$$\begin{aligned} a_1(S_t, t)\frac{\partial^2 F}{\partial S^2} + a_2(S_t, t)\frac{\partial^2 F}{\partial S \partial t} + a_3(S_t, t)\frac{\partial^2 F}{\partial t^2} \\ + a_4(S, t)\frac{\partial F}{\partial S} + a_5(S_t, t)\frac{\partial F}{\partial t} + a_6(S_t, t)F = f(S_t, t) \end{aligned}$$

where the a_i and f are functions of the independent variables S_t and t only. Here F and all its derivatives enter the PDE with exponents of one. The Black–Scholes PDE is therefore linear.

If f is zero everywhere, then the linear PDE is also *homogeneous*. The Black–Scholes PDE is homogeneous as the right-hand side of the equation is zero.

Lastly, if $a_2^2 - 4a_1a_3 = 0$, then the PDE is *parabolic*. In the Black–Scholes equation a_2 and a_3 are both zero, so the Black–Scholes equation is parabolic.

These properties mean that techniques are available to simplify the PDE and construct solutions.

Here is a quick overview of the approach. The Black–Scholes equation can be transformed to a very simple parabolic equation.

$$\frac{\partial U}{\partial \tau} = \frac{1}{2}\sigma^2 \frac{\partial^2 U}{\partial x^2}$$

by letting $F(S_t, t) = e^{-r\tau}U(x, \tau)$ where

$$\tau = T - t \text{ and } x = \log S_t + (r - \frac{1}{2}\sigma^2)\tau$$

Parabolic equations have been studied extensively in physics – originally by Joseph Fourier in his *Théorie Analytique de la Chaleur* (1822) to model the flow of heat through a thin rod and later in other diffusion-type applications such as modeling the spread of dye through a liquid. A technique known as separation of variables can be used to solve this simpler equation. The solution is assumed to be separable into a product of functions of x and τ:

$$U(x,\tau) = X(x)\mathcal{T}(\tau)$$

Substituting this expression for U in the PDE above results in two ordinary differential equations. These are then easily solved using conventional approaches.[3]

In the next chapter, we are going to take a very different approach to finding the derivative price – risk-neutral pricing – which gives us an expectation that can be evaluated explicitly in some cases. This is the continuous analogue of the approach we used in the binomial model. We will show that, just as in the binomial approach, the method is consistent with the Black- Scholes PDE.

The assumption at the start of this section that payoff can only occur at option maturity implies that we have only been considering European-style options. However, the analytical techniques described above cannot be used to give a price for the majority of American-style options. Instead, approximate solutions are found using numerical methods. The method of finite differences is most commonly used. In this method, the derivatives appearing in the PDE and the boundary conditions are replaced by their finite difference approximations. Then, this system of equations is solved iteratively.

9.5 The Black–Scholes formula

The function $F(S_t, t)$ that satisfies the Black–Scholes equation and exactly meets the liability $P(S_T) = [S_T - K]^+$ for a European call option on a non-dividend paying stock is

$$F(S_t, t) = S_t\Phi(d_1) - e^{-r(T-t)}K\Phi(d_2) \tag{9.12}$$

where $\Phi(\cdot)$ is the cumulative distribution function of the standard normal distribution, and

$$d_1 = \frac{\log(S_t/K) + (r + \frac{1}{2}\sigma^2)(T-t)}{\sigma\sqrt{T-t}}, \quad d_2 = d_1 - \sigma\sqrt{T-t}$$

This is the Black–Scholes formula for the European call option price. It can be derived by the analytical techniques discussed in the previous section or by risk-neutral pricing approaches as shown in the next chapter. It can be shown that it satisfies the Black–Scholes equation by calculating all the relevant partial derivatives and substituting as was done before in Exercise 9.1. This is given as a student problem in the questions at the end of this chapter.

Exercise 9.2. *Consider a European call option in which the exercise price of the option is \$45 and the time to expiry is 4 months. The current price of the underlying asset is \$40 and the volatility is 20% per annum. Given that the risk-free interest rate is 5% per annum, find the value of this option.*

Proof.

$$
\begin{aligned}
d_1 &= \frac{\log(S_t/K) + (r + \frac{1}{2}\sigma^2)(T-t)}{\sigma\sqrt{T-t}} \\
&= \frac{\log(40/45) + (0.05 + \frac{1}{2}0.2^2)(1/3)}{0.2\sqrt{1/3}} = -0.8180 \implies \Phi(d_1) = 0.2067 \\
d_2 &= d_1 - \sigma\sqrt{T-t} \\
&= -0.8180 - 0.2\sqrt{1/3} = -0.9334 \implies \Phi(d_2) = 0.1753
\end{aligned}
$$

$$
\begin{aligned}
c_t &= S_t\Phi(d_1) - e^{-r(T-t)}K\Phi(d_2) \\
&= 40(0.2067) - e^{-0.05/3}45(0.1753) = \$0.51
\end{aligned}
$$

□

9.6 Summary

Before the pioneering work by Black, Scholes and Merton, option valuation was theoretically unsound. Their methods provided finance with the first widely used model for calculating the theoretical value of options. The methodology has further generated new types of financial instruments and facilitated more efficient risk management. When awarding Scholes and Merton the 1997 Economic Sciences prize in memory of Alfred Nobel, the Royal Swedish Academy of Sciences stated that their approach 'stands out among the foremost contributions to economic sciences over the last 25 years.'

In this chapter, the Black–Scholes equation was derived using no-arbitrage arguments. The derivative value was replicated by constructing a self-financing portfolio of the underlying asset and a riskless asset. The self-financing and no-arbitrage properties lead to the Black–Scholes equation. Any function representing the price of a derivative on the underlying asset must satisfy this equation. Variants of the equation were then developed to allow for dividend-paying assets and commodities.

Once we have specified the derivative payoff at maturity the Black–Scholes equation can be solved for the derivative price using well-known techniques from physics. This is not the approach that we will take. Instead, in the next chapter, we will use risk-neutral pricing to find the derivative price. This approach follows the logic and conceptual framework of the binomial model.

Key things to remember

- Derivation of Black–Scholes equation: Section 9.2.
- The form of the Black–Scholes equation depends on the assumptions. Change the assumption and the equation changes: Section 9.3.
- In derivative pricing we are basically trying to solve a partial differential equation with boundary conditions given by the payoff: Section 9.4.

Further reading

Shreve (2005) and Klebaner (2012) provide the mathematical finance foundations for understanding the Black–Scholes equation. Mazzoni (2018) considers a number of variants of the equation with different underlying assets. Wilmott (2013) explains how to solve this partial differential equation using techniques from physics. Numerical methods to price American options are also explained.

Notes

1. From Chapter 3, the value of a forward contract is $(K_t - K_0)e^{-r(T-t)}$ where K_0 is the original forward price and K_t is the forward price if negotiated at time t. Also from Chapter 3, $K_t = S_t e^{r(T-t)}$.
2. 'Storage' costs include security. More expensive commodities such as gold would require greater security.
3. An alternative approach to solving the heat equation would be to use integral transforms. See Mazzoni (2018) section 13.4 or Wilmott (2013) section 7.2 for details.

Questions

1 Consider a situation in which we have a risky asset with price S_t and a riskless asset with price B_t. The first asset is a stock which is assumed to pay dividends continuously at a rate q. Show that, in this case, the value of a self-financing portfolio must satisfy

$$dV = \phi_t(dS + qS_t dt) + \psi_t dB$$

Solution

We calculate the increase in value due to dividends from t to $t + \Delta t$ in stages.

From Chapter 3, the dividend rate per unit time is

$$q = \frac{\text{Dividends per share}}{\text{Price per share}}$$

Total dividends per unit time are then

$$q \times \text{price per share}(S_t) \times \text{no. of shares}(\phi_t) = q\phi_t S_t$$

And total dividends in the time interval Δt are $q\phi_t S_t \Delta t$.

Then, the portfolio value at time t is:

$$V_t = \phi_t S_t + \psi_t B_t,$$

while at time $t + \Delta t$, the portfolio value is

$$V_{t+\Delta t} = \phi_t S_{t+\Delta t} + q\phi_t S_t \Delta t + \psi_t B_{t+\Delta t}$$

So,

$$V_{t+\Delta t} - V_t = \phi_t(S_{t+\Delta t} - S_t) + q\phi_t S_t \Delta t + \psi_t(B_{t+\Delta t} - B_t)$$

In the limit,

$$\begin{aligned} dV &= \phi_t dS + q\phi_t S_t dt + \psi_t dB \\ &= \phi_t(dS + qS_t dt) + \psi_t dB \end{aligned}$$

2 An investor in the Black–Scholes world chooses a portfolio strategy such that the value of risky assets is constant after discounting. In other words, the amount ϕ_t of risky assets held is chosen in such a way that

$$\phi_t S_t = aB_t$$

where a is a constant. Show that the self-financing condition for this portfolio strategy can be written as

$$dV = B_t\Big((\mu a + \psi_t r)dt + a\sigma dW\Big)$$

where $\{W_t\}_{t\geq 0}$ is standard Brownian motion.

Solution
From the self-financing property

$$\begin{aligned} dV &= \phi_t dS + \psi_t dB \\ &= \phi_t(\mu S_t dt + \sigma S_t dW) + \psi_t r B_t dt \\ &= (\phi_t \mu S_t + \psi_t r B_t)dt + \phi_t \sigma S_t dW \\ &= (\mu a B_t + \psi_t r B_t)dt + a\sigma B_t dW \\ &= B_t\Big((\mu a + \psi_t r)dt + a\sigma dW\Big) \end{aligned}$$

3 Complete the derivation of the version of the Black–Scholes equation where the underlying asset is a stock paying dividends continuously at rate q (Equation 9.10).

Solution We have three equations

$$\frac{\partial F}{\partial t} + \frac{\partial F}{\partial S}(\mu - q)S_t + \frac{1}{2}\frac{\partial^2 F}{\partial S^2}\sigma^2 S_t^2 = \phi_t \mu S_t + \psi_t r B_t$$

$$\phi_t \sigma S_t = \frac{\partial F}{\partial S}\sigma S_t$$

$$F = \phi_t S_t + \psi_t B_t$$

From the last two equations

$$\phi_t = \frac{\partial F}{\partial S}$$

and $\psi_t B_t = F - \phi_t S_t$

Substituting in the first equation

$$\frac{\partial F}{\partial t} - \frac{\partial F}{\partial S} q S_t + \frac{1}{2}\frac{\partial^2 F}{\partial S^2}\sigma^2 S_t^2 = rF - r\frac{\partial F}{\partial S} S_t$$

$$\Longrightarrow \frac{\partial F}{\partial t} + \frac{1}{2}\sigma^2 S_t^2 \frac{\partial^2 F}{\partial S^2} + (r - q) S_t \frac{\partial F}{\partial S} - rF = 0$$

4 Complete the derivation of the version of the Black–Scholes equation where the underlying asset is a commodity with associated storage costs at rate c (Equation 9.11).

Solution We have three equations

$$\frac{\partial F}{\partial t} + \frac{\partial F}{\partial S}\mu S_t + \frac{1}{2}\frac{\partial^2 F}{\partial S^2}\sigma^2 S_t^2 = \phi_t(\mu - c) S_t + \psi_t r B_t$$

$$\phi_t \sigma S_t = \frac{\partial F}{\partial S}\sigma S_t$$

$$F = \phi_t S_t + \psi_t B_t$$

From the last two equations

$$\phi_t = \frac{\partial F}{\partial S}$$

and $\psi_t B_t = F - \phi_t S_t$

Substituting in the first equation

$$\frac{\partial F}{\partial t} + \frac{1}{2}\frac{\partial^2 F}{\partial S^2}\sigma^2 S_t^2 = rF - (r + c)\frac{\partial F}{\partial S} S_t$$

$$\Longrightarrow \frac{\partial F}{\partial t} + \frac{1}{2}\sigma^2 S_t^2 \frac{\partial^2 F}{\partial S^2} + (r + c) S_t \frac{\partial F}{\partial S} - rF = 0$$

5 Rewrite the Black–Scholes equation in terms of the forward price for an asset with no income. Assume the underlying asset follows geometric Brownian motion.

Solution

The Black–Scholes Equation (9.8) is:

$$\frac{\partial F}{\partial t} + \frac{1}{2}\sigma^2 S_t^2 \frac{\partial^2 F}{\partial S^2} + rS_t \frac{\partial F}{\partial S} - rF = 0$$

The forward price at time t for an asset with no income is $K_t = S_t e^{r(T-t)}$ from Theorem 3.1. Rewrite the price function for any derivative on the underlying asset in terms of the forward price

$$F(S_t, t) = G(K_t, t)$$

Then

$$\frac{\partial F}{\partial t} = \frac{\partial G}{\partial t} + \frac{\partial G}{\partial K}\frac{\partial K}{\partial t} = \frac{\partial G}{\partial t} - rS_t e^{r(T-t)} \frac{\partial G}{\partial K}$$

$$\frac{\partial F}{\partial S} = \frac{\partial G}{\partial K}\frac{\partial K}{\partial S} = e^{r(T-t)} \frac{\partial G}{\partial K}$$

and $$\frac{\partial^2 F}{\partial S^2} = \frac{\partial}{\partial K}\left(e^{r(T-t)} \frac{\partial G}{\partial K}\right)\frac{\partial K}{\partial S} = \frac{\partial^2 G}{\partial K^2}\left(e^{r(T-t)}\right)^2$$

Substituting

$$\left(\frac{\partial G}{\partial t} - rS_t e^{r(T-t)} \frac{\partial G}{\partial K}\right) + \frac{1}{2}\sigma^2 S_t^2 \frac{\partial^2 G}{\partial K^2}\left(e^{r(T-t)}\right)^2 + rS_t\left(e^{r(T-t)} \frac{\partial G}{\partial K}\right) - rG = 0$$

$$\left(\frac{\partial G}{\partial t} - rK_t \frac{\partial G}{\partial K}\right) + \frac{1}{2}\sigma^2 K_t^2 \frac{\partial^2 G}{\partial K^2} + rK_t \frac{\partial G}{\partial K} - rG = 0$$

$$\frac{\partial G}{\partial t} + \frac{1}{2}\sigma^2 K_t^2 \frac{\partial^2 G}{\partial K^2} - rG = 0$$

Note: This version of the Black–Scholes equation is known as the *Black 76 model* which is used in Chapter 13 to price bond options, interest rate caps and floors and swaptions.

6 Determine the values of α for which the function $F(S_t, t) = S_t^{\alpha}$ is a solution of the Black–Scholes equation.

Solution

When $F(S_t, t) = S_t^\alpha$,

$$\frac{\partial F}{\partial t} = 0, \ \frac{\partial F}{\partial S} = \alpha S_t^{\alpha-1}, \ \frac{\partial^2 F}{\partial S^2} = \alpha(\alpha - 1)S_t^{\alpha-2}$$

Substituting these into the Black–Scholes PDE results in

$$\frac{1}{2}\alpha(\alpha - 1)S_t^{\alpha-2}\sigma^2 S_t^2 + r\alpha S_t^{\alpha-1}S_t - rS_t^\alpha = 0.$$

Or

$$\frac{1}{2}\sigma^2\alpha^2 + (r - \frac{1}{2}\sigma^2)\alpha - r = 0$$

Solving the quadratic equation

$$\alpha = \frac{-(r - \frac{1}{2}\sigma^2) \pm \sqrt{(r - \frac{1}{2}\sigma^2)^2 + 2\sigma^2 r}}{\sigma^2}$$
$$= \frac{-(r - \frac{1}{2}\sigma^2) \pm (r + \frac{1}{2}\sigma^2)}{\sigma^2} = 1 \text{ or } \frac{-2r}{\sigma^2}$$

7 Given the pricing function $F(S_t, t)$ for the European call option in 10.9 show that

$$\frac{\partial F}{\partial S} = \Phi(d_1)$$

where $\Phi(\cdot)$ is the cumulative distribution function of the standard normal distribution.

Solution

First

$$\frac{\partial d_1}{\partial S} = \frac{1}{\sigma\sqrt{T-t}}\left[\frac{\partial}{\partial S}\log\frac{S_t}{K}\right] = \frac{1}{S_t\sigma\sqrt{T-t}} = \frac{\partial d_2}{\partial S}.$$

Second, note that

$$\Phi'(x) = \frac{1}{\sqrt{2\pi}}e^{-\frac{x^2}{2}}.$$

Then

$$\frac{\Phi'(d_2)}{\Phi'(d_1)} = \frac{e^{-\frac{d_2^2}{2}}}{e^{-\frac{d_1^2}{2}}} = e^{\frac{1}{2}(d_1^2-d_2^2)} = e^{\frac{1}{2}(d_1+d_2)(d_1-d_2)}.$$

And

$$\begin{aligned}\frac{1}{2}(d_1+d_2)(d_1-d_2) &= \frac{1}{2}(2\frac{\log(S_t/K)+r(T-t)}{\sigma\sqrt{T-t}})(\sigma\sqrt{T-t}) \\ &= \log(S_t/K)+r(T-t).\end{aligned}$$

Therefore

$$\begin{aligned}\frac{\Phi'(d_2)}{\Phi'(d_1)} &= e^{\log(S_t/K)+r(T-t)} = \frac{S_t}{K}e^{r(T-t)} \\ \implies \Phi'(d_2)Ke^{-r(T-t)} &= S_t\Phi'(d_1).\end{aligned}$$

Finally

$$\begin{aligned}\frac{\partial F}{\partial S} &= S_t\Phi'(d_1)\frac{\partial d_1}{\partial S} + \Phi(d_1) - e^{-r(T-t)}K\Phi'(d_2)\frac{\partial d_2}{\partial S} \\ &= S_t\Phi'(d_1)\left(\frac{\partial d_1}{\partial S} - \frac{\partial d_2}{\partial S}\right) + \Phi(d_1) \\ &= \Phi(d_1)\end{aligned}$$

8 Verify that $F(S_t, t)$ as given by the Black–Scholes call option pricing formula is a solution of the Black–Scholes equation.

Solution
First

$$\begin{aligned}\frac{\partial F}{\partial t} &= S_t\Phi'(d_1)\frac{\partial d_1}{\partial t} - e^{-r(T-t)}K\Phi'(d_2)\frac{\partial d_2}{\partial t} - re^{-r(T-t)}K\Phi(d_2) \\ &= S_t\Phi'(d_1)\frac{\partial d_1}{\partial t} - S_t\Phi'(d_1)\frac{\partial d_2}{\partial t} - re^{-r(T-t)}K\Phi(d_2) \\ &\qquad\qquad\qquad\qquad\qquad\qquad \text{from Question 7} \\ &= S_t\Phi'(d_1)\left(\frac{\partial d_1}{\partial t} - \frac{\partial d_2}{\partial t}\right) - re^{-r(T-t)}K\Phi(d_2) \\ &= -\frac{S_t\Phi'(d_1)\sigma}{2\sqrt{T-t}} - re^{-r(T-t)}K\Phi(d_2)\end{aligned}$$

From before, we have $\frac{\partial F}{\partial S} = \Phi(d_1)$ then

$$\frac{\partial^2 F}{\partial S^2} = \Phi'(d_1)\frac{\partial d_1}{\partial S} = \frac{\Phi'(d_1)}{S_t\sigma\sqrt{T-t}}.$$

So

$$\frac{\partial F}{\partial t} + \frac{1}{2}\frac{\partial^2 F}{\partial S^2}\sigma^2 S_t^2 + r\frac{\partial F}{\partial S}S_t - rF$$
$$= -\frac{S_t\Phi'(d_1)\sigma}{2\sqrt{T-t}} - re^{-r(T-t)}K\Phi(d_2) + \frac{1}{2}\frac{\Phi'(d_1)}{S_t\sigma\sqrt{T-t}}\sigma^2 S_t^2$$
$$+ r\Phi(d_1)S_t - r\left(S_t\Phi(d_1) - e^{-r(T-t)}K\Phi(d_2)\right)$$
$$= 0$$

9 (a) Consider a European call option in which the exercise price of the option is \$100 and the time to expiry is 6 months. The current price of the underlying asset is \$80 and volatility is 30% per annum. Given that the risk-free interest rate is 4% per annum find the value of this option.

(b) Determine the corresponding replicating portfolio in shares and cash for 10000 call options.

Solution

(a)

$$d_1 = -0.8516 \implies \Phi(d_1) = 0.1972$$
$$d_2 = -1.0637 \implies \Phi(d_2) = 0.1437$$
$$c_t = 80(0.1972) - e^{-0.04/2}100(0.1437) = \$1.69$$

(b) From Equation (9.9), the number of shares in the replicating portfolio is defined by

$$\phi_t = \frac{\partial F}{\partial S}$$

Question 7 above states that, for a European call option,

$$\frac{\partial F}{\partial S} = \Phi(d_1)$$

So the corresponding replicating portfolio should contain $10000 \times 0.1972 = 1972$ shares for 10000 call options.
The value of the derivatives must equal the value of shares and cash in the replicating portfolio. Therefore, cash in the replicating portfolio would be $10000(c_t - \phi_t S_t) = 10000(1.69 - 0.1972 \times 80) = -140888$. That is, borrow \$ 140888.

10 Option pricing in continuous time

Introduction

Financial markets are inherently dynamic, with asset prices evolving in a continuous and uncertain manner. Continuous-time option pricing theory provides both the mathematical tools and conceptual framework to assess the fair value of contingent claims.

The central idea is that option prices are determined not by subjective forecasts of future returns, but by the absence of arbitrage opportunities in well-functioning markets. In this framework, the value of an option is expressed as the discounted expected payoff under risk-neutral dynamics. Many of the ideas we have already met in the binomial model.

Recall, in the binomial approach, we found the initial derivative price was

$$C_0 = e^{-rn} E^{\mathbb{Q}} C_n$$

where C_n is the payoff from the derivative after n-periods. We will show in the continuous case

$$C_t = e^{-r(T-t)} E_t^{\mathbb{Q}} C_T$$

where C_T is the payoff from the derivative at T. This is the *risk-neutral derivative pricing formula*.

We will then highlight that our solution is consistent with the Black–Scholes equation.

How do we derive the risk-neutral derivative pricing formula? This is not easy ! We will need the *Martingale Representation theorem*. How do we then price derivatives with this result? Also not easy. We will build on ideas encountered before in the binomial model

- equivalent probability measures
- Radon-Nikodým derivatives

as well as a very useful result known as *the theorem of Girsanov*.

DOI: 10.4324/9781003477624-13

10.1 Martingale representation theorem

In the binomial model approach to pricing derivatives, we showed that it is always possible to create a replicating portfolio for any contingent claim by solving two simple simultaneous equations. The derivatives market was then said to be complete meaning that the payoff from any derivative can be replicated by a portfolio of the underlying stock and cash.

In the continuous case, it is also true but we need a result known as the Martingale Representation theorem.

Recall from Definition 4.2, we have that a stochastic process $\{X_t\}$ is a martingale if $\mathbb{E}X_t$ finite for all t and $\mathbb{E}_t X_T = X_t$ for all $0 \leq t \leq T$.

In Theorem 6.5, we showed that every Itô integral is a martingale. The converse is also true. Every martingale can be written as an Itô integral.

The Martingale Representation Theorem in continuous time states the following:

Theorem 10.1. *If $\{X_t\}$ is a martingale with respect to the filtration generated by standard Brownian motion, then there exists a unique adapted process ξ_t such that*

$$X_t = X_0 + \int_0^t \xi_s dW$$

Using differential shorthand, we can write this result as $dX = \xi_t dW$.

A proof of this result is given in Oksendal (2013).

The relevance of the specification of the filtration here is that the only source of uncertainty in this model is due to standard Brownian motion as in the Black–Scholes set-up of the previous chapter. There we had a risky asset dependent on standard Brownian motion along with a risk-free asset. The Martingale Representation Theorem states that any martingale in that set-up can be represented as an Itô integral.

The Martingale Representation Theorem is important for two reasons. First, it will be seen that it implies that every derivative security in the Black–Scholes model can be replicated, that is, the market is complete. This is explained in Section 10.4. Second, it allows us to conclude that martingales are driftless processes.

Theorem 10.2. *Given an Itô process X_t in differential form:*

$$dX = \mu(t, X_t)dt + \sigma(t, X_t)dW$$

then X_t is a martingale is equivalent to the process having no drift term, that is, $\mu(t, X_t) = 0$.

Proof. The first part that $\mu(t, X_t) = 0$ implies the process is a martingale has already been shown in Section 6.6.3.

The second part that X_t a martingale implies that $\mu(t, X_t) = 0$ relies on the Martingale Representation Theorem. Since X_t a martingale, it can be written in differential form as $dX = \xi_t dW$. There is no drift term in this expression. □

10.2 Change of probability measure

In the proof of the risk-neutral derivative pricing formula in continuous time, we will change the probability measure from the real-world probability measure $\mathbb{P}$ to the risk-neutral world probability measure $\mathbb{Q}$ just as we did before in the discrete case described in Chapter 8.

Recall the idea of equivalent probability measures from Section 8.6. Two probability measures $\mathbb{P}$ and $\mathbb{Q}$ are *equivalent* means that each event that has positive probability under one measure must also have positive probability under the other one, and vice versa. In the binomial model, this meant that the probability tree looked the same for both probability measures, that is, the same stock price paths were possible. However, the probability that the stock price followed any particular path was different for each measure. In the continuous time model, the same meaning of equivalence applies. Therefore, $\mathbb{P}$ and $\mathbb{Q}$ equivalent means the probability measures agree on which paths are not possible.

For example, let $f(x)$ be the probability density function of a variable distributed $N(0, \sigma)$. The probability that X is between u and v is given by the dashed area under this curve in Figure 10.1

$$P(u \leq X \leq v) = \int_u^v f(x)dx$$

We can then change the probability measure so that each point x is associated with a new probability density function $g(x)$. In the example given in the figure, the probability density function $g(x)$ is $f(x)$ shifted to the right so that it becomes the density function of a variable distributed $N(\mu, \sigma)$. The probability that X is between u and v is now the grey area under the $y = g(x)$ curve

$$P(u \leq X \leq v) = \int_u^v g(x)dx$$

The probability measures $\mathbb{P}$ and $\mathbb{Q}$ associated with $f(x)$ and $g(x)$ respectively are equivalent since these probabilities are positive for all u and v.

In Section 8.6, the Radon-Nikodým derivative ($\Theta = \frac{d\mathbb{Q}}{d\mathbb{P}}$) allowed us to change from real-world probabilities $\mathbb{P}$ to risk-neutral probabilities $\mathbb{Q}$. For each path along the tree diagram, it was simply the ratio of the $\mathbb{Q}$ risk-neutral probability of this path to the $\mathbb{P}$ real-world probability of this path.

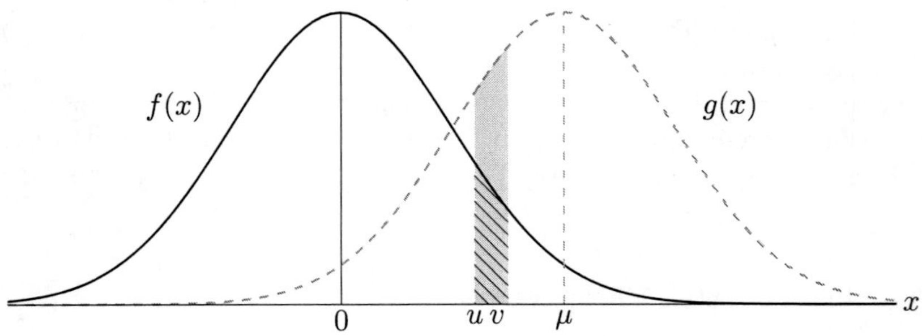

Figure 10.1 Change of measure

In the continuous case, the Radon-Nikodým derivative becomes the $\theta(x)$ such that

$$\theta(x) = \frac{g(x)}{f(x)}$$

$P^{\mathbb{Q}}(A)$, the probability of A under $\mathbb{Q}$, then becomes:

$$P^{\mathbb{Q}}(A) = \int_A g(x)dx = \int_A \frac{g(x)}{f(x)} f(x)dx = \int_A \theta(x)f(x)dx$$

So we can simply replace the risk-neutral probability density, $g(x)$, by the product of the Radon-Nikodým derivative, $\theta(x)$, and the real-world probability density, $f(x)$.

The expectation of a random variable Y under $\mathbb{Q}$ is then

$$\mathbb{E}^{\mathbb{Q}} Y = \int_{\Omega} Y(x)g(x)dx = \int_{\Omega} Y(x)\theta(x)f(x)dx = \mathbb{E}^{\mathbb{P}}\Theta Y \tag{10.1}$$

where $f(x)$, $g(x)$ and $\theta(x)$ are the real-world probability density, risk-neutral probability density and the Radon-Nikodým derivative associated with every possible x.

To continue the example above let $f(x)$ be the probability density function of a variable distributed $N(0, \sigma)$.

$$f(x) = \frac{1}{\sigma\sqrt{2\pi}} e^{-\frac{x^2}{2\sigma^2}}$$

and let $g(x)$ be the probability density function of a variable distributed $N(\mu, \sigma)$.

$$g(x) = \frac{1}{\sigma\sqrt{2\pi}} e^{-\frac{(x-\mu)^2}{2\sigma^2}}$$

then the Radon-Nikodým derivative is

$$\theta(x) = \frac{g(x)}{f(x)} = e^{\frac{x\mu - \mu^2/2}{\sigma^2}}$$

10.3 The theorem of Girsanov

While the preceding section illustrated how a change of measure can be used to shift the mean of a single normal random variable, the same principle extends naturally to continuous-time stochastic processes. In this setting, a change of measure alters the drift term of an Itô process without affecting its volatility. This result, formalized in Girsanov's theorem, provides the foundation for much of modern mathematical finance, as it enables us to recast asset price dynamics under a risk-neutral measure and thereby evaluate option prices expressed in terms of risk-neutral expectations.

The following theorem specifies how to transform an Itô process into a new Itô process with a different drift by a change in probability measure. The Radon-Nikodým derivative now becomes a stochastic process evolving over time.

Theorem 10.3. *Let γ be an adapted process and $\mathcal{F}$ be the filtration generated by standard Brownian motion W. Suppose that we change measure according to the Radon-Nikodým process Θ defined by*

$$d\Theta = -\gamma_t \Theta_t dW, \quad \Theta_0 = 1.$$

The following assumption ensures Θ is well-defined (the Novikov condition)

$$\mathbb{E}\int_0^T \gamma_s^2 \Theta_s^2 ds < \infty$$

Under the new measure obtained in this way, the stochastic process $\widetilde{W}$ defined by

$$d\widetilde{W} = \gamma_t dt + dW, \quad \widetilde{W}_0 = 0,$$

is a standard Brownian motion.

The converse is also true. If we can find a γ_t to make $d\widetilde{W}$ a standard Brownian motion under a new probability measure, then $d\Theta = -\gamma_t \Theta_t dW$ is the Radon-Nikodým derivative to change measure.

Shreve (2004) Section 5.2 has a proof of Girsanov's theorem. Here we will just indicate the logic behind this result.

For simplicity, take γ to be a constant.

Note that

$$d\Theta = -\gamma\Theta_t dW \implies \Theta_t = e^{-\gamma W_t - \frac{\gamma^2 t}{2}}$$

from the solution to geometric Brownian motion stochastic differential equation in Exercise 6.4. Note that Θ_t has a similar form to the Radon-Nikodým derivative to shift the mean of a normal random variable found previously.

And also we have that $d\widetilde{W} = \gamma dt + dW \implies \widetilde{W}_t = \gamma t + W_t$.

We will show that using the new measure $\widetilde{W}_t$ has the same moment generating function as standard Brownian motion. Let's call the new probability measure $\mathbb{Q}$.

Recall the following from Section 2.7:

- The moment-generating function of a random variable X is $\mathbb{E}(e^{rX})$ for some real number r (Equation 2.7).
- The moment-generating function of $X \sim N(\mu, \sigma^2)$ is $e^{r\mu + \frac{1}{2}\sigma^2 r^2}$ (Theorem 2.6).
- If two random variables have the same moment-generating function, then they have the same probability distribution.

The moment-generating function of $\widetilde{W}_t$ is

$$\begin{aligned}
\mathbb{E}^{\mathbb{Q}}(e^{r\widetilde{W}_t}) &= E^{\mathbb{P}}(\Theta_t e^{r\widetilde{W}_t}) \\
&= \mathbb{E}^{\mathbb{P}}(e^{-\gamma W_t - \frac{\gamma^2 t}{2}} e^{r(\gamma t + W_t)}) \\
&= e^{r\gamma t - \frac{1}{2}\gamma^2 t}\mathbb{E}^{\mathbb{P}}(e^{(r-\gamma)W_t}) \\
&= e^{r\gamma t - \frac{1}{2}\gamma^2 t} e^{\frac{1}{2}(r-\gamma)^2 t} && \text{from Equation (2.9)} \\
&= e^{\frac{1}{2}r^2 t}
\end{aligned}$$

That is, $\widetilde{W}_t$ under $\mathbb{Q}$ has the same moment generating function as a normally distributed variable with mean 0 and variance t just as we would expect if it followed standard Brownian motion.

Exercise 10.1. *Given the asset price evolutions:*

$$\begin{aligned}
dS &= \mu S_t dt + \sigma S_t dW \\
dB &= r B_t d_t
\end{aligned}$$

where W is standard Brownian motion under a probability measure $\mathbb{P}$ and μ and σ are both constant, find an equivalent measure $\mathbb{Q}$ under which $\frac{S_t}{B_t}$ is a martingale.

Proof. It was shown in Exercise 8 of Chapter 5 that

$$d\,\frac{S_t}{B_t} = (\mu - r)\frac{S_t}{B_t}dt + \sigma\frac{S_t}{B_t}dW$$

When $\mu \neq r$ the process $\frac{S_t}{B_t}$ is not a martingale under the original measure $\mathbb{P}$ since it has a trend term (see Theorem 10.1).

Note that we can rewrite the expression above as

$$d\frac{S_t}{B_t} = \sigma\frac{S_t}{B_t}\underline{(\frac{\mu - r}{\sigma}dt + dW)} = \sigma\frac{S_t}{B_t}\underline{d\widetilde{W}} \tag{10.2}$$

The process $\frac{S_t}{B_t}$ is trendless and hence a martingale under a new measure $\mathbb{Q}$ if $\widetilde{W}$ is a standard Brownian motion under $\mathbb{Q}$. Girsanov's theorem tell us how to find such a $\mathbb{Q}$. Letting $\gamma = \frac{\mu-r}{\sigma}$ we have the stochastic process $\widetilde{W}$ defined by

$$d\widetilde{W} = \frac{\mu - r}{\sigma}dt + dW$$

is a standard Brownian motion under a new probability measure $\mathbb{Q}$ where the Radon-Nikodým process Θ to change from $\mathbb{P}$ to $\mathbb{Q}$ is defined by

$$d\Theta = -\left(\frac{\mu - r}{\sigma}\right)\Theta_t dW$$

Given the process S_t defined by

$$dS = \mu S_t dt + \sigma S_t dW$$

the expression $\frac{\mu-r}{\sigma}$ is known as the *market price of risk* (or *Sharpe ratio*) of S_t under $\mathbb{P}$ and this defines the Radon-Nikodým process.

The market price of risk indicates how much extra return an investor receives per unit of risk measured by σ for holding this asset. □

It is straightforward to find the risky asset price process under $\mathbb{Q}$. Substituting $dW = d\widetilde{W} - \gamma dt$ where $\gamma = \frac{\mu-r}{\sigma}$ we have

$$\begin{aligned} dS &= \mu S_t dt + \sigma S_t dW \\ &= \mu S_t dt + \sigma S_t (d\widetilde{W} - \gamma dt) \\ &= r S_t dt + \sigma S_t d\widetilde{W} \end{aligned} \tag{10.3}$$

Therefore, changing from probability measure $\mathbb{P}$ to probability measure $\mathbb{Q}$ changes the asset price process from

$$dS = \mu S_t dt + \sigma S_t dW$$

to

$$dS = rS_t dt + \sigma S_t d\widetilde{W}$$

where $\widetilde{W}_t$ is a Brownian motion under $\mathbb{Q}$. Under the risk-neutral probability measure $\mathbb{Q}$ the asset price trends at the risk-free rate. That is, the mean rate of return of this risky asset has become the risk-free rate. Investors therefore no longer require compensation for risk justifying the use of the expression *risk-neutral pricing*. The market price of risk for S_t under $\mathbb{Q}$ is zero since

$$\frac{r-r}{\sigma} = 0$$

Exercise 10.2. *Consider a model with two risky assets driven by the same standard Brownian process:*

$$dS_1 = \mu_1 S_{1,t}\, dt + \sigma_1 S_{1,t}\, dW$$

$$dS_2 = \mu_2 S_{2,t}\, dt + \sigma_2 S_{2,t}\, dW$$

where $\{\mu_1, \mu_2, \sigma_1, \sigma_2\}$ are all constants, find an equivalent probability measure $\mathbb{S}$ under which $\frac{S_{1,t}}{S_{2,t}}$ is a martingale.

Proof.

$$\begin{aligned}
d\frac{S_1}{S_2} &= S_{2,t}^{-1} dS_1 - S_{1,t} S_{2,t}^{-2} dS_2 + S_{1,t} S_{2,t}^{-3} \sigma_2^2 S_{2,t}^2 dt - S_{2,t}^{-2} \sigma_1 \sigma_2 S_{1,t} S_{2,t} dt \\
&= \left(\frac{\mu_1 S_{1,t}}{S_{2,t}} dt + \frac{\sigma_1 S_{1,t}}{S_{2,t}} dW \right) - \left(\frac{\mu_2 S_{1,t}}{S_{2,t}} dt + \frac{\sigma_2 S_{1,t}}{S_{2,t}} dW \right) \\
&\quad + \frac{S_{1,t}}{S_{2,t}} (\sigma_2^2 - \sigma_1 \sigma_2) dt \\
&= \frac{S_{1,t}}{S_{2,t}} \left\{ \left(\mu_1 - \mu_2 + \sigma_2^2 - \sigma_1 \sigma_2 \right) dt + (\sigma_1 - \sigma_2) dW \right\} \\
&= \frac{S_{1,t}(\sigma_1 - \sigma_2)}{S_{2,t}} \left\{ \frac{\mu_1 - \mu_2 + \sigma_2^2 - \sigma_1 \sigma_2}{\sigma_1 - \sigma_2} dt + dW \right\} \\
&= \frac{S_{1,t}(\sigma_1 - \sigma_2)}{S_{2,t}} d\tilde{W}
\end{aligned}$$

The Radon-Nikodým process Θ to change from $\mathbb{P}$ to $\mathbb{S}$ is defined by

$$d\Theta = -\gamma \Theta_t dW$$

where $\gamma = \frac{\mu_1 - \mu_2 + \sigma_2^2 - \sigma_1 \sigma_2}{\sigma_1 - \sigma_2}$. □

In Examples 10.1 and 10.2, an equivalent probability measure was found such that a risky asset divided by either a bond or a second risky asset was a martingale. By dividing in this way we are re-expressing the risky asset in units of the bond or in units of the second risky asset. The denominator is called the *numéraire* and the process is called *changing the numéraire*. This is akin to changing the currency an asset is expressed in from one currency to another. The risk-neutral probability measure found by using the bond as numéraire is key to pricing equity derivatives.

Also later we will see that the process of changing the numéraire turns out to greatly simplify the analysis of interest rate derivatives. The following theorem will provide the theoretical basis. It allows us to assume that if we have a risk-neutral measure, then we can change the numéraire and then price contingent claims with the new measure associated with this numéraire. Moreover it tells us how to find this new measure.

Theorem 10.4. *A market consists of traded assets with prices $S_{1,t}, \ldots, S_{k,t}$.*

Assume that we already have a measure $\mathbb{Q}^N$ equivalent to the real-world probability $\mathbb{P}$ such that $S_{i,t}/N_t$ is a martingale for any i where N_t is a non-dividend paying asset.

Let X_t be another non-dividend paying asset then there exists a probability measure $\mathbb{Q}^X$ such that all $S_{i,t}/X_t$ are martingales as are all C_t/X_t where C_T is a contingent claim at T. Furthermore $\mathbb{Q}^X$ is defined by the Radon-Nikodým derivative

$$\frac{d\mathbb{Q}^X}{d\mathbb{Q}^N} = \frac{X_T/X_t}{N_T/N_t}$$

Proof. The first part can be proved using conditional expectations. See Geman et al. (1995) for details.

Choose an arbitrary asset S_i. Since $S_{i,t}/X_t$ is a martingale under $\mathbb{Q}^X$

$$\mathbb{E}_t^X \frac{S_{i,T}}{X_T} = \frac{S_{i,t}}{X_t}$$

Since $S_{i,t}/N_t$ is a martingale under $\mathbb{Q}^N$

$$\mathbb{E}_t^N \left(\frac{N_t}{X_t} \frac{S_{i,T}}{N_T} \right) = \frac{N_t}{X_t} \mathbb{E}_t^N \left(\frac{S_{i,T}}{N_T} \right) = \frac{N_t}{X_t} \frac{S_{i,t}}{N_t} = \frac{S_{i,t}}{X_t}$$

Therefore

$$\mathbb{E}_t^X \frac{S_{i,T}}{X_T} = \mathbb{E}_t^N \left(\frac{N_t}{X_t} \frac{S_{i,T}}{N_T} \right)$$

Writing the first expectation in terms of the $\mathbb{Q}^N$ measure (see Equation 10.1)

$$\mathbb{E}_t^X \frac{S_{i,T}}{X_T} = \mathbb{E}_t^N \left(\frac{d\mathbb{Q}^X}{d\mathbb{Q}^N} \frac{S_{i,T}}{X_T} \right)$$

Comparing the right-hand sides of the last two equations

$$\frac{d\mathbb{Q}^X}{d\mathbb{Q}^N} = \frac{X_T/X_t}{N_T/N_t} \qquad \square$$

10.4 Risk-neutral derivative pricing formula

The goal of this section is to demonstrate that the risk-neutral derivative pricing formula in the continuous case is

$$C_t = e^{-r(T-t)} \mathbb{E}_t^{\mathbb{Q}} C_T$$

where C_T is the derivative payoff at time T and C_t is its price at time t.

First, we make the same assumptions as in Section 9.2 of the previous chapter:

1. There are no risk-free arbitrage opportunities.
2. No market frictions – no taxes or transaction costs.
3. Borrowing and investing occur at the same constant rate r.
4. Competitive and well-functioning markets
   - No maximum or minimum units of trading.
   - No limits to short selling (e.g., due to credit limits and collaterals).
5. The underlying asset can be traded continuously and its price follows a geometric Brownian motion.

The asset price evolutions are also the same as before in Equation (9.2):

$$dS = \mu S_t dt + \sigma S_t dW$$
$$dB = r B_t dt$$

Construct a portfolio of these assets such that the value of this portfolio at time t is

$$V_t = \phi_t S_t + \psi_t B_t \qquad (10.4)$$

with the self-financing property

$$dV = \phi_t dS + \psi_t dB$$

$\mathbb{Q}$ is chosen as the probability measure such that $\frac{S_t}{B_t}$ is a martingale. The previous section demonstrated how this was done. We will now show that $\frac{V_t}{B_t}$ is also a martingale under $\mathbb{Q}$.

Using the stochastic quotient rule

$$
\begin{aligned}
d\frac{V_t}{B_t} &= \frac{1}{B_t}dV - \frac{V_t}{B_t^2}dB \\
&= \frac{1}{B_t}(\phi_t dS + \psi_t dB) - \frac{\phi_t S_t + \psi_t B_t}{B_t^2}dB \\
&= \phi_t(\frac{1}{B_t}dS - \frac{S_t}{B_t^2}dB) \\
&= \phi_t d\frac{S_t}{B_t} \\
&= \phi_t(\sigma\frac{S_t}{B_t}d\widetilde{W})
\end{aligned}
\tag{10.5}
$$

where the last two lines follow from the previous section. This process has no drift term under $\mathbb{Q}$ therefore $\frac{V_t}{B_t}$ is a martingale.

The martingale relationship holds for any portfolio constructed from the risky and riskless assets. In particular, it holds for the portfolio whose value V_T at time T equals the derivative payoff C_T:

$$
\frac{V_t}{B_t} = \mathbb{E}_t^{\mathbb{Q}}\frac{V_T}{B_T} = \mathbb{E}_t^{\mathbb{Q}}\frac{C_T}{B_T}
$$

By the principle of no arbitrage, the value V_t of this portfolio must equal the derivative price C_t for all time t. Therefore

$$
\frac{C_t}{B_t} = \mathbb{E}_t^{\mathbb{Q}}\frac{C_T}{B_T} \tag{10.6}
$$

Since

$$
dB = rB_t dt \implies B_t = B_0 e^{rt}
$$

we then have our risk-neutral derivative pricing formula in continuous time

$$
C_t = B_t\mathbb{E}_t^{\mathbb{Q}}\frac{C_T}{B_T} = B_0 e^{rt}\mathbb{E}_t^{\mathbb{Q}}\frac{C_T}{B_0 e^{rT}} = e^{-r(T-t)}\mathbb{E}_t^{\mathbb{Q}}C_T
$$

Hence, if there exists a portfolio with value equal to the derivative payoff C_T at time T, then we know its value at all times and this must correspond to the derivative price. Can we be certain that we can construct this portfolio using the assets available? The next steps show that the holdings ϕ_t and ψ_t of risky and riskless assets in the replicating portfolio exist for all t.

If we fix $c_t = e^{-r(T-t)}E^{Q}tC_T$ from the outset then the discounted price process $\frac{C_t}{B_t}$ is a martingale by reversing the argument to Equation (10.6) above. Then, the Martingale Representation Theorem (Theorem 10.1) states there exists ξ_t such that

$$d\frac{C_t}{B_t} = \xi_t d\tilde{W} \tag{10.7}$$

In order that C_t equals the replicating portfolio value V_t for all t then Equations (10.5) and (10.7) must equal.

$$\phi_t \sigma \frac{S_t}{B_t} = \xi_t$$

or

$$\phi_t = \frac{\xi_t}{\sigma \frac{S_t}{B_t}}$$

Since ξ_t exists, the number of risky assets in the replicating portfolio ϕ_t is well-defined. The total riskless investment $\psi_t B_t$ can in turn be found from rearranging Equation (10.4) to give $\psi_t B_t = V_t - \phi_t S_t$ where V_t is set equal to C_t as determined by the pricing formula.

The replicating portfolio can be constructed regardless of how the derivative pay-off is defined. It is therefore possible to replicate every derivative dependent on S_t using some combination of the underlying risky asset and the riskless asset. The market is therefore complete.

Furthermore, the derivative value $C_t = F(S_t, t)$ will be consistent with the Black–Scholes equation since all the assumptions here are the same as in the derivation in the previous chapter. Also the asset price evolutions are specified in the same way, the replicating portfolio is subject to the self-financing condition and we rely on the condition of no arbitrage between a derivative and the portfolio that replicates its payoff at maturity.[1]

The Martingale Representation Theorem tells us that ϕ_t exists but not how to determine it. However, our analysis of the Black–Scholes equation in Equation (9.9) found that once $C_t = F(S_t, t)$ was determined then

$$\phi_t = \frac{\partial F}{\partial S}$$

10.5 Applying the risk-neutral pricing formula

Now that we have our derivative pricing formula $C_t = e^{-r(T-t)}\mathbb{E}_t^{\mathbb{Q}} C_T$ we will determine the price of some simple derivatives.

Exercise 10.3. *The price movements of a non-dividend paying share S_t are governed by geometric Brownian motion. Find the fair price to pay at time t for a derivative paying S_T^n at time T where n is a real number.*

Proof. Apply the risk-neutral derivative pricing formula:

$$C_t = e^{-r(T-t)}\mathbb{E}_t^{\mathbb{Q}} C_T = e^{-r(T-t)}\mathbb{E}_t^{\mathbb{Q}} S_T^n$$

Since under $\mathbb{Q}$

$$dS = rS_t dt + \sigma S_t d\widetilde{W}$$

substituting r for μ in Exercise 6.4

$$S_T = S_t e^{(r-\frac{1}{2}\sigma^2)(T-t)+\sigma W_{T-t}}$$

and

$$S_T^n = S_t^n e^{n(r-\frac{1}{2}\sigma^2)(T-t)+n\sigma W_{T-t}}$$

The exponent here is a normally distributed variable with mean $n(r-\frac{1}{2}\sigma^2)(T-t)$ and variance $n\sigma^2(T-t)$. Applying the result from Equation (2.9)

$$\mathbb{E}_t^{\mathbb{Q}} S_T^n = S_t^n e^{n(r+\frac{1}{2}(n-1)\sigma^2)(T-t)} \tag{10.8}$$

and the price of the derivative at time t is

$$C_t = e^{-r(T-t)} S_t^n e^{n(r+\frac{1}{2}(n-1)\sigma^2)(T-t)}$$ □

Exercise 10.4. *A digital option has payoff $C_T = 1_{S_T \geq K}$; that is to say, the payoff is 1 if $S_T \geq K$ and 0 otherwise. Find the price of the digital option at time t.*

Proof. Apply the risk-neutral derivative pricing formula:

$$C_t = e^{-r(T-t)}\mathbb{E}_t^{\mathbb{Q}} C_T = e^{-r(T-t)}\mathbb{E}_t^{\mathbb{Q}}(1_{S_T \geq K})$$

It is not as easy to find an expression for the expectation here as in the previous example since the pay-off cannot be expressed in the form e^X for X a normally distributed random variable.

Instead we have to calculate the expectation from first principles.

$$C_t = e^{-r(T-t)}\mathbb{E}_t^{\mathbb{Q}}(1_{S_T \geq K})$$

$$= \frac{e^{-r(T-t)}}{\sqrt{2\pi}} \int_{z^*}^{\infty} 1 \times e^{-\frac{1}{2}z^2} dz \qquad \text{where } S_T \geq K \text{ when } z \geq z^*$$

$$= \frac{e^{-r(T-t)}}{\sqrt{2\pi}} \int_{-\infty}^{-z^*} e^{-\frac{1}{2}z^2} dz \qquad \text{since } e^{-\frac{1}{2}z^2} \text{ symmetric about } Y\text{-axis}$$

$$= e^{-r(T-t)}\Phi(d) \quad \text{for } d = -z^*$$

where Φ is the cumulative normal distribution.

z^* is easily determined by letting $\widetilde{W}_{T-t} = \sqrt{T-t}Z$ where Z is a standard normal variable:

$$S_T \geq K \implies S_t e^{(r-\frac{1}{2}\sigma^2)(T-t)+\sigma\sqrt{T-t}z} \geq K$$

$$\implies z \geq \frac{\log(K/S_t) - (r - \frac{1}{2}\sigma^2)(T-t)}{\sigma\sqrt{T-t}}$$

The expression on the right-hand side is z^* hence

$$d = -z^* = \frac{\log(S_t/K) + (r - \frac{1}{2}\sigma^2)(T-t)}{\sigma\sqrt{T-t}} \qquad \square$$

10.6 European call and put options

The approach to finding the Black–Scholes European call option pricing formula is similar to the digital option example above where the payoff becomes $C_T = [S_T - K]^+$ with strike price K. This is left as an end-of-chapter exercise.

The result for a non-dividend paying stock is

$$c_t = S_t\Phi(d_1) - e^{-r(T-t)}K\Phi(d_2) \tag{10.9}$$

where $d_1 = \frac{\log(S_t/K)+(r+\frac{1}{2}\sigma^2)(T-t)}{\sigma\sqrt{T-t}}$, $d_2 = d_1 - \sigma\sqrt{T-t}$

The price of a European put option can be found similarly or simply by using the call formula with put-call parity.

Exercise 10.5. *The principle of put-call parity states that a portfolio of a European call option on a non-dividend paying share and cash worth $Ke^{-r(T-t)}$ should have the same value as a European put option and a share at $t < T$, that is,*

$$c_t + Ke^{-r(T-t)} = p_t + S_t$$

Use put-call parity to derive the price of a put option.

Proof. Put-call parity states

$$c_t + Ke^{-r(T-t)} = p_t + S_t$$

So

$$\begin{aligned}
p_t &= c_t + Ke^{-r(T-t)} - S_t \\
&= S_t\Phi(d_1) - e^{-r(T-t)}K\Phi(d_2) + Ke^{-r(T-t)} - S_t \\
&= Ke^{-r(T-t)}(1 - \Phi(d_2)) - S_t(1 - \Phi(d_1)) \\
&= Ke^{-r(T-t)}\Phi(-d_2) - S_t\Phi(-d_1) \qquad (10.10)
\end{aligned}$$

□

These formulae are in turn easily amended to accommodate dividend-paying shares. Suppose dividends are payable continuously at a constant rate of q per annum per share. As before in subsection 9.3.1, the share price process when no dividend is paid is

$$dS = \mu S_t dt + \sigma S_t dW$$

When the dividend is paid at rate q, the new share price process becomes

$$dS = (\mu - q)S_t dt + \sigma S_t dW$$

Solving we get

$$\begin{aligned}
S_T &= S_t e^{(\mu - q - \frac{1}{2}\sigma^2)(T-t) + \sigma W_{T-t}} \\
&= \underbrace{e^{-q(T-t)} S_t}\, e^{(\mu - \frac{1}{2}\sigma^2)(T-t) + \sigma W_{T-t}}
\end{aligned}$$

Therefore, the price of a dividend paying share starting from value S_t evolves like the price of a non-dividend paying share starting from value $e^{-q(T-t)}S_t$.

So we can simply amend the call and put option formulae above for dividend paying shares.

Substituting $e^{-q(T-t)}S_t$ for S_t in Equation (10.9), the price of a European call option on a dividend paying asset is

$$c_t = S_t e^{-q(T-t)}\Phi(d_1) - e^{-r(T-t)}K\Phi(d_2)$$

And noting that $\log \frac{S_t e^{-q(T-t)}}{K} = \log(\frac{S_t}{K}) + \log(e^{-q(T-t)}) = \log(\frac{S_t}{K}) - q(T-t)$

$$d_1 = \frac{\log(S_t/K) + (r - q + \frac{1}{2}\sigma^2)(T-t)}{\sigma\sqrt{T-t}}$$

with $d_2 = d_1 - \sigma\sqrt{T-t}$ as before.

Similarly substituting $e^{-q(T-t)}S_t$ for S_t in Equation (10.10), the price of a European put option on a dividend paying asset is

$$p_t = Ke^{-r(T-t)}\Phi(-d_2) - S_t e^{-q(T-t)}\Phi(-d_1)$$

10.7 Extensions of the risk-neutral derivative pricing formula

In Example 10.1 where there was one risky asset S_t with one source of uncertainty W_t the market price of risk for the risky asset determined the Radon-Nikodým process to change from the real-world probability measure $\mathbb{P}$ to the risk-neutral probability measure $\mathbb{Q}$. Then $\frac{S_t}{B_t}$ was a martingale under this measure. That is, given

$$dS = \mu S_t dt + \sigma S_t dW$$

then the Radon-Nikodým process to change from $\mathbb{P}$ to $\mathbb{Q}$ is

$$d\Theta = -\gamma\Theta_t dW$$

where γ is equal to the market price of risk for S_t, that is, $\gamma = \frac{\mu - r}{\sigma}$.

What would happen if there were multiple risky assets and may be also multiple sources of uncertainty ?

Let's deal with the situation of two assets, $S_{1,t}$ and $S_{2,t}$, with one source of uncertainty W first.

Consider the model below with two risky assets that are driven by the same standard Brownian motion:

$$dS_1 = \mu_1 S_{1,t}dt + \sigma_1 S_{1,t}dW$$

$$dS_2 = \mu_2 S_{2,t}dt + \sigma_2 S_{2,t}dW$$

and a riskless asset B_t where $B_t = B_0 e^{rt}$.

We want to find a probability measure $\mathbb{Q}$ such that $\frac{S_{1,t}}{B_t}$ and $\frac{S_{2,t}}{B_t}$ are martingales.

In order for $\frac{S_{1,t}}{B_t}$ to be a martingale under $\mathbb{Q}$ the Radon-Nikodým process to change from $\mathbb{P}$ to $\mathbb{Q}$ is

$$d\Theta = -\gamma\Theta_t dW$$

where γ is equal to the market price of risk for $S_{1,t}$, that is, $\gamma = \frac{\mu_1 - r}{\sigma_1}$.

Similarly, for $\frac{S_{2,t}}{B_t}$ to be a martingale, γ must be equal to the market price of risk for $S_{2,t}$, that is, $\gamma = \frac{\mu_2 - r}{\sigma_2}$.

Therefore, we need to find a γ that satisfies the simple system of equations

$$\frac{\mu_1 - r}{\sigma_1} = \gamma$$
$$\frac{\mu_2 - r}{\sigma_2} = \gamma$$

If γ can be found, then $\mathbb{Q}$ exists. In which case, both risky assets have the same market price of risk.

$$\gamma = \frac{\mu_1 - r}{\sigma_1} = \frac{\mu_2 - r}{\sigma_2}$$

If both risky assets do not have the same market price of risk, then there is an arbitrage opportunity.

To see this let

$$\mu = \frac{\mu_1 - r}{\sigma_1} - \frac{\mu_2 - r}{\sigma_2} > 0$$

Then set up a portfolio holding

$$\phi_{1,t} = \frac{1}{S_{1,t}\sigma_1} \text{ of the first risky asset}$$
$$\phi_{2,t} = -\frac{1}{S_{2,t}\sigma_2} \text{ of the second risky asset}$$

and cash of value

$$\psi_t = V_t - \phi_{1,t}S_{1,t} - \phi_{2,t}S_{2,t} = V_t - \frac{1}{\sigma_1} + \frac{1}{\sigma_2}$$

Then the change in value of this portfolio is

$$\begin{aligned}
dV &= \phi_{1,t}\, dS_1 + \phi_{2,t}\, dS_2 + r\psi_t dt \\
&= \frac{1}{S_{1,t}\sigma_1}(\mu_1 S_{1,t}dt + \sigma_1 S_{1,t}dW) - \frac{1}{S_{2,t}\sigma_2}(\mu_2 S_{2,t}dt + \sigma_2 S_{2,t}dW) \\
&\quad + r(V_t - \frac{1}{\sigma_1} + \frac{1}{\sigma_2})dt \\
&= \left(\frac{\mu_1 - r}{\sigma_1} - \frac{\mu_2 - r}{\sigma_2} + rV_t\right)dt \\
&= \left(\mu + rV_t\right)dt
\end{aligned}$$

As r and μ are strictly positive, this portfolio is growing in value with certainty.

Also, the portfolio can be made initially costless by setting

$$\psi_0 = -\phi_{1,0}S_{1,0} - \phi_{2,0}S_{2,0} = -\frac{1}{\sigma_1} + \frac{1}{\sigma_2}$$

There is therefore an arbitrage opportunity.

In the example given with two assets $S_{1,t}$ and $S_{2,t}$ and one source of uncertainty W_t, finding a probability measure where $\frac{S_{1,t}}{B_t}$ and $\frac{S_{2,t}}{B_t}$ were martingales led to a simple system of two equations linked by one market price of risk parameter γ.

More generally, it can be shown that given a market model with m assets and d sources of uncertainty, finding a probability measure where all discounted asset prices are martingales leads to a system of m equations linked by d market price of risk parameters $\gamma_{j,t}$, $j = 1, ..., d$. If $\gamma_{j,t}$ can be found to satisfy this system of equations, then the risk-neutral probability measure $\mathbb{Q}$ exists. It can also be shown that if $\mathbb{Q}$ exists, then the market model is arbitrage-free. This result is known as the *first fundamental theorem of asset pricing*.

Definition 10.1. *A probability measure is risk-neutral if*

1. *it is equivalent to the real-world measure* $\mathbb{P}$ *and*
2. *for all assets* S_i *the discounted asset price* $\frac{S_{i,t}}{B_t}$ *is a martingale*

Theorem 10.5. *(First fundamental theorem of asset pricing)*
A market model has a risk-neutral probability measure if and only if arbitrage is not possible.

A system of equations can possibly have multiple solutions. If the system of m market price of risk equations described above is satisfied by more than one set of d market price of risk parameters $\gamma_{j,t}$, then there are multiple possible risk-neutral probability measures. This is not possible if the market is complete.

To see why this is the case, consider a model with two risk-neutral measures $\mathbb{Q}$ and $\mathbb{Q}'$ and a digital option with payoff $C_T = 1_X$, that is, paying \$1 if any events in set X occur at time T. If the market is complete, then a replicating portfolio can be set up with value at maturity equal to the payoff value. As $\mathbb{Q}$ is a risk-neutral measure, the discounted portfolio value process $\frac{V_t}{B_t}$ is a martingale

$$\frac{V_t}{B_t} = \mathbb{E}_t^{\mathbb{Q}} \frac{V_T}{B_T} = \mathbb{E}_t^{\mathbb{Q}} \frac{C_T}{B_T} = \frac{1}{B_T} P^{\mathbb{Q}}(X)$$

As $\mathbb{Q}'$ is also a risk-neutral measure, the discounted portfolio value process $\frac{V_t}{B_t}$ is also a martingale under $\mathbb{Q}'$ so

$$\frac{V_t}{B_t} = \frac{1}{B_T} P^{\mathbb{Q}'}(X)$$

In which case, the probabilities $P^{\mathbb{Q}}(X)$ and $P^{\mathbb{Q}'}(X)$ are the same for any set of events X and therefore the probability measures are identical.

The converse that the uniqueness of the risk-neutral probability measure implies completeness can also be shown to be true. We then have a result known as the *second fundamental theorem of asset pricing.*

Theorem 10.6. *(Second fundamental theorem of asset pricing)*
A market model is complete if and only if the risk-neutral probability measure is unique.

According to the Fundamental Theorems of Asset Pricing, three scenarios arise when we model a financial market.

(1) **(Arbitrage)** There is no risk-neutral measure since there are arbitrage opportunities. This model is useless for option pricing.
(2) **(No arbitrage but incomplete)** There are multiple risk-neutral measures.
    Albin et al. (2024) show that option prices will be the same across these measures for those derivatives where we can set up a self-financing replicating portfolio. For non-replicable claims, the price depends on which risk-neutral measure you pick.
(3) **(No arbitrage and complete)** There is only one risk-neutral measure and hence a unique price for all options.

As has been demonstrated, the Black–Scholes model falls under the third scenario. The unique risk-neutral measure was found using Girsanov's theorem. All options can then be priced and this price is unique.

However, stochastic volatility models and credit risk models fall under the second scenario. These market models are incomplete leading to multiple risk-neutral measures yielding different fair values for some derivatives. In practice, such models are calibrated to observed market spreads in order to pin down a unique price.

10.8 Summary

This chapter showed how to determine the fair value of derivative securities whose payoff depends on the future value of an underlying asset. The central principle is that we can construct a self-financing portfolio that matches this payoff. Then, in a well-functioning market with no-arbitrage opportunities, the value of this portfolio should equate to the derivative price at all times.

The outcome is the risk-neutral pricing formula where the derivative price is the discounted expected payoff under the risk-neutral measure $\mathbb{Q}$.

The risk-neutral measure is the measure under which the discounted asset price is a martingale. Girsanov's theorem tells us how to find this probability measure. Then the discounted derivative price $(\frac{C_t}{B_t})$ is also a martingale under $\mathbb{Q}$ and the risk-neutral pricing formula follows.

To apply the risk-neutral pricing formula, remember that

- changing from the real-world probability measure $\mathbb{P}$ to the risk-neutral probability measure $\mathbb{Q}$ implies that geometric Brownian model of asset price evolution assumed in the Black–Scholes model changes from

$$dS = \mu S dt + \sigma S dW$$

to

$$dS = rS dt + \sigma S d\widetilde{W}$$

where $\widetilde{W}_t$ is a Brownian motion under $\mathbb{Q}$
- When the payoff is of the form e^X for X a normally distributed random variable, we can apply Equation (2.9) to evaluate the expectation. Forwards are of this type.
- Otherwise we have to evaluate the expectation from first principles. European options are of this type.

In the Black–Scholes model, there is a unique risk-neutral measure and hence one unique price for all options. This is not true in all mathematical models for option pricing. There may be multiple risk-neutral measures and hence multiple option prices for some derivatives. Another possibility is that there is no risk-neutral measure and no option price.

Key things to remember

- Derivative price is given by the risk-neutral pricing formula

$$C_t = e^{-r(T-t)} E^{\mathbb{Q}} C_T$$

where C_T is the payoff from the derivative at T.
- Theorem of Girsanov: Section 10.3.
- Derivation of risk-neutral pricing formula: Section 10.4.
- Applying the risk-neutral pricing formula: Section 10.5 and questions at end of the chapter.

Further reading

The material in this chapter is quite challenging. Re-reading Chapter 8 on the binomial model will help build the intuition about the key concepts of risk-neutral pricing, change of measure and the Radon-Nikodým derivative. McDonald (2013) is an excellent text to develop this intuition that is not overly reliant on stochastic calculus. Klebaner (2012); Shreve (2005) and Oksendal (2013) take the material to a more rigorous mathematical level and deal with the more complicated payoff structures in exotic and American options.

Notes

1. The relationship between the Black-Scholes equation and risk-neutral pricing is a special case of a more general result known as the *Feynman–Kac theorem*, which finds the solution to parabolic partial differential equations in terms of the risk-neutral expectation of a stochastic process. The risk-neutral pricing formula found here in this section can be found by directly applying the theorem to the Black-Scholes equation. This is the approach taken in some texts.

Questions

1 Find the risk-neutral probability measure for the following models where it exists.

a.
$$S_t = e^{2t+W_t}$$
$$B_t = e^t$$

b.
$$S_t = e^{2t}$$
$$B_t = e^t$$

c.
$$S_t = e^{2W_t}$$
$$B_t = e^{W_t}$$

d. (Bachelier model)
$$S_t = S_0 + \mu t + \sigma W_t$$
$$B_t = 1$$

[Albin et al. (2024) Exercise 11.18(i)(h)(j) and Exercise 11.34(a)]

Solution

a.
$$dS = \frac{5}{2}S_t dt + S_t dW \qquad \text{from Exercise 1 in Chapter 5}$$
$$dB = B_t dt$$

This is of the form considered in Example 10.1. Hence the Radon-Nikodým process Θ to change from $\mathbb{P}$ to the risk-neutral probability measure $\mathbb{Q}$ is defined by

$$d\Theta = -\left(\frac{\frac{5}{2}-1}{1}\right)\Theta_t dW = -\frac{3}{2}\Theta_t dW$$

b.
$$dS = 2S_t dt$$
$$dB = B_t dt$$

Comparing with Example 10.1 the σ term is now zero hence Θ and $\mathbb{Q}$ do not exist.

c.
$$dS = 2S_t dt + 2S_t dW \qquad \text{from Exercise 1 in Chapter 5}$$
$$dB = \frac{1}{2}B_t dt + B_t dW \qquad \text{from Exercise 1 in Chapter 5}$$

Comparing with Example 10.2 the Radon-Nikodým process is

$$d\Theta = -\frac{2-\frac{1}{2}+1-2}{2-1}\Theta_t dW = -\frac{1}{2}\Theta_t dW$$

d. $d\frac{S_t}{B_t} = dS = \mu dt + \sigma dW = \sigma(\frac{\mu}{\sigma}dt + dW) = \sigma d\widetilde{W}_t$

From Girsanov's theorem

$$d\Theta = -\frac{\mu}{\sigma}\Theta_t dW$$

2 Given the asset price processes in the Black–Scholes model, show that the fair price to pay at time t for a forward on S_t with forward price K and time to expiry $T-t$ is $C_t = S_t - Ke^{-r(T-t)}$.

Solution

The pay-off of a forward on a non-dividend paying share is:

$$\text{Payoff} = S_T - K$$

Apply the risk-neutral derivative pricing formula:

$$C_t = e^{-r(T-t)}\mathbb{E}_t^{\mathbb{Q}} C_T = e^{-r(T-t)}\mathbb{E}_t^{\mathbb{Q}}(S_T - K)$$

Setting $n = 1$ in Equation (10.8)

$$\mathbb{E}_t^{\mathbb{Q}}(S_T - K) = \mathbb{E}_t^{\mathbb{Q}} S_T - K = S_t e^{r(T-t)} - K$$

and the price of the derivative at time t is

$$C_t = e^{-r(T-t)}(S_t e^{r(T-t)} - K) = S_t - Ke^{-r(T-t)}$$

Note that this is the same as the expression given in Theorem 3.4 substituting the forward price from Theorem 3.1.

3 A contract provides a pay-off equal to the value of the square of the share price at maturity time T less the square of the forward price K. Find the fair price to pay at time t for this derivative. Assume that the share price follows geometric Brownian motion with a constant force of dividend yield q.

Solution The pay-off is:

$$\text{Payoff} = S_T^2 - K^2$$

Setting $n = 2$ in Equation (10.8) and amending for the dividend yield

$$\mathbb{E}_t^{\mathbb{Q}}(S_T^2 - K^2) = S_t^2 e^{2(r-q+\frac{1}{2}\sigma^2)(T-t)} - K^2 = S_t^2 e^{(2r-2q+\sigma^2)(T-t)} - K^2$$

Then the price of the derivative at time t is

$$C_t = S_t^2 e^{(r-2q+\sigma^2)(T-t)} - K^2 e^{-r(T-t)}$$

4 A non-dividend paying share S_t is currently priced at $S_0 = \$10$.

(a) Find the fair price to pay now for a derivative paying S^3 in six months given that the volatility of S_t is 30% p.a. and the risk-free interest rate is 5% p.a.

(b) Determine the holdings of shares and cash in a portfolio used to replicate this position.

Solution

(a) Setting $n = 3$ the price of the derivative at time t is

$$\begin{aligned} C_t &= e^{-r(T-t)} S_t^3 e^{(3r+3\sigma^2)(T-t)} \\ &= e^{-0.05\times 0.5} 10^3 e^{(3\times 0.05+3\times 0.3^2)0.5} \\ &= \$1203.21 \end{aligned}$$

(b)

$$\phi_t = \frac{\partial F}{\partial S} = e^{-r(T-t)} 3S_t^2 e^{(3r+3\sigma^2)(T-t)} = 360.97 \text{ i.e. } 360.97 \text{ shares}$$

and

$$\psi_t B_t = F - \phi_t S_t = -\$2406.44 \text{ i.e. borrowing } \$\ 2406.44 \text{ in cash}$$

5 Let W be a standard Brownian motion. Suppose that the risk-free rate is r and that S_t is a risky asset such that:

$$dS = rS_t dt + \sigma S_t dW, \quad o \leq r, S_0 < \infty$$

where $\sigma > 0$ is a constant.

You are investigating the discounted payoff process $Y_t^{(n)} = e^{-rt}S_t^n$ for positive integer n and $0 < t < T$.

(a) Derive the stochastic dynamics of $Y_t^{(n)}$ in the form

$$dY^{(n)} = \mu(t, S_t)dt + \sigma(t, S_t)dW$$

(b) Determine, using Girsanov theorem, an equivalent martingale measure $\mathbb{Q}^A$ such that $Y_t^{(n)}$ is a martingale with respect to $\mathbb{Q}^A$.

Now assume $\sigma = 1, r = 0$.

(c) Determine the Radon-Nikodym process $Z = \frac{d\mathbb{Q}^A}{d\mathbb{Q}}$.

You are interested in pricing a derivative whose payoff is given by the expression:

$$V_T = e^{-\frac{49}{8}T - \frac{7}{2}W_T} S_T^8$$

(d) Determine a closed-form expression for today's price of V_T.

[QFIQF – Spring 2024 Q2]

Solution

(a) From the time-dependent Itô formula

$$\begin{aligned} dY^{(n)} = d(e^{-rt}S_t^n) &= \Big(- re^{-rt}S_t^n + (ne^{-rt}S_t^{n-1})rS_t \\ &\quad + \frac{1}{2}(n(n-1)e^{-rt}S_t^{n-2})\sigma^2 S_t^2 \Big)dt + (ne^{-rt}S_t^{n-1})\sigma S_t dW \\ &= e^{-rt}S_t^n\Big(- r + nr + \frac{1}{2}n(n-1)\sigma^2\Big)dt + e^{-rt}S_t^n n\sigma dW \\ &= e^{-rt}S_t^n\Big((n-1)(r + \frac{1}{2}n\sigma^2)\Big)dt + e^{-rt}S_t^n n\sigma dW \end{aligned}$$

(b)

$$\begin{aligned} d(e^{-rt}S_t^n) &= e^{-rt}S_t^n\Big((n-1)(r + \frac{1}{2}n\sigma^2)\Big)dt + e^{-rt}S_t^n n\sigma dW \\ &= n\sigma e^{-rt}S_t^n\left(\frac{(n-1)(r + \frac{1}{2}n\sigma^2)}{n\sigma}dt + dW\right) \end{aligned}$$

The Radon-Nikodým process $Z = \frac{d\mathbb{Q}^A}{d\mathbb{Q}}$ to change from $\mathbb{Q}$ to $\mathbb{Q}^A$ is defined by

$$dZ = -\gamma Z_t dW$$

where $\gamma = \frac{(n-1)(r+\frac{1}{2}n\sigma^2)}{n\sigma}$

Now assume $\sigma = 1, r = 0$.

(c) The Radon-Nikodým process $Z = \frac{d\mathbb{Q}^A}{d\mathbb{Q}}$ is

$$\begin{aligned} dZ &= -\gamma Z_t dW \\ \implies Z_t &= e^{-\gamma W_t - \frac{1}{2}\gamma^2 t} \\ &= e^{-\frac{1}{2}(n-1)W_t - \frac{1}{8}(n-1)^2 t} \end{aligned}$$

(d)

$$S_t = S_0 e^{(r-\frac{1}{2}\sigma^2)t+\sigma W_t} = S_0 e^{-\frac{1}{2}t+W_t}$$

Apply the risk-neutral derivative pricing formula:

$$\begin{aligned} C_0 &= e^{-rT}\mathbb{E}^{\mathbb{Q}} C_T \\ &= \mathbb{E}^{\mathbb{Q}}(e^{-\frac{49}{8}T-\frac{7}{2}W_T} S_T^8) \\ &= \mathbb{E}^{\mathbb{Q}}(e^{-\frac{49}{8}T-\frac{7}{2}W_T}(S_0^8 e^{-4T+8W_T})) \\ &= S_0^8 \mathbb{E}^{\mathbb{Q}} e^{-\frac{81}{8}T+\frac{9}{2}W_T} \\ &= S_0^8 \qquad \text{from Equation (2.9)} \end{aligned}$$

Alternatively using $Z_T = e^{-\frac{1}{2}(n-1)W_T-\frac{1}{8}(n-1)^2T} = e^{-\frac{49}{8}T-\frac{7}{2}W_T}$ for $n = 8$. Then using Equation (10.1)

$$C_0 = \mathbb{E}^{\mathbb{Q}} C_T = \mathbb{E}^{\mathbb{Q}} e^{-\frac{49}{8}-\frac{7}{2}W_T} S_T^8 = \mathbb{E}^{\mathbb{Q}} Z_T S_T^8 = \mathbb{E}^{\mathbb{Q}^A} S_T^8 = S_0^8$$

Since $Y_t^{(n)} = e^{-rt}S_t^n$ is a martingale with respect to $\mathbb{Q}^A$ for all n and all r.

6 (a) Suppose a derivative pays $D_2 = £100x^2$ at maturity after 2 years where x is uniformly distributed $U(0, 1)$ under the risk-neutral measure. Determine the fair price of this derivative given the risk-free interest rate is 2% p.a.

(b) Suppose that another derivative security pays the amount D_2 at time $t = 2$, but only if D_2 is at least £25. Determine the fair price of this derivative.

Solution

(a) Apply the risk-neutral derivative pricing formula:

$$\begin{aligned} C_t &= e^{-r(T-t)}\mathbb{E}_t^{\mathbb{Q}} C_T \\ &= e^{-0.02\times 2}\mathbb{E}_t^{\mathbb{Q}}(100x^2) \end{aligned}$$

$$= 100e^{-0.02\times 2}\int_0^1 x^2 dx \qquad \text{the pdf is 1 for } 0 \le x \le 1$$
$$= 100e^{-0.02\times 2}(1/3) = £32.03$$

(b)
$$C_t = 100e^{-0.02\times 2}\int_{0.5}^1 x^2 dx$$
$$= 100e^{-0.02\times 2}(1/3 - 1/3(0.5)^3) = £28.02$$

7 Show that the pay-off function for the call option is

$$F(S_t, t) = S_t\Phi(d_1) - e^{-r(T-t)}K\Phi(d_2)$$

where $\Phi(\cdot)$ is the CDF of the standard normal distribution, and

$$d_1 = \frac{\log(S_t/K) + (r + \frac{1}{2}\sigma^2)(T-t)}{\sigma\sqrt{T-t}}, \; d_2 = d_1 - \sigma\sqrt{T-t}$$

Solution

$$C_t = e^{-r(T-t)}\mathbb{E}_t^{\mathbb{Q}} C_T$$
$$= e^{-r(T-t)}\mathbb{E}_t^{\mathbb{Q}}[S_T - K]^+$$
$$= \frac{e^{-r(T-t)}}{\sqrt{2\pi}}\int_{z^*}^{\infty}(S_T - K)e^{-\frac{1}{2}z^2}dz \qquad \text{where } S_T \ge K \text{ when } z \ge z^*$$
$$= \frac{e^{-r(T-t)}}{\sqrt{2\pi}}\int_{z^*}^{\infty}(S_t e^{(r-\frac{1}{2}\sigma^2)(T-t)+\sigma\sqrt{T-t}z} - K)e^{-\frac{1}{2}z^2}dz$$
$$= S_t\int_{z^*}^{\infty}\frac{1}{\sqrt{2\pi}}e^{-\frac{1}{2}(z^2 - 2\sigma\sqrt{T-t}z + \sigma^2(T-t))}dz$$
$$- e^{-r(T-t)}K\int_{z^*}^{\infty}\frac{1}{\sqrt{2\pi}}e^{-\frac{1}{2}z^2}dz$$

In the first integral, let
$y = z - \sigma\sqrt{T-t} \implies y^2 = z^2 - 2\sigma\sqrt{T-t}z + \sigma^2(T-t)$.
Then integrate between y^* and ∞ where

$$y^* = z^* - \sigma\sqrt{T-t}$$
$$= \frac{\log(K/S_t) - (r - \frac{1}{2}\sigma^2)(T-t)}{\sigma\sqrt{T-t}} - \sigma\sqrt{T-t}$$

$$= \frac{\log(K/S_t) - (r + \frac{1}{2}\sigma^2)(T-t)}{\sigma\sqrt{T-t}}$$

So

$$\begin{aligned} C_t &= S_t \int_{y^*}^{\infty} \frac{1}{\sqrt{2\pi}} e^{-\frac{1}{2}y^2} dy - e^{-r(T-t)} K \int_{z^*}^{\infty} \frac{1}{\sqrt{2\pi}} e^{-\frac{1}{2}z^2} dz \\ &= S_t \int_{-\infty}^{-y^*} \frac{1}{\sqrt{2\pi}} e^{-\frac{1}{2}y^2} dy - e^{-r(T-t)} K \int_{-\infty}^{-z^*} \frac{1}{\sqrt{2\pi}} e^{-\frac{1}{2}z^2} dz \\ &= S\Phi(d_1) - e^{-r(T-t)} K\Phi(d_2) \end{aligned}$$

With

$$d_1 = -y^* = \frac{\log(S_t/K) + (r + \frac{1}{2}\sigma^2)(T-t)}{\sigma\sqrt{T-t}} \text{ and}$$

$$d_2 = -z^* = -y^* - \sigma\sqrt{T-t} = d_1 - \sigma\sqrt{T-t}$$

8 Show that the pay-off function for the put option is

$$F(S_t, t) = Ke^{-r(T-t)}\Phi(-d_2) - S_t\Phi(-d_1)$$

where $\Phi(\cdot)$ is the CDF of the standard normal distribution, and

$$d_1 = \frac{\log(S_t/K) + (r + \frac{1}{2}\sigma^2)(T-t)}{\sigma\sqrt{T-t}}, \; d_2 = d_1 - \sigma\sqrt{T-t}$$

Solution

$$\begin{aligned} C_t &= e^{-r(T-t)}\mathbb{E}_t^{\mathbb{Q}} C_T \\ &= e^{-r(T-t)}\mathbb{E}_t^{\mathbb{Q}}[K - S_T]^+ \\ &= \frac{e^{-r(T-t)}}{\sqrt{2\pi}} \int_{-\infty}^{z^*} [K - S_T]^+ e^{-\frac{1}{2}z^2} dz \\ &= \frac{e^{-r(T-t)}}{\sqrt{2\pi}} \int_{-\infty}^{z^*} (K - S_t e^{(r-\frac{1}{2}\sigma^2)(T-t)+\sigma\sqrt{T-t}z}) e^{-\frac{1}{2}z^2} dz \\ &= e^{-r(T-t)} K \int_{-\infty}^{z^*} \frac{1}{\sqrt{2\pi}} e^{-\frac{1}{2}z^2} dz \\ &\quad - S_t \int_{-\infty}^{z^*} \frac{1}{\sqrt{2\pi}} e^{-\frac{1}{2}(z^2 - 2\sigma\sqrt{T-t}z + \sigma^2(T-t))} dz \end{aligned}$$

In the second integral, let $y = z - \sigma\sqrt{T-t}$ as was done in the previous question.

So

$$\begin{aligned} p_t &= e^{-r(T-t)}K \int_{-\infty}^{z^*} \frac{1}{\sqrt{2\pi}} e^{-\frac{1}{2}z^2}\, dz - S_t \int_{-\infty}^{y^*} \frac{1}{\sqrt{2\pi}} e^{-\frac{1}{2}y^2}\, dy \\ &= e^{-r(T-t)}K\Phi(-d_2) - S_t\Phi(-d_1) \end{aligned}$$

9 Use Monte Carlo simulation to estimate the price of a European call option with the following parameters:

- Initial stock price: $S_0 = 40$
- Strike price: $K = 45$
- Risk-free interest rate: $r = 5\%$ per annum
- Volatility: $\sigma = 20\%$ per annum
- Time to maturity: $T = 4$ months

Instructions:

- Amend the code in Section 6.7.2 to model geometric Brownian motion model under the risk-neutral measure. That is, change *mu* to *r*.
- Using the parameters given, simulate 500 independent paths of the stock price at maturity.
- For each simulated path, compute the call payoff:

$$\text{Payoff} = [S_T - K]^+$$

- Compute the average payoff over all simulated paths.

$$\mathbb{E}^{\mathbb{Q}}[S_T - K]^+$$

- Find the discounted average payoff. This is the price of the option.

$$e^{-rT}\mathbb{E}^{\mathbb{Q}}[S_T - K]^+$$

- Compare your answer to Exercise 9.2 where the Black–Scholes formula for the European call price was applied.

11 Option Greeks

Overview

In options trading and risk management, understanding the sensitivity of an option's value to changes in underlying variables – *market risk* – is just as important as knowing the option's price itself. These sensitivities, collectively known as the Greeks, quantify how option values respond to shifts in factors such as the underlying asset price, volatility and time. Practitioners seldom hold options in isolation. Portfolios are dynamically hedged, and risk must be managed continuously in changing markets. The Greeks provide the essential toolkit for this purpose: they allow traders to measure exposures, design hedging strategies, and evaluate the robustness of portfolios against different sources of market risk.

This chapter develops the theory and application of option Greeks. We begin with their formal definitions and illustrate their role in hedging strategies. *Delta hedging* involves neutralizing exposure to the underlying asset price by taking offsetting positions. In *Gamma hedging*, positions are adjusted to reduce the risk of large, nonlinear moves while *Vega hedging* means using options to manage exposure to volatility changes.

There are other strategies derivative traders use to manage market risk not considered in this chapter. Traders can offset positions across maturities, strikes, or asset classes ; manage their position size or they can trade derivatives in combination to reduce outright exposure.

The focus here is on managing exposure to market risk. There are also other risks in managing derivatives that we do not consider – counterparty risk, liquidity risk, regulatory risk, operational risks and so on. The risk management of derivatives is a much broader topic worthy of an entire textbook.

Introduction

Derivatives can be traded in two main ways through organized exchanges or through privately negotiated contracts *over-the-counter*. A financial institution selling an over-the-counter option must manage its exposure. The underlying asset price could move adversely creating a significant payoff for the option

DOI: 10.4324/9781003477624-14

holder and significant losses for the financial institution. This chapter addresses risk managing options to reduce or offset potential losses from price movements in the underlying asset. In particular, it deals with the technique of *hedging* which means taking opposite positions in the underlying asset or other options in order to neutralize such risks.

For example, consider the situation where a bank sells a European call option on 1000 shares of XYZ stock. The strike price is \$110 and the premium received is \$1.46 per share (or \$1,460 in total). If the stock price is \$120 at maturity, the bank must pay out

$$1000 \times (\$120 - \$110) = \$10,000$$

This is much greater than the \$1,460 the bank received for selling the option. The bank has made a loss of \$8,540 on the trade.

In this case, the bank has done nothing to mitigate their risk exposure. This is known as a *naked position*.

Instead, we will consider risk managing the option by hedging. We will see that it does not eliminate the risk entirely but will reduce potential losses from adverse price movements.

11.1 Delta hedging

Having bought an option we can now calculate the quantity of the underlying asset we need to purchase in order to minimize losses from price movements.

Let the option price be $C_t = F(S_t, t)$ then using the Taylor expansion in two variables (Section 1.10)

$$\Delta C = \frac{\partial F}{\partial S}\Delta S + \frac{\partial F}{\partial t}\Delta t + \frac{1}{2}\left(\frac{\partial^2 F}{\partial S^2}(\Delta S)^2 + 2\frac{\partial^2 F}{\partial S \partial t}\Delta S \Delta t + \frac{\partial^2 F}{\partial t^2}(\Delta t)^2\right) + \ldots \quad (11.1)$$

As we are only concerned with the effect of price movements on the option price we focus on the $\Delta S, (\Delta S)^2, (\Delta S)^3 \ldots$ terms.

$$\Delta C = \frac{\partial F}{\partial S}\Delta S + \frac{1}{2}\frac{\partial^2 F}{\partial S^2}(\Delta S)^2 + \ldots \quad (11.2)$$

If ΔS small, then:

$$\Delta C \approx \frac{\partial F}{\partial S}\Delta S$$

We will create a portfolio containing this option and an amount n of the underlying asset so that the portfolio value V does not change as the price moves. The portfolio value is

$$V = C + nS$$

and the change in value due to a change in the asset price is

$$\begin{aligned}\Delta V &= \Delta C + n\Delta S \\ &\approx \frac{\partial F}{\partial S}\Delta S + n\Delta S \\ &= \left(\frac{\partial F}{\partial S} + n\right)\Delta S\end{aligned}$$

Choosing $n = -\frac{\partial F}{\partial S}$ means that the change in the portfolio value ΔV will be zero. The derivative $\frac{\partial F}{\partial S}$ was called the *delta* of the option in Chapter 10 hence the technique of *delta hedging*. The combination of the long position in the option and the short position in the underlying asset is said to be *delta neutral*.

As an example, consider being long in a 6-month European call option with strike price \$50 and current stock price \$45. Assume the share pays no dividends, volatility is 20% and the risk-free rate is 3%.

From Equation (3.1), the current price of the European call option c_t is

$$\begin{aligned}c_t &= S_t\Phi(d_1) - e^{-r(T-t)}K\Phi(d_2) \\ &= 45\,\Phi(-0.568) - e^{-0.03\times 0.5}\,50\,\Phi(-0.710) = 1.05\end{aligned}$$

Question 7 Chapter 9 gives the delta of the European call option

$$\text{delta} = \frac{\partial F}{\partial S} = \Phi(d_1) = \Phi(-0.568) = 0.285$$

Therefore we create a portfolio long in the call option and short 0.285 shares. The initial value of this portfolio is

$$V = 1.05 - 0.285 \times 45 = -11.77$$

In Table 11.1, we consider the effect of price movements to the portfolio value. Say the stock price moves instantaneously to 44. The recalculated derivative

Table 11.1 Delta hedging

	S	c	ΔV
Initial	45	1.05	
ΔS small	44	0.79	0.03
	46	1.36	0.03
ΔS large	40	0.20	0.57
	50	3.19	0.71

price is 0.79. And the portfolio value is now

$$V = 0.79 - 0.285 \times 44 = -11.74$$

The derivative price has changed substantially but the portfolio value has hardly changed ($\Delta V = 0.03$). The effect of price movements has not been eliminated but has been significantly reduced.

The table also shows that the hedge is not quite as effective with larger price movements. The changes in value when the share price moves instantaneously to 40 or 50 are \$0.57 and \$0.71 respectively. This is because we assumed in our calculations of n above that ΔS was small. The next section shows how we can improve upon the delta hedge when price movements are large.

If instead we had taken a short position in the option then, following the logic above, we would take the opposite position in the underlying asset. That is, we would go long delta shares.

Delta hedging can be applied to any option. The delta of a European put option on a share can be found by differentiating the pricing function $p_t = F(S_t, t)$ with respect to S_t or by differentiating both sides of the put-call parity relationship with respect to S_t.

$$\begin{aligned} c_t + Ke^{-r(T-t)} &= p_t + S_t \\ \Longrightarrow \text{ delta}_{\text{ call}} &= \text{delta}_{\text{ put}} + 1 \\ \Longrightarrow \text{ delta}_{\text{ put}} &= \text{delta}_{\text{ call}} - 1 \\ &= \Phi(d_1) - 1 \end{aligned} \tag{11.3}$$

We then proceed as before offsetting our option position with a delta position in the share.

11.2 Gamma hedging

Looking again at the Taylor expansion in Equation (11.2)

$$\Delta C = \frac{\partial F}{\partial S}\Delta S + \frac{1}{2}\frac{\partial^2 F}{\partial S^2}(\Delta S)^2 + \ldots$$

we see that the approximation for the change in derivative value used in delta-hedging could be improved by retaining the second-order term, that is,

$$\Delta C \approx \frac{\partial F}{\partial S}\Delta S + \frac{1}{2}\frac{\partial^2 F}{\partial S^2}(\Delta S)^2$$

The second-order partial derivative $\frac{\partial^2 F}{\partial S^2}$ is known as the *gamma* or Γ of the option.

To gamma hedge in addition to delta hedging we will create a portfolio containing the option with price $C_{1,t} = F_1(S_t, t)$, an amount n_1 of the underlying asset and an amount n_2 of another option on the underlying asset with price $C_{2,t} = F_2(S_t, t)$ so that the portfolio value is neutral to price movements. The change in value due to a change in the asset price is now

$$\begin{aligned}
\Delta V &= \Delta C_1 + n_1 \Delta S + n_2 \Delta C_2 \\
&\approx \left(\frac{\partial F_1}{\partial S} \Delta S + \frac{1}{2} \frac{\partial^2 F_1}{\partial S^2} (\Delta S)^2 \right) + n_1 \Delta S + n_2 \left(\frac{\partial F_2}{\partial S} \Delta S + \frac{1}{2} \frac{\partial^2 F_2}{\partial S^2} (\Delta S)^2 \right) \\
&= \left(\frac{\partial F_1}{\partial S} + n_2 \frac{\partial F_2}{\partial S} + n_1 \right) \Delta S + \frac{1}{2} \left(\frac{\partial^2 F_1}{\partial S^2} + n_2 \frac{\partial^2 F_2}{\partial S^2} \right) (\Delta S)^2
\end{aligned}$$

Choosing n_1 and n_2 so that the two terms in brackets are zero means that the change in the portfolio value ΔV will be zero. Starting with the $(\Delta S)^2$ term

$$n_2 = -\frac{\partial^2 F_1}{\partial S^2} \Big/ \frac{\partial^2 F_2}{\partial S^2} = -\frac{\Gamma_1}{\Gamma_2}$$

and substituting in the ΔS term

$$n_1 = -\frac{\partial F_1}{\partial S} - n_2 \frac{\partial F_2}{\partial S} = -\text{delta}_1 + \frac{\Gamma_1}{\Gamma_2} \text{delta}_2$$

Does this improve the hedge in the previous section? We create the portfolio containing the European call option, n_1 shares and an amount n_2 of another option. Let this other option be a European call with strike 55 maturing in 9 months. The initial price of this option is calculated to be 0.67.

Question 8 Chapter 9 gives the gamma of the European call option

$$\Gamma = \frac{\partial^2 F}{\partial S^2} = \frac{\Phi'(d_1)}{S_t \sigma \sqrt{T-t}}$$

Remember from the first fundamental theorem of calculus $\Phi'(x) = \phi(x)$ (Section 2.6.4). Therefore,

$$\begin{aligned}
\Gamma_1 &= \frac{\phi(-0.568)}{45 \times 0.2 \times \sqrt{0.5}} = 0.053 \\
\Gamma_2 &= \frac{\phi(-0.942)}{45 \times 0.2 \times \sqrt{0.75}} = 0.033 \\
\Longrightarrow n_2 &= -\frac{\Gamma_1}{\Gamma_2} = -\frac{0.053}{0.033} = -1.624
\end{aligned}$$

Table 11.2 Gamma hedging

	S	c_1	c_2	ΔV
Initial	45	1.05	0.67	
ΔS small	44	0.79	0.51	0.00
	46	1.36	0.86	0.00
ΔS large	40	0.20	0.14	0.02
	50	3.19	2.00	-0.05

And

$$\begin{aligned}
\text{delta}_1 &= \Phi(-0.568) = 0.285 \\
\text{delta}_2 &= \Phi(-0.942) = 0.173 \\
\Longrightarrow n_1 &= -\text{delta}_1 + \frac{\Gamma_1}{\Gamma_2}\,\text{delta}_2 = -0.285 + \frac{0.053}{0.033} \times 0.173 = -0.004
\end{aligned}$$

Then we create a portfolio long in the call option and short 0.004 shares and 1.624 options. The initial value of this portfolio is

$$V = 1.05 - 0.004 \times 45 - 1.624 \times 0.67 = -0.2085$$

In Table 11.2, we consider the effect of price movements to the portfolio value. The effect of price movements on the portfolio value is much less than when just delta hedging in Table 11.1.

The gamma of a European put option will be the same as the gamma of a European call option since

$$\Gamma_{\text{put}} = \frac{\partial^2 F}{\partial S^2} = \frac{\partial}{\partial S}\Big(\Phi(d_1) - 1\Big) = \frac{\partial}{\partial S}\Phi(d_1) = \Gamma_{\text{call}}$$

11.3 Time effects

In this section, change is no longer instantaneous and we must take account of the passage of time. Time can affect

1. Interest that accumulates on any borrowing or investment.
2. The derivative price

11.3.1 Interest

Let's consider the accumulation of interest. Using our earlier notation, we create a portfolio long in the option and short a combination of cash and shares with the same initial value. Then with ψ units of cash and φ shares the value of this portfolio Π_0 at $t = 0$ is

$$\Pi_0 = C_0 - (\varphi S_0 + \psi) = 0$$

If the combination of cash and shares is not rebalanced, the portfolio value at time t becomes

$$\Pi_t = C_t - (\varphi S_t + e^{rt}\psi) \tag{11.4}$$

Exercise 11.1. *Several months ago, an investor sold 100 units of a one-year European call option on a non-dividend-paying stock. She immediately delta-hedged the commitment with shares of the stock, but has not ever re-balanced her portfolio. She now decides to close out all positions. You are given the following information:*

(i) The risk-free interest rate is constant.

(ii)

	Several months ago	*Now*
Stock price	*\$40.00*	*\$50.00*
Call option price	*\$ 8.88*	*\$14.42*
Put option price	*\$ 1.63*	*\$ 0.26*
Call option delta	*0.794*	

The put option in the table above is a European option on the same stock and with the same strike price and expiration date as the call option.

Calculate her profit.

[IFM - Sample Q47]

Proof. We can use put-call parity to find e^{rt}. Let time at 'several months ago' be zero, current time be t and expiration date be T.

$$c_t + Ke^{-r(T-t)} = p_t + S_t$$

$$\text{At time } 0,\ Ke^{-rT} = 1.63 + 40 - 8.88 = 32.75$$

$$\text{At time } t,\ Ke^{-r(T-t)} = 0.26 + 50 - 14.42 = 35.84$$

$$\text{Dividing gives } e^{rt} = \frac{Ke^{-r(T-t)}}{Ke^{-rT}} = 1.09$$

The investor is short the option and long in shares. Having closed out all positions, the change in value is

$$\begin{aligned} -C_t + (\varphi S_t + e^{rt}\psi) &= -Ct + \varphi St + e^{rt}(C_0 - \varphi S_0) \text{ where } \varphi \text{ is the option delta} \\ &= -14.42 + 0.794 \times 50 + 1.09(8.88 - 0.794 \times 40) \\ &= 0.24 \end{aligned}$$

Her profit is $100 \times 0.24 = \$24.12$.

□

The previous example did not consider rebalancing. An option's delta will change as the underlying price moves. If the trader hedges only once, she will soon be exposed again. The trader instead will continuously adjust her hedge to maintain her position delta neutral. This is called *dynamic hedging* because the hedge evolves dynamically as market conditions change. The next example describes this process.

Exercise 11.2. *An investor sells 1000 one-year European call options on a non dividend-paying stock for \$2.08 each. She immediately delta-hedges the commitment with shares of the stock and rebalances every month. Given the stock price evolutions in the table below and parameters r = 5%;K = 45; σ = 20%, determine the hedging portfolio at each point in time.*

Evolution	*Month*	*0*	*1*	*2*	*3*	*4*	*5*	*6*
1	*Stock price (\$)*	*40.00*	*38.94*	*38.02*	*43.91*	*45.28*	*44.83*	*47.49*
2	*Stock price (\$)*	*40.00*	*40.24*	*40.17*	*41.09*	*38.76*	*36.13*	*37.92*

Evolution	*Month*	*7*	*8*	*9*	*10*	*11*	*12*
1	*Stock price (\$)*	*44.23*	*47.18*	*51.45*	*53.26*	*51.61*	*54.60*
2	*Stock price (\$)*	*39.94*	*41.86*	*41.54*	*42.89*	*44.24*	*42.93*

Table 11.3 Dynamic hedging

	In-the-money			Out-of-the-money		
t	S_t	*delta*	*hedge*	S_t	*delta*	*hedge*
0	40.00	0.406	406	40.00	0.406	406
1	38.94	0.337	337	40.24	0.402	402
2	38.02	0.273	273	40.17	0.381	381
3	43.91	0.564	564	41.09	0.412	412
4	45.28	0.627	627	38.76	0.265	265
5	44.83	0.596	596	36.13	0.121	121
6	47.49	0.735	735	37.92	0.168	168
7	44.23	0.537	537	39.94	0.243	243
8	47.18	0.730	730	41.86	0.336	336
9	51.45	0.935	935	41.54	0.266	266
10	53.26	0.986	986	42.89	0.328	328
11	51.61	0.993	993	44.24	0.423	423
12	54.60	1.000	1000	42.93	0.000	0

Proof. The option ends in-the-money for the first stock price evolution.
Initially

$$d_1 = \frac{\log(S_t/K) + (r + \frac{1}{2}\sigma^2)(T-t)}{\sigma\sqrt{T-t}} = \frac{\log(40/45) + (0.05 + \frac{1}{2}0.2^2)}{0.2} = -0.239$$

The call delta is

$$\text{delta} = \Phi(-0.239) = 0.406$$

She is short in calls so buys $1000 \times 0.406 = 406$ shares.

This result is summarized in Table 11.3 with calculations repeated for subsequent months.

At month 1, the delta is $\Phi(-0.420) = 0.337$. The hedge is then long 337 shares.

And so on for the remaining months until maturity after one year.

At maturity, the option is in-the-money. She sells the shares she owns and collects the strike price.

The option ends out-of-the-money for the second stock price evolution. The calculations above are repeated and are given in the table. There are no shares remaining in the hedge position at expiration. □

Increasing the frequency of rebalancing will keep your portfolio close to delta-neutral at all times. The investor then has reduced risk exposure to small price moves.

More frequent rebalancing however implies higher transaction costs. When trading shares, there are brokerage commissions as well as the hidden cost of differences in bid-ask spreads. These costs will accumulate with frequent rebalancing. The trader may then decide on fixed intervals of rebalancing to limit transaction costs. Alternatively, the trader may decide how often to rebalance based on

(a) volatility – if low then share price moves little and hence delta changes little
(b) how deep in or out-of-the-money – delta changes little when far from the strike price
(c) time to expiry – delta can change quickly close to expiration especially if close to the strike price

Observe how little delta changes weeks 9-12 for the in-the-money scenario in Table 11.3 far from the strike price \$45. Contrast this with the same time period for the out-of-the-money scenario as the share price gets close to the strike price.

11.3.2 Theta

Time also affects the option price. Assuming Δt and ΔS small, the Taylor expansion in Equation (11.1) becomes

$$\Delta C \approx \text{delta}\,\Delta S + \frac{\partial F}{\partial t}\Delta t + \frac{1}{2}\Gamma(\Delta S)^2 \tag{11.5}$$

The derivative $\frac{\partial F}{\partial t}$ is known as the option *theta* or Θ. The only uncertainty in the Black–Scholes model is due to asset price movements. There is no uncertainty about the passage of time so theta is not a risk parameter as such. Theta tells us how the price of an option changes as the option approaches maturity.

$$\Delta C \approx \text{delta}\,\Delta S + \Theta\Delta t + \frac{1}{2}\Gamma(\Delta S)^2 \tag{11.6}$$

Exercise 11.3. *The table below lists the Greek measures of two long European options on Stock XYZ.*

Option	*Delta*	*Gamma*	*Theta (per day)*
A	*0.7890*	*0.0180*	*–0.0118*
B	*0.6795*	*0.0222*	*–0.0129*

Your company has a portfolio consisting of a long position in 100 Option A and a short position in 200 Option B.

Estimate the change in the value of your company's portfolio by using the Taylor series expansion of the option prices if the price of the underlying stock decreases by 10 after 5 days.

[QFIQF Spring 2022 Q12(d)]

Proof. Using

$$\Delta C \approx \text{delta} \times (-10) + \Theta \times 5 + \frac{1}{2}\Gamma \times (-10)^2$$

Option	*Delta*	*Gamma*,Γ	*Theta*,Θ	ΔC
A	0.7890	0.0180	–0.0118	–7.0490
B	0.6795	0.0222	–0.0129	–5.7495

$$\Delta V = 100 \times -7.0490 - 200 \times -5.7495 = 445$$

□

11.4 Vega

The price of an option is very dependent on the measure of future volatility which by its nature is unknowable. A trader may therefore be concerned about minimizing her exposure to volatility shifts.

In the Black–Scholes model, volatility is regarded as constant. The discussion in Chapter 7 showed that volatility in fact clearly changes over time. The Black–Scholes model is therefore inappropriate as a guide to minimizing the risk of changing volatility. A model incorporating stochastic volatility is a more suitable framework.

However the Black–Scholes model is in practice used this way for simplicity. The sensitivity of the option price to changes in volatility is called the *vega* (ν)

$$\nu = \frac{\partial C}{\partial \sigma}$$

For European call and put options, the vega is

$$\nu = S_t\sqrt{T - t}\Phi'(d_1)$$

which is always positive. This is what we expected in Section 3.2. The higher the volatility, the greater the chance that the asset price can move in favor of the holder of the option before expiry. Notice also that vega weakens with time to expiration $T - t$.

Delta, gamma and vega hedging requires a portfolio with the underlying asset and two other different options on the same asset. Option traders will delta hedge routinely by adjusting their position in the underlying asset. Gamma and vega are not generally managed in this way as option gamma and vega are interdependent. It can be difficult to either find the right combination of options or purchase enough options in the market to achieve neutrality in all three Greeks. These risks are then managed within limits rather than eliminated by adding options that reduce gamma and vega as the opportunity arises.

Some traders trade on volatility by finding option implied volatilities that they believe are mistaken. They achieve this by structuring option strategies that profit mainly from changes in volatility. Volatility traders will then rely heavily on calculating vegas.

Exercise 11.4. *You manage a portfolio of variable annuities invested in the Russell 2000 Index. You construct a hedge for the portfolio. You are given the following for the hedge position:*

- *Yesterday's market price was \$6.72.*
- *Today's market price is \$6.93.*

Also

- *Yesterday's implied volatility was 24% per annum.*
- *Yesterday's Vega was 30.11.*

Estimate today's implied volatility of the Russell 2000 Index underlying the hedge position based on the information given above.

[QFIQF – Fall 2018 q5(d)]

Proof. From the definition of the vega

$$\begin{aligned}
\nu &= \frac{\partial C}{\partial \sigma} \\
&\approx \frac{C_{t+1} - C_t}{\sigma_{t+1} - \sigma_t} \\
\implies 30.11 &= \frac{6.93 - 6.72}{\sigma_{t+1} - 0.24} \\
\implies \sigma_{t+1} &= 24.7\%
\end{aligned}$$

□

11.5 Greeks for European calls and puts

Table 11.4 summarizes the Greeks for European calls and put options on a dividend-paying stock under the Black–Scholes model. The signs of the Greeks are also added as plus (+) or minus (–). For example, $\Phi(\cdot)$ is always positive so the delta for a European call is positive and $\Phi(\cdot)$ is also always less than one so the delta for a European put should always be negative. The sign of Θ for a European put is ambiguous.

Table 11.4 Greeks for European calls and puts

	Call		*Put*	
Delta	$e^{-q(T-t)}\Phi(d_1)$	(+)	$e^{-q(T-t)}(\Phi(d_1)-1)$	(–)
Γ	$\dfrac{\phi(d_1)e^{-q(T-t)}}{S_t\sigma\sqrt{T-t}}$	(+)	Γ_{call}	(+)
Θ	$-\dfrac{S_t\phi(d_1)\sigma e^{-q(T-t)}}{2\sqrt{T-t}} + qS_t\Phi(d_1)e^{-q(T-t)} - rKe^{-r(T-t)}\Phi(d_2)$	(–)	$\Theta_{\text{call}} - qS_te^{-q(T-t)} + rKe^{-r(T-t)}$	(?)
ν	$S_t\sqrt{T-t}\,\phi(d_1)e^{-q(T-t)}$	(+)	ν_{call}	(+)

As before

$$d_1 = \frac{\log(S_t/K) + (r - q + \frac{1}{2}\sigma^2)(T-t)}{\sigma\sqrt{T-t}} \text{ and } d_2 = d_1 - \sigma\sqrt{T-t}$$

and the derivative of the cumulative normal probability distribution function $\Phi(x)$ is the normal probability density function $\phi(x)$

$$\Phi'(x) = \phi(x)$$

The direction of the response of European option prices to changes in the share price, time and volatility is discussed in Section 3.2. The signs in Table 3.1 are seen to be consistent with Table 11.4.

There we argued that an increase in the share price meant the price of a European call should rise and the price of a European put should fall. This agrees with the sign of the delta. Higher volatility was presumed to increase call and put option prices agreeing with the sign of the vega above.

The results for time are also consistent though they appear at first glance contradictory. In Section 3.2, expiration date T was increasing while the current time t was fixed, that is, time to maturity was getting longer. Theta is calculated on the basis that expiration date is fixed and that time is progressing toward that date, that is, time to maturity is getting shorter. Therefore negative Θ for a European call is what we would have expected from the earlier discussion. The mathematical ambiguity about the sign of Θ for a European put also agrees with Section 3.2.

Exercise 11.5. *You are given the following for a European call option on a non-dividend-paying stock:*

- *The implied volatility = 20%*
- *The time-to-maturity = 6 months*
- *The stock price = $90*
- *Vega = 0.207, expressed as the sensitivity of option price change with respect to 1 percentage point change in the level of the volatility (from 20% to 21%)*

Calculate the option's Gamma.

[QFIQF – Fall 2019 q8(c)]

Proof. Adjust the vega for a unit change in the level of the volatility

$$\nu = 0.01 \times S_t\sqrt{T-t}\,\phi(d_1)$$

$$\Longrightarrow\ 0.207 = 0.01 \times 90 \times \sqrt{0.5} \times \phi(d_1) \ \Longrightarrow\ \phi(d_1) = 0.325$$

Then gamma is

$$\begin{aligned}\Gamma &= \frac{\phi(d_1)}{S_t\sigma\sqrt{T-t}} \\ &= \frac{0.325}{90 \times 0.20 \times \sqrt{0.5}} = 0.026\end{aligned}$$

□

There are other Greeks of less significance used occasionally by traders. Given option price $C_t = F(S_t, t)$

- Rho (ρ) measures the sensitivity of the option to changes in the risk-free interest rate.

$$\rho := \frac{\partial F}{\partial r}$$

- Lambda (λ) measures the sensitivity of the option to changes in the dividend yield.

$$\lambda := \frac{\partial F}{\partial q}$$

In theory both of these parameters could vary over the term of the option contract although assumed constant in the Black–Scholes model. Once again other models that allow these parameters to vary stochastically should be used to assess these risks.

11.6 Summary

This chapter explored the role of option Greeks in understanding and managing the market risk of option positions. The Greeks measure the sensitivities of option values to changes in the factors affecting option prices and serve as the foundation of hedging strategies in derivatives markets.

Through delta hedging, traders construct positions that are insensitive to small price changes by taking offsetting positions in the underlying asset. Delta hedging only accounts for small moves in the underlying asset price, larger shifts can still introduce risk. To address this, gamma hedging adds a second option to neutralize curvature leading to a more robust hedge. Vega is the sensitivity of the option price to volatility. Since volatility is unobservable, managing vega is crucial for traders exposed to volatility risk. Vega hedging involves taking positions in other options to offset volatility exposure. While delta hedging is routine, gamma and vega hedges are often managed within limits rather than eliminated completely due to practical constraints.

Not only does the passage of time affect interest accrual and cash balances in a hedged portfolio, but option values also erode through theta. Continuous rebalancing is theoretically ideal but, in practice, traders balance hedge frequency against transaction costs.

Hedging with Greeks allows traders to isolate, control and manage exposures, but always with practical limitations such as transaction costs, liquidity constraints and model assumptions. Together, these tools form the backbone of modern option risk management. Traders also manage market risk by combining derivatives, diversification and risk controls.

Managing derivatives also involves other non-market risks – model and parameter risk, hedging limitations, counterparty exposure and operational/systemic risks. Effective management requires robust risk systems, stress testing and regulatory compliance.

Key things to remember

- Given a derivative price, calculate the Greeks.
- The process of dynamic hedging.

Further reading

Pirie (2021) provides a practitioner-friendly introduction to derivatives. It covers Greeks conceptually with a strong emphasis on their role in risk management. This is an excellent text for building intuition for students from a finance/economics background. McDonald (2013) develops option pricing theory and links it to hedging applications. The discussion of Greeks is systematic and well-suited for students aiming to connect pricing models to practical hedging strategies. Hull (2021) explains the Greeks in a highly applied way including their use in delta and gamma hedging and their importance in managing derivative portfolios. Derivative trading strategies to manage market risk are also comprehensively treated. Jarrow and Chatterjea (2024) is a rigorous yet accessible treatment of derivatives. It situates Greeks within arbitrage pricing and risk-neutral valuation. Hedging is combined with consideration of other risks to build a broader unified risk management framework.

Questions

1 For two European call options, Call-I and Call-II, on a stock, you are given:

Greek	*Call-I*	*Call-II*
Delta	0.5825	0.7773
Gamma	0.0651	0.0746
Vega	0.0781	0.0596

Suppose you just sold 1000 units of Call-I.

Determine the numbers of units of Call-II and stock you should buy or sell in order to both delta-hedge and gamma-hedge your position in Call-I.

[MFE - Spring 2007 Q10]

Solution From Section 11.2

$$n_2 = -\frac{\Gamma_1}{\Gamma_2} = -0.0651/0.0746 = -0.8727$$

and

$$n_1 = -\text{delta}_1 + \frac{\Gamma_1}{\Gamma_2}\,\text{delta}_2 = 0.0958$$

The signs are then reversed since we have a short position in the call. So sell 95.8 units of stock and buy 872.7 units of Call-II.

2 Assume you have purchased European put options for 100,000 shares of a non dividend-paying stock and you are given the following information.

- Price of stock = \$49.16
- Strike price = \$50.00
- Continuously-compounded risk-free interest rate = 5% per annum
- Volatility = 20% per annum
- There are 20 weeks remaining until maturity

(a) Determine the initial position you should take in the underlying stock to implement a delta-hedging strategy.

(b) You now have the following information

	Stock price	d_1
1	\$49.33	0.10
2	\$49.09	0.05

You decide to readjust the delta hedging strategy on a weekly basis. Calculate the cumulative costs, including interest, of the hedge at the end of week 2.

Proof. (a) Initially

$$d_1 = \frac{\log(S_t/K) + (r + \frac{1}{2}\sigma^2)(T-t)}{\sigma\sqrt{T-t}} = \frac{\log(49.16/50) + (0.05 + \frac{1}{2}0.2^2)\frac{20}{52}}{0.2\sqrt{\frac{20}{52}}} = 0.080$$

Using the delta of a put in Equation (11.3)

$$\text{delta} = \Phi(0.080) - 1 = -0.4679$$

We are long in puts so we buy $100000 \times 0.4679 = 46,793$ shares.

(b) The initial hedge costs $46,793 \times \$49.17 = \$2,300,364$

Using the given d_1 values and Equation (11.3), we can calculate the delta values for weeks 1 and 2.

	Stock price	d_1	*delta*	*hedge*
0	\$49.16	0.08	-0.4679	46,793
1	\$49.33	0.10	-0.4602	46,017
2	\$49.09	0.05	-0.4801	48,006

At week 1 we must sell $46,793 - 46,017 = 776$ shares and in week 2 we must buy $48,006 - 46,017 = 1989$ shares.

So the cumulative cost at 5% interest is

$$2300364e^{0.05\times\frac{2}{52}} - (777 \times 49.33)e^{0.05\times\frac{1}{52}} + (1989 \times 49.09) = \$2,364,101$$

□

3 You are a portfolio manager at a small financial institution. You have a long position in a vanilla call option which is worth \$10 today when the underlying stock price is $S = \$100$.

The change in the call option price $C(S,t)$ when the underlying stock price changes by a small amount dS from time t to time $t + dt$ is:

$$C(S + dS, t + dt) - C(S,t) = \Theta dt + \Delta dS + \frac{1}{2}\Gamma(dS)^2$$

where

$$\Theta = \frac{\partial C}{\partial t}, \Delta = \frac{\partial C}{\partial S} \text{ and } \Gamma = \frac{\partial^2 C}{\partial S^2}$$

And the change in the value of the call over a calendar day (assuming 365 days in a year) is:

- \$5.99, when the stock price moves up by 10%, and
- −\$4.01, when the stock price moves down by 10%

Assume the option's theta Θ is -3.65 and the risk-free rate of return is 0%.

(a) Determine the option's delta and gamma using the Taylor series expansion.
(b) Show that the payoff of a delta-hedged portfolio is

$$\Theta\,dt + \frac{1}{2}\Gamma(dS)^2$$

when the stock price moves in either direction.

[QFIQF – Spring 2019 q7(a)(b)]

Solution

(a) Using

$$C(S + dS, t + dt) - C(S, t) = \Theta dt + \Delta dS + \frac{1}{2}\Gamma(dS)^2$$

we have

$$\begin{aligned} 5.99 &= -\,3.65 \times \frac{1}{365} + 10\Delta + 50\Gamma \\ -4.01 &= -\,3.65 \times \frac{1}{365} - 10\Delta + 50\Gamma \\ &\Longrightarrow \Delta = 0.5 \\ &\text{and } \Gamma = 0.02 \end{aligned}$$

(b) $dV = dC - \Delta dS \implies dV = \Theta dt + \frac{1}{2}\Gamma(dS)^2$

4 Your company owns 100 European call options with a strike of \$100 on a stock. You are given:

- The stock pays no dividends.
- The continuously compounded risk-free interest rate is 0%.
- Initially, the options are at-the-money with one year to expiration with the price of \$7.97 per option.
- One week later, the price of the stock is \$104, the delta of the call option is 0.62, and the price of the call option is \$10.20.

- Two weeks later, the price of the stock is \$100, and the price of the call option is \$7.81.
- The implied volatility is 20% for both valuation and hedging.

You are asked to hedge your option portfolio using the underlying shares. Assume that you can only buy or sell whole shares and all options or shares are liquidated at the end of two weeks.

(a) Calculate the profit or loss at the end of two weeks from delta hedging (ignore any cash withdrawal or infusion that may be needed for rebalancing) if:
  (i) You rebalance your hedge only at the end of each week.
  (ii) You never rebalanced your hedge.

(b) Explain why the hedge profits or losses are different between the two hedging strategies in part (a).

[QFIQF – Spring 2019 Q6(c)(d)]

Solution

(a) Initially

$$\text{delta} = \Phi(0.100) = 0.54$$

				(i)		(ii)	
Week	*S*	*delta*	*C*	ΔV	V	ΔV	V
0	100	0.54	7.97		–46.03		–46.03
1	104	0.62	10.20	0.07	–45.96		–45.96
2	100		7.81	0.09	–54.19	–0.16	–46.19
				0.16		–0.16	

(i) $\Delta V = 100 \times 0.16 = 16$

(ii) $\Delta V = 100 \times -0.16 = -16$

(b) From Equation (11.6)

$$\Delta C \approx \text{delta}\, \Delta S + \Theta \Delta t + \frac{1}{2}\Gamma(\Delta S)^2$$

But our hedge is only negating the delta. For negative or positive price movements, the portfolio long in the option benefits from Γ. Over time the option loses value through Θ.

The hedge in (i) benefits from Γ in week 1 and week 2 offsetting the loss in Θ.

The hedge in (ii) gets no benefit from Γ as $\Delta S = 0$ but loses on Θ.

5 Find the delta and gamma of the derivative priced according to the formula:

$$C_t = F(S_t, t) = S_t^3 e^{2S_t}$$

where S_t is the share price at time t.

Solution

$$\text{delta} = \frac{\partial F}{\partial S} = 3S_t^2 e^{2S_t} + 2S_t^3 e^{2S_t} = S_t^2 e^{2S_t}(3 + 2S_t)$$

$$\begin{aligned}\text{gamma} &= \frac{\partial^2 F}{\partial S^2} = 6S_t e^{2S_t} + 6S_t^2 e^{2S_t} + 6S_t^2 e^{2S_t} + 4S_t^3 e^{2S_t} \\ &= 2S_t e^{2S_t}(3 + 6S_t + 2S_t^2)\end{aligned}$$

6 Show that the formula for the vega of a European call option on a non dividend-paying share is $\nu = S_t\sqrt{T-t}\,\Phi'(d_1)$.

Solution

For a European call option

$$c_t = S_t\Phi(d_1) - e^{-r(T-t)}K\Phi(d_2).$$

with

$$d_2 = d_1 - \sigma\sqrt{T-t}$$

Then

$$\frac{\partial d_2}{\partial \sigma} = \frac{\partial d_1}{\partial \sigma} - \sqrt{T-t}$$

So

$$\begin{aligned}\nu &= \frac{\partial F}{\partial \sigma} \\ &= S_t\Phi'(d_1)\frac{\partial d_1}{\partial \sigma} - e^{-r(T-t)}K\Phi'(d_2)\left(\frac{\partial d_1}{\partial \sigma} - \sqrt{T-t}\right) \\ &= S_t\Phi'(d_1)\sqrt{T-t}\end{aligned}$$

where we have used the result from Question 7 in Chapter 9 that

$$e^{-r(T-t)}K\Phi'(d_2) = S_t\Phi'(d_1)$$

7 You have three call options on the same stock with different time-to-maturities: 1 year, 1 month and 1 week. Figures 11.1–11.4 illustrate how your options' Greeks vary with the underlying stock price and time-to-maturity based on Black–Scholes-Merton model. In each figure, the three lines (dotted, solid and dashed) represent the same Greek (Delta, Gamma, Vega or Theta). All three lines in each figure are shown with the same strike price, interest rate, dividend yield and implied volatility.

(i) Determine which figure corresponds to Delta, Gamma, Vega and Theta, respectively. Justify your answers.

(ii) Determine which line (dotted, solid or dashed) corresponds to time-to-maturity of 1 year, 1 month and 1 week, respectively, by analyzing any one of the four figures. Briefly justify your answers based on each Greek's characteristics.

[QFIQF – Fall 2019 q8(a)(b)]

Solution

This question is most easily solved by considering both parts together.

From Table 11.4, theta is always negative for a European call. So Figure 11.4 is theta.

The delta is the cumulative normal probability distribution function $\Phi(\cdot)$ which takes on values in the interval [0,1]. So Figure 11.3 is delta.

For dates close to maturity, changes in S/K should imply large changes in d_1 since $T - t$ is in the denominator. For dates far from maturity, changes in S/K should affect d_1 less. So

X(dotted) = 1 week

Y(solid) = 1 month

Z(dashed) = 1 year

Gamma is the derivative of delta with respect to S so it corresponds to the slope of the curves in Figure 11.3. Gamma should be highest for the dotted line for S close to the strike price and highest for the dashed line far from the strike price. So Figure 11.2 is gamma. Vega is then Figure 11.1.

(i) Figure 11.1 (vega); Figure 11.2 (gamma); Figure 11.3 (delta); Figure 11.4 (theta)

(ii) X(dotted) = 1 week ; Y(solid) = 1 month; Z(dashed) = 1 year

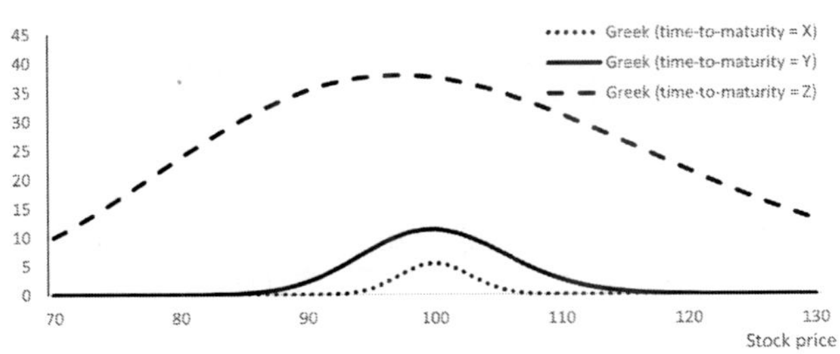

Figure 11.1 Stock option Greek (option strike price = $100)

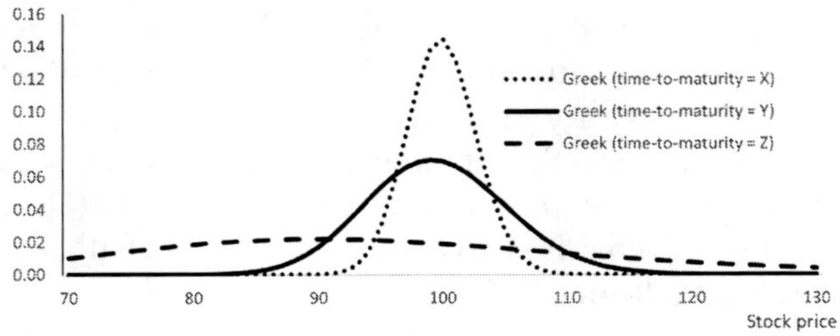

Figure 11.2 Stock option Greek (option strike price = $100)

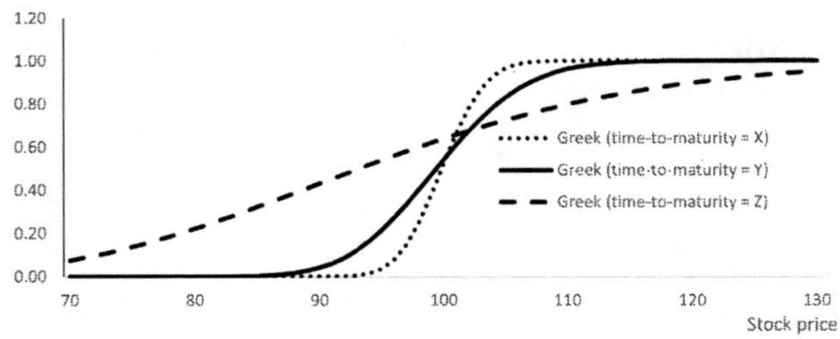

Figure 11.3 Stock option Greek (option strike price = $100)

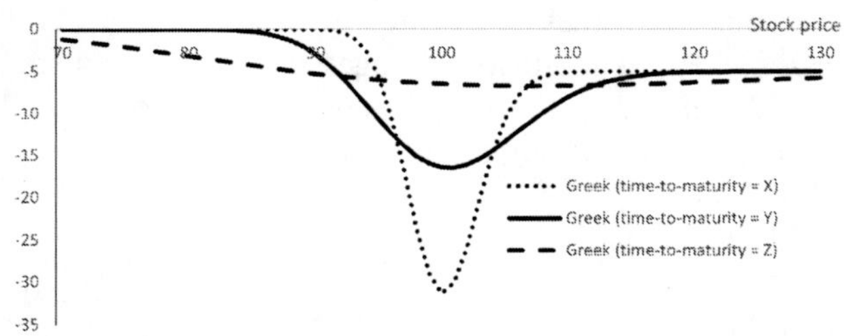

Figure 11.4 Stock option Greek (option strike price = $100)

8 Identify whether each of the following statements is true or false. Briefly justify your answer.

(i) The delta of a European call option that is out-of-the-money will converge to 0 as the expiration date approaches.

(ii) For an out-of-the-money option with the underlying having low volatility, if the volatility increases, the delta of the option increases.

(iii) Gamma approaches 0 for deep-in-the money calls.

(iv) Gamma approaches 1 for deep-out-of-the-money puts.

(v) For an out-of-the-money option with an underlying asset price that is exhibiting low volatility, Gamma is expected to be relatively low.

(vi) For an option that happens to be right at-the-money very near to the expiry date, a stable Gamma is likely to be observed.

[QFIQF – Spring 2019 Q6(b) / Spring 2022 Q12(c)]

Solution

These questions can be answered with reference to the Greeks formulae in Table 11.4 or the graphs in Question 11.6 or as below from an intuitive understanding of the behavior of call and put options.

(i) True. The option price reflects the likelihood of a payoff at expiration. If the option is out-of-the-money changes in the share price will have less and less effect on the likelihood of a payoff approaching expiration.

(ii) True. For an out-of-the-money option with low volatility delta will be low as changes in the share price have little effect on the likelihood of a payoff. Increasing the volatility increases the possibility there will be a payoff. Changes in the share price now have some effect on the likelihood of a payoff.

(iii) True. If the call is deep in the money, then delta will be practically one. Changing the share price will have little effect on delta.

(iv) False. If the put is deep-out-of-the-money, then delta will be practically zero. Changing the share price will have little effect on delta so should be close to zero.

(v) True. For an out-of-the-money option with low volatility delta will be low as changes in the share price have little effect on the likelihood of a payoff at maturity. Changes in the share price should have little effect on delta.

(vi) False. If the option is right at-the-money very near to expiry date small movements in the share price could flip delta between zero and one.

Part IV

Interest rate derivatives

12 Fixed-income instruments

Introduction

Interest rate derivatives are financial instruments whose payoffs are dependent in some way on the level of interest rates. They are among the most widely traded derivatives in global financial markets and are used mainly to manage interest rate exposure and hedge interest rate risks.

The interest rate derivatives market is much larger than equity or foreign exchange derivatives markets. At the end of December 2024, the notional amount of interest rate derivative contracts outstanding was \$548 trillion and the contracts had a gross market value of \$11.5 trillion. By contrast, the notional amount outstanding for equity-linked derivatives was \$8.9 trillion while the notional for foreign exchange derivatives was \$0.13 trillion (BIS, 2024).

The most common type of interest rate derivatives is interest rate swaps – agreements to exchange fixed-rate and floating-rate payments. The mechanics of these derivatives and their pricing are described in this chapter. Options on interest rates (caps, floors and swaptions) – contracts that provide protection against rate increases or decreases – are described in the next chapter.

Interest rate derivatives are not only important because of their vast market size but also in their role as essential tools for risk management and financial stability. They allow firms to protect against fluctuations in borrowing costs, enable banks to manage mismatched balance sheets and serve as reference points for pricing bonds and loans.

12.1 Zero-coupon bonds

A bond is basically a loan. Government, companies or other organizations borrow by issuing bonds. The issuer promises to pay periodic interest (called the *coupon*) during the bond's life and repay the bond's face value (*principal* or *notional amount*) at a set date in the future when the bond matures.

A *zero-coupon bond* is a type of bond where the issuer repays only the principal at maturity. Investment banks and brokers create zero-coupon bonds called STRIPS (Separate Trading of Registered Interest and Principal of

DOI: 10.4324/9781003477624-16

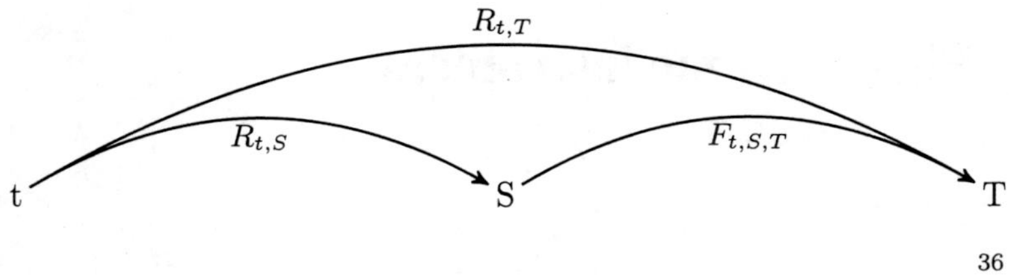

Figure 12.1 Spot and forward rates

Securities) by stripping regular US Treasury bonds into separate principal and interest components.

12.2 Definitions

Let the price at time t of a zero-coupon bond with principal \$1 at maturity T be $P_{t,T}$.

The *spot rate* or *zero rate* or *yield-to-maturity* is the annualized rate of return from holding the bond until maturity.

Using annual compounding the spot rate $R_{t,T}$ is the interest rate that gives a future value of \$1

$$P_{t,T}(1 + R_{t,T})^{T-t} = 1 \implies R_{t,T} = \left(\frac{1}{P_{t,T}}\right)^{\frac{1}{T-t}} - 1 \tag{12.1}$$

Market rates are generally quoted on a discrete interest basis but it is often easier to work with continuous compounding. The continuously compounded spot rate $R_{t,T}$ is then given by

$$P_{t,T}e^{R_{t,T}(T-t)} = 1 \implies R_{t,T} = -\frac{\log P_{t,T}}{T-t} \tag{12.2}$$

The *short rate* r_t is the spot rate that applies to an infinitesimally short period of time at time t, that is,

$$r_t := \lim_{\delta t \to 0} R_{t,t+\delta t}$$

The short rate is an important concept when modeling stochastic interest rates. It will be seen in Chapter 14 that once the short rate is specified, many other important relationships are also defined.

The *forward rate* is the interest rate agreed upon today for a loan or investment that will start at a later date. This rate should be consistent with the corresponding spot rates.

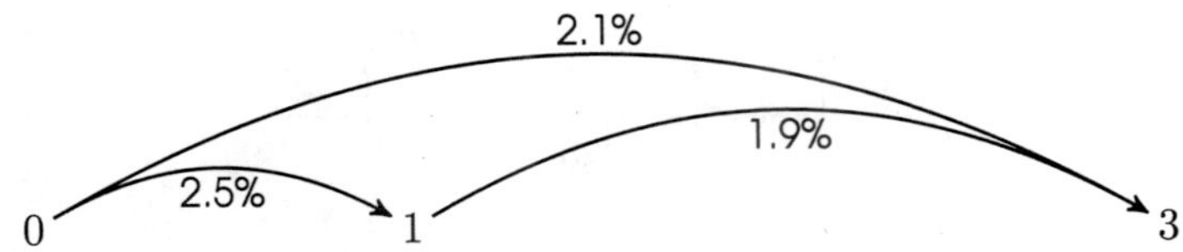

Figure 12.2 Spot and forward rates example

The forward rate $F_{t,S,T}$ from time S in the future to a later time T is given by[1]

$$e^{R_{t,S}(S-t)} \times e^{F_{t,S,T}(T-S)} = e^{R_{t,T}(T-t)} \implies F_{t,S,T} = \frac{R_{t,T}\,(T-t) - R_{t,S}\,(S-t)}{T-S} \tag{12.3}$$

For example, a 1-year spot rate $R_{0,1}$ of 2.5% and a 3-year spot rate $R_{0,3}$ of 2.1% imply a forward rate $F_{0,1,3}$ from year 1 to year 3 of

$$F_{0,1,3} = \frac{0.021 \times 3 - 0.025 \times 1}{3-1} = 1.9\%$$

There is also an instantaneous forward rate. The *instantaneous forward rate* at time t for maturity T written $f_{t,T}$ is the forward rate that applies over an infinitesimally short time interval $[T, T + \delta T]$, that is,

$$f_{t,T} := \lim_{\delta T \to 0} F_{t,T,T+\delta T}$$

We have the following two relationships between the instantaneous forward rate and the other definitions.

The first relationship is found by taking the limit of the definition of the forward rate (12.3)

$$\begin{aligned} F_{t,T,T+\delta T} &= \frac{R_{t,T+\delta T}\,(T+\delta T-t) - R_{t,T}\,(T-t)}{\delta T} \\ &= R_{t,T+\delta T} + (T-t)\frac{R_{t,T+\delta T} - R_{t,T}}{\delta T} \\ \implies f_{t,T} &= R_{t,T} + (T-t)\frac{\partial R_{t,T}}{\partial T} \end{aligned} \tag{12.4}$$

The second relationship is found by rearranging the definition of the spot rate (12.2) and substituting into the definition of the forward rate (12.3)

$$\begin{aligned} F_{t,T,T+\delta T} &= -\frac{1}{\delta T}(\log P_{t,T+\delta T} - \log P_{t,T}) \\ \implies f_{t,T} &= -\frac{\partial \log P_{t,T}}{\partial T} \end{aligned} \tag{12.5}$$

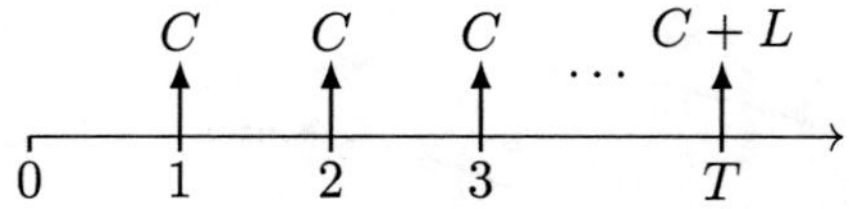

Figure 12.3 Coupon bonds

12.3 Coupon bonds

A bond is a debt instrument issued by a borrower (government, corporation or financial institution) to raise capital. An investor buys the bond and receives periodic interest payments (known as the *coupon*) – typically paid semiannually or annually – and the repayment of the *principal* (the *face value* or *par value*) at maturity.

A coupon bond pays periodic coupon payments C and the principal L at maturity T. The coupon payment is calculated as $C = \frac{Lc}{f}$ where c is the annual coupon rate and f the number of payments per year.

The price of a coupon bond P is its present value:

$$P = \sum_{t=1}^{T} P_{0,t}\, C_t = \sum_{t=1}^{T} e^{-R_{0,t} \times t}\, C_t \tag{12.6}$$

where C_t is the cashflow at t (either C or $C+L$) and using discrete compounding (Equation 12.1). Let the interest rate $R_{0,t}$ used for discounting be the constant y that equates the theoretical price with the market price of the bond. This is known as the *bond yield* and will change as interest rates change.

$$P = \sum_{t=1}^{T} e^{-yt} C_t$$

In order to risk manage bonds, we need to calculate the sensitivity of its price to changes in the yield y. Taking the Taylor expansion (Section 1.10) of $P = f(y)$ gives an expression for the change in bond price as y changes:

$$\Delta P = f(y + \Delta y) - f(y) = f'(y)\Delta y + \frac{f''(y)}{2}(\Delta y)^2 + \ldots$$

Ignoring the second term onwards gives an approximation for the price change

$$\Delta P \approx f'(y)\Delta y$$

Defining the *modified duration* D^* as

$$D^* := -\frac{1}{P} f'(y)$$

implies

$$\frac{\Delta P}{P} \approx -D^{*}\Delta y$$

A portfolio with a modified duration of zero should be insensitive to small interest rate changes. Even with such adjustments modified duration hedging will not eliminate interest rate risk. Jarrow and Chatterjea (2024) show that a portfolio with a modified duration of zero will only be neutral to parallel shifts in the yield curve. It will be seen in Section 12.6 that the yield curve changes in many ways not just through parallel shifts.

If the interest rate changes are large, then the second term in the Taylor expansion term needs to be included when estimating the price change. Defining the *convexity* C as

$$C := \frac{1}{P} f''(y)$$

implies

$$\frac{\Delta P}{P} \approx -D^{*}\Delta y + \frac{1}{2} C(\Delta y)^{2}$$

12.4 Reference rates

Interest rate derivative payoffs are determined using a reference rate for short-term interest rates.

Traditionally, LIBOR, the *London Interbank Offered Rate*, which was the rate at which London banks, could borrow from each other performed this role. Due to abuse of the LIBOR system uncovered in 2012, LIBOR has been gradually phased out and replaced by other benchmarks. LIBOR was expressed in terms of simple interest, not compounded interest since interbank loans were for fixed terms such as 1 month or 3 months. When LIBOR is used to construct discount factors or forward rates, those are typically derived by linking these simple-interest periods together. So the relationship with zero-coupon bond prices for $t < T$ becomes

$$P_{t,T}(1 + R_{t,T}(T - t)) = 1 \tag{12.7}$$

and the relationship with forward rates for $t < S < T$ becomes

$$\begin{aligned} 1 + R_{t,T}(T - t) &= (1 + R_{t,S}(S - t))(1 + F_{t,S,T}(T - S)) \\ \implies 1 + F_{t,S,T}(T - S) &= \frac{1 + R_{t,T}(T - t)}{1 + R_{t,S}(S - t)} = \frac{P_{t,S}}{P_{t,T}} \end{aligned} \tag{12.8}$$

SOFR, or *Secured Overnight Financing Rate*, is the benchmark interest rate now used for US dollar-denominated financial transactions. It represents the cost of borrowing cash overnight using US Treasury securities as collateral and is based on actual transaction data rather than bank estimates as with LIBOR.

For most eurozone interest rate derivatives, €STR or *euro short-term rate* is the standard reference rate. It is calculated in a similar way to SOFR but using unsecured overnight borrowing in euros. SONIA or *Sterling Overnight Index Average* calculated by the Bank of England is similar to €STR but based on transactions involving a narrower set of counterparties.

12.5 Interest rate swaps

Interest rate swaps are the largest segment of the over-the-counter derivatives market. At the end of 2024, $447 trillion out of the total $548 trillion notional outstanding in interest rate contracts was for swaps making these derivatives the main method of managing interest rate risk worldwide (BIS, 2024). They allow companies, banks and governments to stabilize financing costs and manage asset–liability mismatches.

12.5.1 Mechanics of swaps

An interest rate swap is a financial contract where one party pays fixed interest while the other pays floating interest on a notional amount.

In Figure 12.4, company A agrees to pay company B a swap rate of 2.5% per annum on a principal of $10m for three years. While B agrees to pay A the 6-month the floating rate on the same principal. Payments will be made semi-annually beginning in six months.

In six months, time, A will pay

$$\frac{2.5\%}{2} \times 10\text{m} = \$125\text{k}$$

If the floating rate is 1.8% in six months, time, B will pay

$$\frac{1.8\%}{2} \times 10\text{m} = \$90\text{k}$$

Only the net difference between these payments is actually exchanged. So A pays a net settlement of $35k to B.

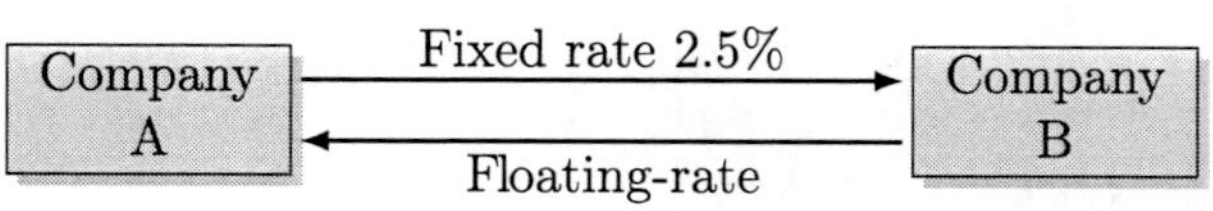

Figure 12.4 Interest rate swaps

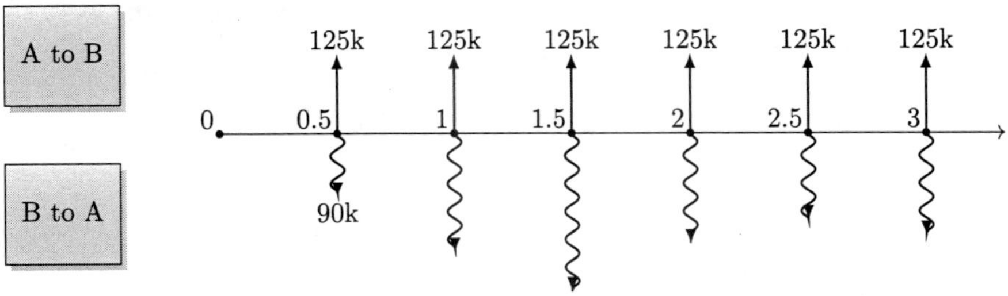

Figure 12.5 Timeline of interest rate swap payments

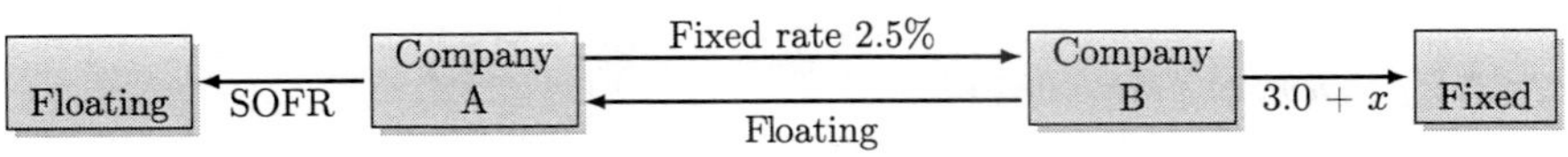

Figure 12.6 Comparative advantage argument for swaps

The next payment occurs at one year. The fixed payment is \$125k as before. The floating payment is re-calculated using the reset 6-month rate. And so on till the contract matures after three years.

Why do companies do this? The comparative advantage argument is the traditional explanation. Suppose before entering the swap described above, the companies wanted to borrow and faced these rates where x and y positive.

	Fixed (%)	*Floating (%)*
A	3.0	SOFR
B	$3.0 + x$	SOFR + y

Company A has an absolute advantage for both ways of borrowing. Let the spread for fixed rate borrowing (x) be less than for floating rate borrowing (y) then A also has a comparative advantage in floating rate borrowing. A should then borrow floating and B should borrow fixed and enter the swap.

After the swap: A is paying and receiving floating. And A pays B a fixed rate of 2.5%.

B is paying a fixed rate of $(3.0 + x)$ and receiving a fixed rate of 2.5%. And B pays A the floating rate.

A would have paid 3.0% in fixed rate borrowing without the swap so fixed borrowing costs are reduced by

$$3.0 - 2.5 = 0.5\%$$

B would have paid floating + y in floating rate borrowing without the swap so floating borrowing costs are reduced by

$$(\text{floating} + y) - (\text{floating} + 0.5 + x) = y - x - 0.5$$

The total gain to both parties is

$$0.5 + (y - x - 0.5) = y - x$$

So the swap happens because a total gain exists due to the difference in spreads. Then the parties have to come together to agree a swap rate to share that gain.

12.5.2 Calculating the swap rate

In the example above, the fixed rate paid by company A called the swap rate was 2.5%. The swap rate for an interest rate swap is determined by finding the fixed interest rate so that the party paying floating and the party paying fixed end up paying the same amount. That is, the fixed rate whereby the present value of the fixed and floating payments at the start of the contract are the same

$$\text{Present value of the fixed leg} = \text{Present value of the floating leg}$$

Let the principal be L, the swap rate s and payment dates $T_1, \ldots, T_N$ with constant reset period δ. Each fixed payment is then

$$\text{Principal} \times \text{time between resets} \times \text{swap rate} = L\delta s$$

The present value of the fixed leg is

$$\sum_{i=1}^{N} P_{0,T_i}(L\delta s)$$

From (Equation 12.8), the floating interest rate that will apply from time T_{i-1} to time T_i is

$$F_{0,T_{i-1},T_i} = \frac{1}{\delta}\left(\frac{P_{0,T_{i-1}}}{P_{0,T_i}} - 1\right)$$

The present value of the floating leg is then

$$\sum_{i=1}^{N} P_{0,T_i} L\delta\left(\frac{1}{\delta}\left(\frac{P_{0,T_{i-1}}}{P_{0,T_i}} - 1\right)\right) = L\sum_{i=1}^{N}(P_{0,T_{i-1}} - P_{0,T_i}) = L(1 - P_{0,T_N}) \quad (12.9)$$

Equating the fixed and floating legs

$$L\delta s \sum_{i=1}^{N} P_{0,T_i} = L(1 - P_{0,T_N})$$

$$\Longrightarrow s = \frac{1 - P_{0,T_N}}{\delta \sum_{i=1}^{N} P_{0,T_i}} \tag{12.10}$$

A forward swap is an interest rate swap that starts at a future date rather than immediately. Until that start date, no cash flows are exchanged.

Consider a forward swap starting at future time T_0 with payment dates $T_1, \ldots, T_N$ and constant reset period δ. The forward breakeven swap rate s_{0,T_0,T_N} can be calculated in a similar manner to above giving

$$s_{0,T_0,T_N} = \frac{P_{0,T_0} - P_{0,T_N}}{\delta \sum_{i=1}^{N} P_{0,T_i}} \tag{12.11}$$

We can show that constant spot rates imply constant swap rates. If spot rates R_{t,T_i} are constant, then all discount terms are of the form $P_{t,T_i} = e^{-R(T_i - t)} = e^{Rt}e^{-RT_i}$ and

$$s_{t,T_0,T_N} = \frac{P_{t,T_0} - P_{t,T_N}}{\delta \sum_{i=1}^{N} P_{t,T_i}} = \frac{e^{-RT_0} - e^{-RT_N}}{\delta \sum_{i=1}^{N} e^{-RT_i}}$$

The right-hand side is independent of t and hence the forward swap rate s_{t,T_0,T_N} is constant for all $t < T_0$.

12.6 Term structure

The *term structure* or zero-coupon *yield curve* or *zero-rate curve* is a graph that shows the relationship between spot rates and the maturity dates for financial instruments of the same credit quality.

There are different term structures depending on the market of interest. Markets regard the yield curve derived from US Treasury securities as a proxy for risk-free borrowing in US dollars over different maturities. Discounting with the Treasury yield curve would therefore be appropriate for pricing government debt or municipal bonds. The Treasury yield curve is constructed by taking market prices of T-bills for maturities less than one-year and using T-notes and T-bonds for longer maturities. Spot rates at the short end of the curve are easily calculated since T-bills are zero coupon while for longer maturities where securities pay coupons, the spot rates can be iteratively found by progressing from shortest to longest maturities using (Equation 12.6). Interpolation is then used for maturities for which securities are unavailable. All yields are nominal and affected by inflation expectations.

For swaps and derivatives pricing, the most common curve used for discounting is the SOFR Overnight Indexed Swap (OIS) curve. For interest-rate derivatives, this is because SOFR is generally the reference rate. And for other derivatives, it is because the collateral used when trading earns the overnight rate.

An overnight indexed swap is a swap as described in the previous section where the floating rate is the compounded average of SOFR over the time between resets. As SOFR is derived from overnight borrowing with US Treasuries as collateral the SOFR curve also reflects risk-free borrowing. At the short end, both the Treasury and SOFR curves will be similar as both are influenced by Federal Reserve policy and short-term funding conditions. For longer maturities, there will be a spread between the curves as Treasury yields incorporate liquidity risk (T-bonds are less liquid than short-term Treasuries) and maturity risk (interest rates and inflation are more likely to change over long time periods).

The SOFR curve is built from overnight and one-week deposits at the very short end of the curve; future contracts for mid-maturities and OIS swaps for longer maturities. Spot rates are found from swaps by iteratively progressing from shortest to longest maturities using (Equation 12.10) – a process known as *bootstrapping*. An example of bootstrapping from swap rates is given in Question 6 at the end of the chapter. Interpolation then provides rates at intermediate maturities.

12.6.1 Yield curve shapes

Pricing of bonds (Equation 12.6) and swaps (Equation 12.10) as well as the pricing of other interest rate derivatives depends on the shape of the curve used for discounting. When the shape of the curve changes prices will change.

The shape of the yield curve for US Treasury securities is driven by expectations of future interest rates; risk associated with longer maturities and supply and demand of securities at different maturities. The Treasury yield curve at 30th September 2025 is plotted in Figure 12.7.

For short maturities less than one year, the curve is inverted and then steepens for longer maturities. The 10-year Treasury yield was 4.73%, while the 2-year Treasury yield stood at 3.60% – creating a spread of 113 basis points (1.13%). A positive spread typically suggests market optimism about future growth and inflation expectations. The inverted yield for shorter maturities was perhaps due to expectations that the Federal Reserve would cut rates in the short to medium term.

The typical yield curves shapes are given in Figure 12.8. An upward-sloping curve means investors demand higher yields for locking money up longer. This is the most common shape when the economy is expanding. An inverted yield curve signals expectations of lower interest rates in the future and is often an early indicator of future recession. A flat yield curve has yields that are the same for all maturities. This is often seen when the economy is in transition between different states of the economy.

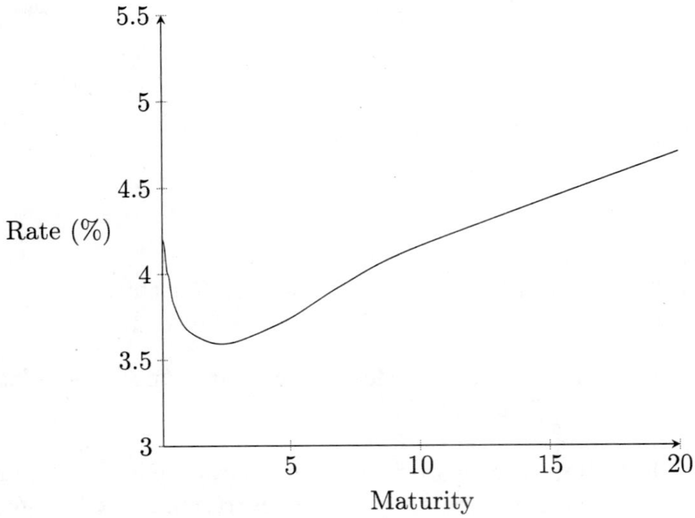

Figure 12.7 Treasury yield curve 30th September 2025. Note: Daily Treasury Par Yield Curve Rates. (*Source*: U.S. Dept. of the Treasury)

12.6.2 *Yield curve changes*

Over time, the yield curve will transition from one shape to another. Changes occur as the market's expectations of future monetary policy change such as

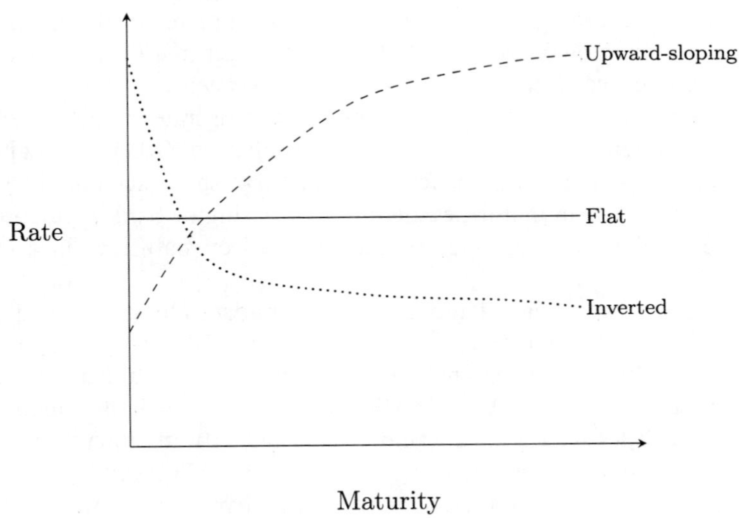

Figure 12.8 Typical yield curves

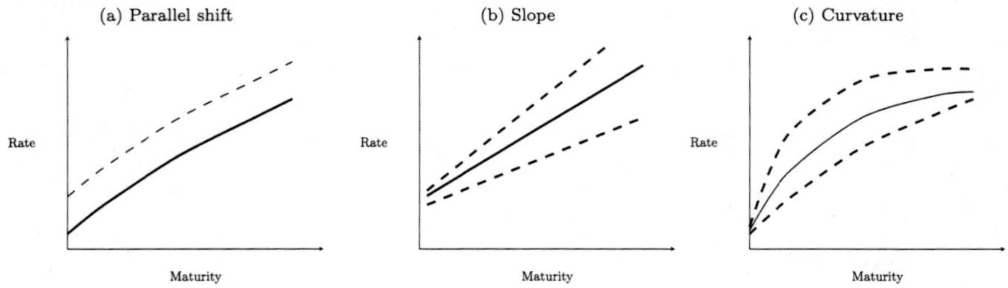

Figure 12.9 Curve changes. (a) Parallel shift (b) Slope (c) Curvature

after major economic releases or Federal Reserve announcements. Changes in yield curves are typically classified as

- parallel shifts – rates at all maturities rise or fall by roughly the same amount.
- changes in slope – (flattening) short-term rates rise relative to long-term rates reducing the curve's slope (steepening) long-term rates rise faster than short-term rates increasing the curve's slope.
- changes in curvature – mid-curve rates move differently than short or long-term rates

Parallel shifts tend to be most common for Treasuries. Yield changes will then be highly positively correlated across the curve. When short rates rise due to a higher federal funds rate, the curve tends to flatten or invert and correlations between changes at the short and long ends are weakly positive. This is because lending rates rise across the curve but this effect is countered by the prospects of slower economic growth and lower inflation at longer maturities. When short rates fall, the curve steepens and correlations between changes at the short and long ends are also weak. This is due to lending rates falling across the curve but countered by expectations of higher inflation. Other factors besides the federal funds rate can influence the yield curve shape such as the supply of securities of certain maturities, quantitative easing or bank regulations on capital requirements. These factors may only affect maturities in a certain portion of the yield curve.

The size of the changes tends not to be uniform along the yield curve. Figure 12.10 shows the volatility term structure of US Treasury yields. Volatility tends to be highest at the short end reflecting uncertainty about Federal Reserve monetary policy decisions. While for longer maturities, volatility tends to be associated with less-changeable liquidity and maturity risk factors reflecting uncertainty about the macroeconomy. The hump in yield volatilities at maturity of two to three years is well-established and is driven by volatile short-rate expectations (Cieslak and Povala, 2016).

Treasury yield volatility is also time-varying. The MOVE Index (Merrill Lynch Option Volatility Estimate) provides a summary measure of the implied volatility of US Treasury yields – just like the VIX index measures the implied volatility of the S&P 500 index (Section 7.4). The MOVE is graphed in Figure 12.11 for the past five years. It peaked in March 2020 due to the COVID

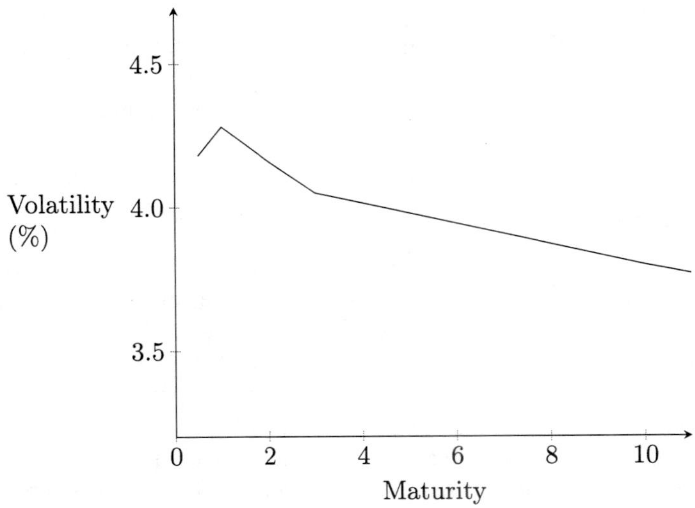

Figure 12.10 Term structure of interest rate volatility. Note: Estimates from US yield curves Q2:1983 to Q1:1995 (Goncalves & Issler 1996)

shock and again in 2022–23 as the Federal Reserve repeatedly raised their target rate to fight inflation.

We will examine interest rate models in later chapters. Potentially, there are many characteristics such a model should capture.

- Can it generate all the yield curve shapes observed in the market?
- What yield curve changes are admissible? – parallel shifts, slope changes and curvature changes
- Is the volatility term structure realistic?
- Is volatility allowed vary stochastically over time? How does the volatility term structure change over time?

Models will become more complicated as these various elements are introduced. There is obviously then a trade-off between analytical tractability and realism.

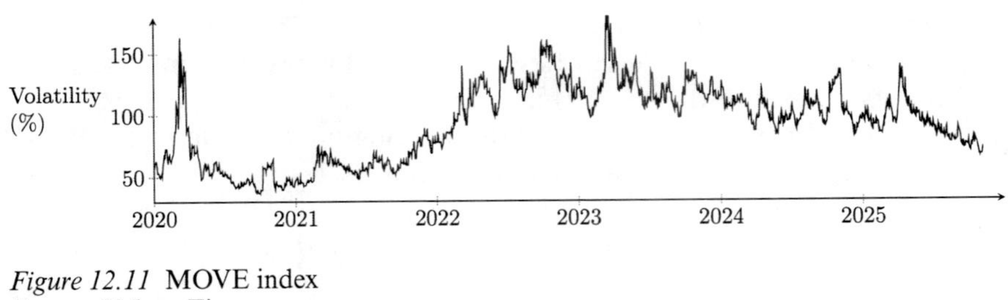

Figure 12.11 MOVE index
Source: Yahoo Finance

Our particular application also guides our choice of model. For example, we could forgo stochastic volatility in our model if the macroeconomic environment was stable. Or we would be more concerned about correlations between yield changes when risk managing derivatives with long tenors and repeated payoffs.

12.7 Summary

This chapter introduces interest rate derivatives. Basic concepts of spot, forward, short rates and zero-coupon bond prices are introduced and defined. Bonds are priced and the mechanics and valuation of swaps are explained through the exchange of fixed and floating cashflows. Finally, the chapter covers term structures – their construction, shapes and dynamics showing how they reflect monetary policy, inflation expectations and economic outlook. Changes in rates at different maturities can be correlated, have different volatilities and can vary over time. Interest rate models covered in later chapters capture some of these features but not all.

Key things to remember

- Basic definitions (Section 12.2).
- Swap rate calculations (Section 12.5.2)
- A model of the yield curve should capture typical yield curve shapes (Section 12.6.1) and the features of yield curve changes (Section 12.6.2).

Further reading

Several complementary texts and articles provide both foundational knowledge and advanced perspectives. Fabozzi and Fabozzi (2021) and Tuckman and Serrat (2022) are essential references on fixed income markets, yield curves and strategies, offering a comprehensive view of bond pricing and risk tools that underpin interest rate derivatives. Jarrow and Chatterjea (2024) also offer an accessible introduction to interest rate derivative securities and risk management.

On the applied side, NY Fed (2021) provides a more thorough understanding of SOFR. Russo and Fabozzi (2025) discuss the practicalities of bootstrapping risk-free rates and constructing SOFR-based curves. Research articles such as Indriawan et al. (2021) and Brace et al. (2024) explore the dynamics of SOFR and its interaction with Federal Reserve policy. Finally, contributions such as Longstaff and Schwartz (1992) and Cieslak and Povala (2016) provide a theoretical and empirical basis for understanding interest rate volatility and the informational content of yield curve dynamics.

Notes

1. This can also be written $e^{F_{t,S,T}(T-S)} = \frac{P_{t,S}}{P_{t,T}}$.

Questions

1 The table below gives the price of a zero-coupon bond with principal \$1 for a range of maturities.

Maturity T	*Price* $P_{0,T}$	*Spot rate* $R_{0,T}$	*Forward rate* $F_{0,T-1,T}$
0	1		
1	0.98	$R_{0,1}$	$F_{0,0,1}$
2	0.95	$R_{0,2}$	$F_{0,1,2}$
3	0.91	$R_{0,3}$	$F_{0,2,3}$
4	0.87	$R_{0,4}$	$F_{0,3,4}$

Complete the table using
(i) discrete compounding
(ii) continuous compounding

Solution
(i) From Equation (12.1)

$$R_{0,1} = \left(\frac{1}{P_{0,1}}\right) - 1 = \left(\frac{1}{0.98}\right) - 1 = 2.04\%$$

$$R_{0,2} = \left(\frac{1}{P_{0,2}}\right)^{\frac{1}{2}} - 1 = \left(\frac{1}{0.95}\right)^{\frac{1}{2}} - 1 = 2.60\% \quad \text{etc.}$$

The discrete forward rate should satisfy

$$\begin{aligned}(1 + R_{0,T})^T(1 + F_{0,T,T+1}) &= (1 + R_{0,T+1})^{T+1} \\ => F_{0,T,T+1} &= P_{0,T}/P_{0,T+1} - 1 \text{ from (Equation 12.1)}\end{aligned}$$

$$F_{0,0,1} = \frac{1}{P_{0,1}} - 1 = \frac{1}{0.98} - 1 = 2.04\%$$

$$F_{0,1,2} = \frac{P_{0,1}}{P_{0,2}} - 1 = \frac{0.98}{0.95} - 1 = 3.16\% \quad \text{etc.}$$

(i) From (Equation 12.2)

$$R_{0,1} = -\log P_{0,1} = -\log 0.98 = 2.02\%$$

$$R_{0,2} = -\frac{\log P_{0,2}}{2} = -\frac{\log 0.95}{2} = 2.56\% \quad \text{etc.}$$

From (Equation 12.3)

$$F_{0,0,1} = R_{0,1} = 2.02\%$$

$$F_{0,1,2} = R_{0,2} \times 2 - R_{0,1} = 2.56 \times 2 - 2.02 = 3.11\% \quad \text{etc.}$$

		Discrete		Continuous	
Maturity	*Price*	*Spot (%)*	*Forward (%)*	*Spot (%)*	*Forward (%)*
0					
1	0.98	2.04	2.04	2.02	2.02
2	0.95	2.60	3.16	2.56	3.11
3	0.91	3.19	4.40	3.14	4.30
4	0.87	3.54	4.60	3.48	4.50

2 You are provided with a continuously compounded spot curve and a one-year continuously compounded forward curve directly derived from the spot curve in the following chart.

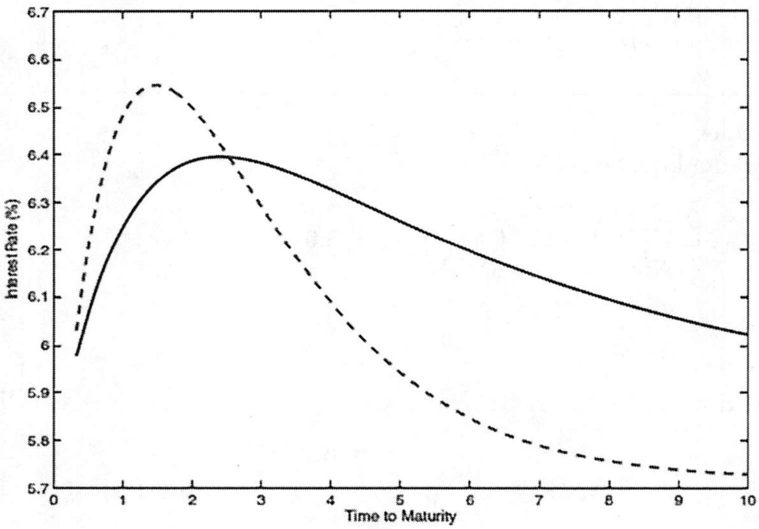

Determine which curve is the spot curve and which curve is the forward curve. Justify your answer by mathematically demonstrating the key relationship between the two curves.

[QFIQF Spring 2019 Q11]

Solution

The solid curve is the spot curve, and the dotted curve is the forward curve derived from the spot curve.

Given the continuously compounded spot rate $R_{0,T}$, the forward rate $F_{0,t,T}$ and the instantaneous forward rate $f_{0,T}$ then from (Equation 12.4)

$$f_{0,T} = R_{0,T} + T\frac{\partial R_{0,T}}{\partial T}$$

When the spot rate is increasing $\frac{\partial R_{0,T}}{\partial T} > 0$, the instantaneous forward rate $f_{0,T}$ should be above the spot rate $R_{0,T}$. This is the case in the first portion of the chart ($T < 2.5$ approximately).
When the spot rate is decreasing $\frac{\partial R_{0,T}}{\partial T} < 0$, the instantaneous forward rate $f_{0,T}$ should be below the spot rate $R_{0,T}$. This is the case in the second portion of the chart ($T > 2.5$ approximately).

3 Given the following data as of Dec 31, 2018:

- A six-month Treasury bill with \$100 face value was priced at \$97.
- A one-year Treasury note with \$100 face value and paying 3% coupons semiannually was priced at \$95.85.

Calculate the discount factors $P_{0,0.5}$ and $P_{0,1}$ as of Dec 31, 2018.

[QFIQF – Spring 2020 Q10]

Solution From the six-month Treasury bill, $P_{0,0.5} = 0.97$
From the one-year Treasury note

$$95.85 = P_{0,0.5} \times (3/2) + P_{0,1}(3/2 + 100) \implies P_{0,1} = 0.93$$

4 At time 0, the prices of three zero-coupon bonds, which mature at time t, $P_{0,t}$, are as follows:

$P_{0,1}$=0.9608
$P_{0,2}$=0.9241
$P_{0,3}$=0.8896
Determine the annual coupon rate for a 3-year bond issued at par at time zero.

[FETE - Spring 2013 Q15(b)]

Solution The coupon is $C = Lc$ as payments are annual. From (Equation 12.6)

$$\text{Price} = P_{0,1}(Lc) + P_{0,2}(Lc) + P_{0,3}(Lc + L)$$

Since the bond is issued at par the initial price is L

$$L = P_{0,1}(Lc) + P_{0,2}(Lc) + P_{0,3}(Lc + L)$$
$$\Longrightarrow 1 = c(0.9608 + 0.9241 + 0.8896) + 0.8896$$
$$\Longrightarrow c = 3.98\%$$

5 Malbec Corp. prefers to pay a floating rate of interest. Valpolicella Corp. prefers to pay a fixed rate of interest.

Current Terms for Borrowing

	Fixed (semiannual compounding)	*Floating*
Malbec	8.0%	6-month LIBOR + 1.30%
Valpolicella	9.7%	6-month LIBOR + 1.91%

(a) Show that a swap rate of 7.36% between Malbec and Valpolicella would lower the borrowing costs for both companies, and calculate the rate reduction for each company.

Malbec and Valpolicella entered into a 3-year interest rate swap. After one year, the current bid and offer rates in the fixed swap market are (with semiannual compounding):

Maturity (years)	*Bid*	*Offer*
2	8.15%	8.19%

Assume:

- The swap has a \$10 million notional, and 4 future swap payments will take place.
- Both sides of the swap pay semiannually.
- Valpolicella has invested in a \$10 million project with the proceeds from their borrowing. The present value of the project's cash flows has decreased in value by \$192,000 due to interest rate moves.
- The 0.5, 1 and 1.5 year continuously compounded LIBOR zero rates are 7.9%, 8.0% and 8.1% respectively.

(b) Evaluate whether or not the swap has proved to be an effective asset liability management tool for Valpolicella.

[FETE - Fall 2010 Q7]

Solution (a) Malbec has a comparative advantage in fixed borrowing so Malbec (M) borrows fixed and Valpolicella (V) borrows floating. They enter a swap with M paying LIBOR and V paying fixed at the swap rate 7.36%.

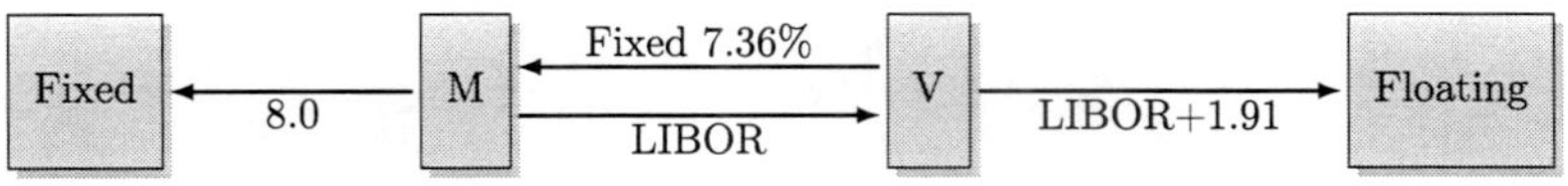

With the swap,
M pays LIBOR + (8.0 - 7.36) = LIBOR+0.64%
V pays 7.36 + 1.91 = 9.27%.

	Without	*With*	*Reduction*
V	9.7	9.27	0.43
M	LIBOR+1.3	LIBOR+0.64	0.66

Note: the total reduction in borrowing costs 0.43 + 0.66 = 1.09 equals the difference in the spreads between fixed (1.7) and floating (0.61).

(b) The swap has been beneficial for V if the gain in value of the swap offsets the loss on the project.

The value of the swap at time zero is 0 to both parties.

The value of the swap to V after one year is the present value of the floating leg minus the present value of the fixed leg. From Section 12.5.2, this is

$$\underbrace{10\text{m}(1 - P_{0,2})}_{\text{floating}} \\ \underbrace{- \, 10\text{m} \times 0.5 \times 0.0736(e^{-0.079\times 0.5} + e^{-0.080\times 1} + e^{-0.081\times 1.5} + P_{0,2})}_{\text{fixed}}$$

Use (Equation 12.10) to calculate $P_{0,2}$. Choose the current swap rate to be the average of bid and ask rates $s = 0.0817$. Then

$$0.0817 = \frac{1 - P_{0,2}}{0.5(e^{-0.079\times 0.5} + e^{-0.080\times 1} + e^{-0.081\times 1.5} + P_{0,2})}$$
$$\Longrightarrow P_{0,2} = 0.852$$

Substituting this into the equation above the value of the swap has increased by \$147,040.

This has offset 77% of the loss on the project. Therefore the swap has proved to be a reasonably effective asset liability management tool.

6 The table below is a slightly simplified version of Exhibit 2 in Russo and Fabozzi (2025) showing SOFR-based swap rates as of September 30,2024.

Year (T_i)	*Swap rate* (s_{T_i})
1	3.852
2	3.436
3	3.309
4	3.261
5	3.247
6	3.254
7	3.261
8	3.280
9	3.299
10	3.319

(a) Bootstrap the swap curve to calculate discount factors for each of the 10 years.
(b) Find the equivalent discretely compounded zero rates and plot the zero-rate curve.

Solution
(a) Rearranging (Equation 12.10)

$$s = \frac{1 - P_{0,T_N}}{\sum_{i=1}^{N} \delta P_{0,T_i}}$$

$$\Longrightarrow s\delta\left(\sum_{i=1}^{N-1} P_{0,T_i} + P_{0,T_N}\right) = 1 - P_{0,T_N}$$

$$\Longrightarrow P_{0,T_N} = \frac{1 - s\delta \sum_{i=1}^{N-1} P_{0,T_i}}{1 + s\delta}$$

$$P_{0,1} = \frac{1}{1 + 0.03852} = 0.9629$$

$$P_{0,2} = \frac{1 - 0.03436 \times 0.9629}{1 + 0.03436} = 0.9348$$

$$P_{0,3} = \frac{1 - 0.03309(0.9629 + 0.9348)}{1 + 0.03309} = 0.9072 \quad \text{etc.}$$

(b) From (Equation 12.1)

$$R_{0,T} = \left(\frac{1}{P_{0,T}}\right)^{\frac{1}{T}} - 1$$

$$R_{0,1} = \left(\frac{1}{0.9629}\right) - 1 = 3.85\%$$

$$R_{0,2} = \left(\frac{1}{0.9348}\right)^{\frac{1}{2}} - 1 = 3.43\% \quad \text{etc.}$$

Year (T_i)	*Swap rate* (s_{T_i})	P_{0,T_i}	R_{0,T_i} (%)
1	3.852	0.9629	3.85
2	3.436	0.9348	3.43
3	3.309	0.9072	3.30
4	3.261	0.8798	3.25
5	3.247	0.8527	3.24
6	3.254	0.8255	3.25
7	3.261	0.7991	3.26
8	3.280	0.7725	3.28
9	3.299	0.7466	3.30
10	3.319	0.7211	3.32

Zero rate curve

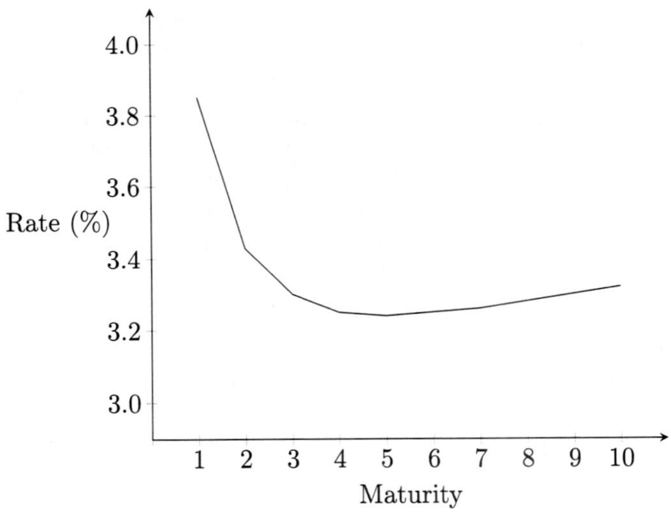

13 Black's model

Introduction

Caps and swaptions rank among the most actively traded interest rate options, and their accurate pricing and hedging are therefore of central importance in financial markets. Black's model has become one of the standard tools in the valuation of these instruments.

The model itself traces its origins to the work of Fischer Black, who initially introduced it to price options on commodity futures building on the Black–Scholes option framework. Over time, the formula was adopted by practitioners from its original setting to price interest rate derivatives without a solid theoretical justification. Subsequent work, however, demonstrated that Black model's can be formally derived from no-arbitrage principles. Extensions of Black's formula into a full arbitrage-free framework are known as *market models*. These models can be divided into two distinct approaches. The first focuses on forward interest rates, which makes the framework well suited to the pricing of caps. The second approach models forward swap rates, which allows for consistent valuation of swaptions. Despite their popularity, these two approaches are not fully compatible. This incompatibility highlights a key limitation of the Black model.

13.1 Black's model

Black (1976) developed the Black–Scholes theory further to price options on commodity futures. For example, the European call option price for commodity futures was derived as

$$c_t = e^{-r(T-t)}\Big(F\Phi(d_1) - K\Phi(d_2)\Big) \tag{13.1}$$

where F is the forward price at t and

$$d_1 = \frac{\log(F/K) + \frac{1}{2}\sigma^2(T-t)}{\sigma\sqrt{T-t}}, d_2 = d_1 - \sigma\sqrt{T-t}$$

DOI: 10.4324/9781003477624-17

This formula can be seen to closely align with the Black–Scholes call option formula (Equation 3.1).

From Theorem 3.1, we have the forward price S_t for an asset with no income at t is $F = S_t e^{r(T-t)}$. Substitute this into the Black–Scholes European call option formula with dividend yield set to zero

$$\begin{aligned} c_t &= S_t\Phi(d_1) - e^{-r(T-t)}K\Phi(d_2) \\ &= Fe^{-r(T-t)}\Phi(d_1) - e^{-r(T-t)}K\Phi(d_2) \\ &= e^{-r(T-t)}\Big(F\Phi(d_1) - K\Phi(d_2)\Big) \end{aligned}$$

where

$$\begin{aligned} d_1 &= \frac{\log(S_t/K) + (r - q + \frac{1}{2}\sigma^2)(T-t)}{\sigma\sqrt{T-t}} \\ &= \frac{\log(Fe^{-r(T-t)}/K) + (r + \frac{1}{2}\sigma^2)(T-t)}{\sigma\sqrt{T-t}} \\ &= \frac{\log(F/K) + \frac{1}{2}\sigma^2(T-t)}{\sigma\sqrt{T-t}} \quad \text{and } d_2 = d_1 - \sigma\sqrt{T-t} \end{aligned}$$

This formula was then adopted by traders to price interest rate derivatives without any real theoretical justification. Applications of Black's model to bond options, caps and swaptions are demonstrated in the next sections.

The payoff from a European call option on a stock has the form $[S_T - K]^+$. We will compare the payoffs in each application to the call payoff to see clearly how the formula is applied.

Although the Black's model approach to pricing interest rate derivatives originated as a simple way for traders to quote prices of caps, floors and swaptions, later work established that the formula follows from risk-neutral pricing with a change of numéraire. Even with this framework, there remain theoretical inconsistencies in the application of the model to pricing different types of derivatives.

13.2 Bond options

A bond option is an option to buy or sell a particular bond by a particular date for a particular price. Consider an option on a coupon-paying bond with expiry at time T. This bond matures at a later time T^*.

To apply Black's model, we need to first determine the forward price of a coupon-paying bond. Let the forward mature at T. With a slight change of notation, Theorem 3.2 shows that the forward price for a bond will be

$$F = (BP_t - I)e^{r(T-t)}$$

where BP_t is the current bond price, and I is the present value of the coupons paid until T.

The payoff from a European call option on a bond has the form $[BP_T - K]^+$. Then pricing a bond option is a simple substitution of the forward price for a bond into Black's formula (Equation 13.1) along with a measure of the forward bond price volatility for σ.

Exercise 13.1. *A trader is pricing a 9-month European call option on a bond with par value \$1000. The current bond price is \$975, the strike price is \$1000 and the volatility of the 9-month forward bond price is 10% per annum.*

The bond pays semiannual coupons at a coupon rate of 8% per year. Next coupon payments are expected in 2 months and 8 months.

Suppose that the 2-month, 8-month and 9-month risk-free interest rates are all 4.0% per annum. Find the price of the option.

Proof. The present value of the coupons paid over the 9 months is

$$I = 40e^{-0.04\times\frac{2}{12}} + 40e^{-0.04\times\frac{8}{12}} = 78.68$$

The forward bond price for a contract maturing in 9 months is

$$F = (BP_t - I)e^{r(T-t)} = (975 - 78.68)e^{0.04\times\frac{9}{12}} = 923.62$$

Now apply Black's formula.

$$d_1 = \frac{\log(F/K) + \frac{1}{2}\sigma^2(T-t)}{\sigma\sqrt{T-t}} = \frac{\log(923.62/1000) + \frac{1}{2}0.1^2\frac{9}{12}}{0.1\sqrt{\frac{9}{12}}} = -0.874$$

$$d_2 = d_1 - \sigma\sqrt{T-t} = -0.874 - 0.1\sqrt{\frac{9}{12}} = -0.961$$

And the bond option price BO_t is

$$\begin{aligned} BO_t &= e^{-r(T-t)}\Big(F\Phi(d_1) - K\Phi(d_2)\Big) \\ &= e^{-0.04\times\frac{9}{12}}\Big(923.62\,\Phi(-0.874) - 1000\,\Phi(-0.961)\Big) \\ &= \$7.85 \end{aligned}$$

□

13.3 Interest rate caps and floors

An interest rate cap protects the holder from rising interest rates by setting an upper limit or *cap rate* on the interest rate that will apply to a floating-rate loan. In this way, borrowers protect themselves from rising rates.

An interest rate floor protects the holder from falling interest rates. It sets a lower limit or *floor rate* below which the holder will receive a payout to compensate for the lower rates. Lenders with floating-rate investments buy floors to protect their income when rates fall.

A cap provides insurance against the interest on a floating rate note rising above a cap rate. Consider a cap with cap rate R_K where the floating interest rate R_t is reset on dates T_1, T_2, T_3, T_4. Figure 13.1 demonstrates the payoffs to the holder. At T_1 and T_2, the floating rate is below the cap and there is no payoff. At T_3 the floating rate is above the cap and the holder will receive the spot rate minus the cap rate R_K. The payment is made relative to a notional amount L and is made at the next reset T_4. At T_4, there is no payoff.

For example, suppose the notional amount is \$100m, the time between resets is 3 months and the cap rate is 5%. If the floating rate is 6% at T_3, then the payoff 3 months later is

$$100\text{m} \times \frac{3}{12} \times (6\% - 5\%) = \$250k$$

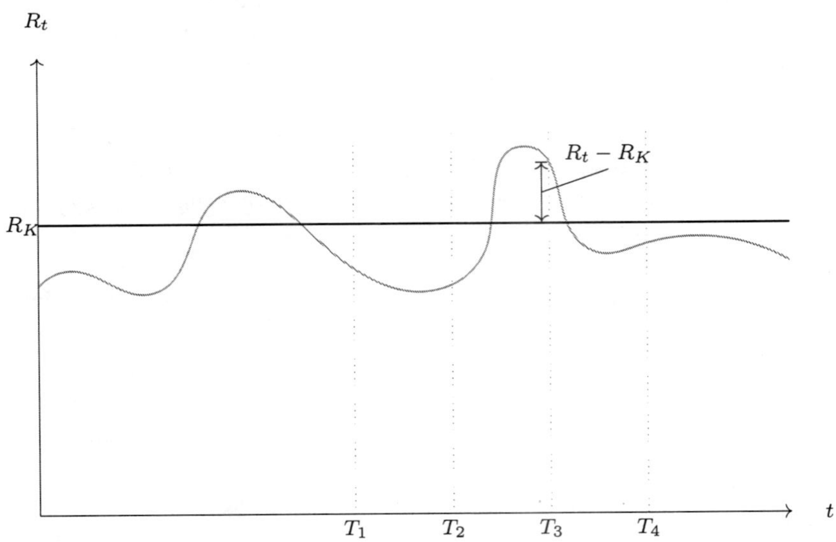

Figure 13.1 Cap payoffs

The payoff from a reset at any time T resembles a call option on the floating rate

$$\text{Payoff received at next reset} = L\delta[R_T - R_K]^+$$

where δ is the length of the reset period. The cap is a portfolio of such call options each of which is known as a *caplet*.

Each of the caplets can then be priced using Black's model. If the next reset is at T^*, then discounting the payoff back to T

$$\text{Payoff discounted to } T = L\delta P_{T,T^*}[R_T - R_K]^+$$

Then pricing a caplet requires multiplying Black's model by $L\delta P_{T,T^*}$ and substituting for the forward interest rate along with a measure of the forward rate volatility. Applying Black's model and writing $e^{-r(T-t)}$ as $P_{t,T}$, the caplet price Cap_t is

$$\text{Cap}_t = L\delta P_{t,T} P_{T,T^*}\Big(F\Phi(d_1) - R_K\Phi(d_2)\Big) \tag{13.2}$$

$$= L\delta P_{t,T^*}\Big(F\Phi(d_1) - R_K\Phi(d_2)\Big) \tag{13.3}$$

where F is the forward interest rate at time t for the time period between T and T^*, that is, F_{t,T,T^*}. The d_1 and d_2 parameters are calculated using a measure of the forward rate volatility for σ.

Exercise 13.2. *An investor purchases a contract that caps a floating interest rate on \$100 million at 5% per annum (with quarterly compounding) for 3 months starting in 1 year.*

The yield curve is flat at 4% per annum with quarterly compounding and the volatility of the 3-month forward rate is 30% per annum.

What is the contract price ?

Proof. This is a caplet and Black's model can be applied.

The yield curve information serves two purposes. First, the discount factor is calculated taking account of discrete compounding

$$P_{0,1.25} = (1 + 0.04/4)^{-5} = 0.951$$

Second, a flat yield curve where spot rates are constant implies forward rates are constant from Section 12.2. Therefore, $F = 4\%$.

Now apply Black's formula.

$$d_1 = \frac{\log(0.04/0.05) + \frac{1}{2}0.3^2}{0.3} = -0.594 \text{ and } d_2 = -0.594 - 0.3 = -0.894$$

And the caplet price is

$$\begin{aligned}\text{Cap}_t &= L\delta P_{t,T^*}\Big(F\Phi(d_1) - R_K\Phi(d_2)\Big)\\ &= 100\text{m} \times 0.25 \times 0.951\Big(0.04\Phi(-0.594) - 0.05\Phi(-0.894)\Big)\\ &= \$42036\end{aligned}$$

□

A floor provides a payoff when the floating rate falls below the cap rate. The payoff from a reset at any time T resembles a put option on the floating rate

$$\text{Payoff received at next reset} = L\delta[R_K - R_T]^+$$

where δ is the length of the reset period. The floor is a portfolio of such put options each of which is known as a *floorlet*. Each of the floorlets can then be priced using the put version of Black's model.

A *collar* is an instrument designed to guarantee that the interest rate on the underlying floating-rate note always lies between two levels. Collars can be priced as a combination of caplets and floorlets.

13.4 Swaptions

Swap options or *swaptions* give the holder the right to enter into an interest rate swap at a specified time in the future. A *payer swaption* is an option to enter into a swap where you pay a fixed rate and receive a floating rate while a *receiver swaption* is an option to enter into a swap where you pay a floating rate and receive a fixed rate.

For example, suppose a payer swaption on a swap beginning in T years and lasting until T_N has a strike fixed rate of 3%. At option maturity in T years, if the market swap rate for a swap until T_N is above 3%, the holder exercises the swaption and enters the swap paying a swap rate of 3%. If the market swap rate is below 3%, then the holder lets the option expire.

Let the fixed rate for the swaption be s_K. Figure 13.2 demonstrates the potential payoffs to the holder. At option maturity T, the swap rate turns out to be s_T which is larger than s_K. The holder enters the swap paying s_K instead of s_T thus saving the difference $s_T - s_K$ whenever the swap payments are made. The amount saved is calculated relative to the notional amount of the swap.

Swaptions can also be priced using Black's model. Consider a payer swaption where the holder has the right to pay a fixed rate s_K and receive a floating rate

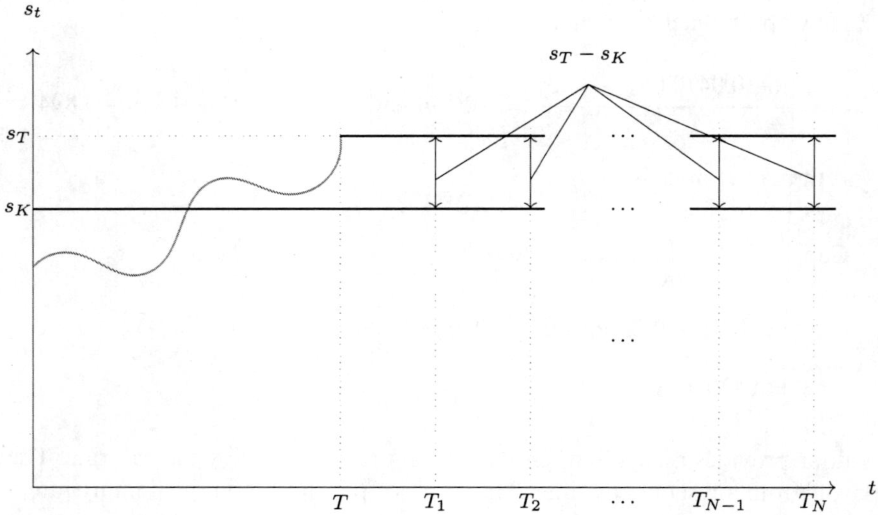

Figure 13.2 Swaption payoffs

on a swap starting in T years with payment dates $T_1, \ldots T_N$ and constant reset period δ. Suppose a notional amount of L.

Each payment on the swap is the fixed rate times $L\delta$. Suppose the swap rate for this swap proves to be s_T at time T. The payoff consists of a series of cash flows equal to

$$\text{Payoff at each reset} = L\delta[s_T - s_K]^+$$

With discounting to time T, the payoff is

$$\text{Payoff discounted to } T = L\delta \sum_{i=1}^{N} P_{T,T_i}[s_T - s_K]^+$$

Then pricing a swaption requires multiplying Black's model by

$$L\delta \sum_{i=1}^{N} P_{T,T_i}$$

and substituting for the forward swap rate along with a measure of the forward swap rate volatility. Applying Black's model and writing $e^{-r(T-t)}$ as $P_{t,T}$, the payer swaption price, PS_t, is

$$\begin{aligned}
PS_t &= L\delta P_{t,T} \sum_{i=1}^{N} P_{T,T_i}\Big(F\Phi(d_1) - s_K\Phi(d_2)\Big) \\
&= L\delta \sum_{i=1}^{N} P_{t,T_i}\Big(F\Phi(d_1) - s_K\Phi(d_2)\Big) \\
&= LA_t\Big(F\Phi(d_1) - s_K\Phi(d_2)\Big)
\end{aligned}$$

where $A_t = \delta \sum_{i=1}^{N} P_{t,T_i}$ and F is the forward swap rate. The d_1 and d_2 parameters are calculated using a measure of the forward swap rate volatility for σ.

> **Exercise 13.3.** *Find the value of a swaption that gives the holder the right to pay 2.6% in a 3-year swap starting in 5 years. The forward swap rate is 2.52% with volatility 10%. Payments are made semiannually and the notional principal is $10 million. The yield curve is flat at 2.5% per annum with continuous compounding.*

Proof. Calculating the annuity term

$$\begin{aligned}
A_t = \delta \sum_{i=1}^{N} P_{t,T_i} &= 0.5(e^{-0.025\times 5.5} + e^{-0.025\times 6} + e^{-0.025\times 6.5} + e^{-0.025\times 7} \\
&\quad + e^{-0.025\times 7.5} + e^{-0.025\times 8}) \\
&= 2.5347
\end{aligned}$$

Applying Black's formula

$$\begin{aligned}
d_1 &= \frac{\log(0.0252/0.026) + \frac{1}{2}0.1^2 \times 5}{0.1\sqrt{5}} = -0.0356 \text{ and} \\
d_2 &= -0.0356 - 0.1\sqrt{5} = -0.2592
\end{aligned}$$

And the payer swaption price is

$$\begin{aligned}
\text{PS}_t &= LA_t\Big(F\Phi(d_1) - s_K\Phi(d_2)\Big) \\
&= 10\text{m} \times 2.5347\Big(0.0252\,\Phi(-0.0356) - 0.026\,\Phi(-0.2592)\Big) \\
&= \$47654
\end{aligned}$$

□

13.5 Intuition behind Black's model

The first problem with using Black's model is that the Black–Scholes framework assumed interest rates are constant. As we are trying to price interest rate derivatives presumably we are uncertain about how interest rates evolve. Indeed, replacing stochastic S_t by Fe^{-rt} treated the interest rate as if it were stochastic.

Black's model can be theoretically justified using no-arbitrage arguments separately for bond options, interest rate caps and swaptions. In this way, interest rates can be considered as stochastic throughout and the resulting formulae are the same as Black's model.

In the Black–Scholes derivations, we considered a market with only two assets – a risky asset and a risk-free asset. Our market here will be zero-coupon bonds of price $P_{t,T}$ for every T up till a given time horizon. Also, interest rates are now stochastic and an investment of cash evolves according to the process

$$dB_t = r_t B_t dt$$

In the Black–Scholes world, we found a risk-neutral measure such that $\frac{S_t}{B_t}$ was a martingale. Here we find a risk-neutral measure such that $\frac{P_{t,T}}{B_t}$ is a martingale for all T.

Assuming the construction of a replicating portfolio of zero-coupon bonds, $\frac{C_t}{B_t}$ is then a martingale for any contingent claim C_t on a bond. Therefore we have

$$\begin{aligned}
\frac{C_t}{B_t} &= \mathbb{E}_t^{\mathbb{Q}}\frac{C_T}{B_T} \\
\implies C_t &= B_t\mathbb{E}_t^{\mathbb{Q}}\frac{C_T}{B_T} = \mathbb{E}_t^{\mathbb{Q}}\left(B_t\frac{C_T}{B_T}\right)
\end{aligned}$$

where C_T is the derivative payoff at time T and C_t is its price at time t.

In the next applications, we use the Radon-Nikodým derivative for changing the numéraire. Under the risk-neutral measure, we had

$$C_t = \mathbb{E}_t^{\mathbb{Q}}\left(B_t\frac{C_T}{B_T}\right)$$

Changing from the risk-free asset B_t to a new numéraire N_t we have from Theorem 10.4 that the derivative price can also be written

$$C_t = \mathbb{E}_t^{\mathbb{Q}}\left(B_t \frac{C_T}{B_T}\right) = \mathbb{E}_t^{\mathbb{N}}\left(\frac{d\mathbb{Q}}{d\mathbb{Q}^N} B_t \frac{C_T}{B_T}\right) = \mathbb{E}_t^{\mathbb{N}}\left(\frac{B_T/B_t}{N_T/N_t} B_t \frac{C_T}{B_T}\right)$$
$$= \mathbb{E}_t^{\mathbb{N}}\left(N_t \frac{C_T}{N_T}\right) \tag{13.4}$$

13.5.1 Option on a bond

In this case, choose $P_{t,T}$ the price at time t of a zero-coupon bond maturing at T as the numéraire N_t. The probability measure associated with this numéraire is known as the *T-forward measure*. Let $\mathbb{E}^T$ be the expectation with respect to this measure. Then the price of an option on a bond maturing at a date T^* after the option maturity date T is

$$\mathrm{BO}_t = \mathbb{E}_t^T\left(P_{t,T} \frac{\mathrm{BO}_T}{P_{T,T}}\right) = P_{t,T}\, \mathbb{E}_t^T(\mathrm{BO}_T)$$

Consider a call option on the bond with strike price K. Letting the bond price at time t be BP_t, the call option price BC_t is

$$\mathrm{BC}_t = P_{t,T}\, \mathbb{E}_t^T[\mathrm{BP}_T - K]^+$$

Assume that BP_T, the bond price at time, T, has a lognormal conditional distribution with volatility σ^2. Then from Theorem 13.1 in the appendix, we have

$$\mathrm{BC}_t = P_{t,T}(\mathbb{E}_t^T \mathrm{BP}_T\, \Phi(d_1) - K\Phi(d_2))$$

where

$$d_1 = \frac{\log(\mathbb{E}_t^T \mathrm{BP}_T / K) + \frac{1}{2}\sigma^2(T-t)}{\sigma\sqrt{T-t}}, d_2 = d_1 - \sigma$$

It remains to show that $\mathbb{E}_t^T \mathrm{BP}_T$ is the forward price of the bond at time t.

Consider a forward on the bond with price F negotiated at time t and maturing at time T. The payoff from the forward at T is $\mathrm{BP}_T - F$. The price f_t of this forward at t is then given by the pricing formula 13.4 as

$$f_t = P_{t,T}\, \mathbb{E}_t^T(\mathrm{BP}_T - F) = P_{t,T}\, (\mathbb{E}_t^T \mathrm{BP}_T - F)$$

But the value of the forward contract when initially negotiated is zero. Therefore $\mathbb{E}_t^T \mathrm{BP}_T = F$.

13.5.2 Caplet

It will be seen in the next chapter that the caplet price can be written in terms of the price of a European put on a zero-coupon bond. The caplet can therefore be priced using the risk-neutral measure described at the start of this section. Changing the numéraire makes pricing simpler.

Choose P_{t,T^*} the price at time t of a zero-coupon bond maturing at T^* as the numéraire N_t. The probability measure associated with this numéraire is known as the T^*-*forward measure*. Let $\mathbb{E}^{T^*}$ be the expectation with respect to this measure.

The price of a caplet resetting at T and maturing at T^* is

$$\text{Cap}_t = \mathbb{E}_t^{T^*}\left(P_{t,T^*}\frac{\text{Cap}_T}{P_{T,T^*}}\right)$$

Given notional principal L; $\delta = T^* - T$; floating rate R_T and caplet strike rate R_K, the caplet payoff at T is $P_{T,T^*}L\delta[R_T - R_K]^+$ and the caplet price is

$$\text{Cap}_t = \mathbb{E}_t^{T^*}\left(P_{t,T^*}\frac{P_{T,T^*}L\delta[R_T - R_K]^+}{P_{T,T^*}}\right) = L\delta P_{t,T^*}\,\mathbb{E}_t^{T^*}[R_T - R_K]^+$$

Assume that R_T, the spot rate at time T, has a lognormal conditional distribution with volatility σ. Then from Theorem 13.1 in the appendix, we have

$$\text{Cap}_t = L\delta P_{t,T^*}(\mathbb{E}_t^{T^*}R_T\,\Phi(d_1) - R_K\Phi(d_2))$$

where

$$d_1 = \frac{\log(\mathbb{E}_t^{T^*}R_T/R_K) + \frac{1}{2}\sigma^2(T-t)}{\sigma\sqrt{T-t}}, d_2 = d_1 - \sigma$$

It remains to show that $\mathbb{E}_t^{T^*}R_T$ is the forward interest rate at time t for the time period between T and T^*, that is, F_{t,T,T^*}

From the definition of the forward rate in (Equation 12.8)

$$\frac{1}{1+\delta F_{t,T,T^*}} = \frac{P_{t,T^*}}{P_{t,T}} \implies F_{t,T,T^*} = \frac{1}{\delta}\left(\frac{P_{t,T} - P_{t,T^*}}{P_{t,T^*}}\right)$$

As the numerator is the price of a combination of zero-coupon bonds, the fraction is a martingale under the measure with P_{t,T^*} as numéraire. Hence

$$F_{t,T,T^*} = \mathbb{E}_t^{T^*}F_{T,T,T^*} = \mathbb{E}_t^{T^*}R_T$$

since the forward rate over the time interval from T to T^* becomes the spot rate when we reach time T.

13.5.3 Swaption

It will be seen in the next chapter that the swaption price can be written as a coupon bond put option. The swaption can therefore be priced using the risk-neutral measure described at the start of this section. Changing the numéraire makes pricing simpler.

Consider a payer swaption giving the right to pay s_K and receive a floating rate beginning T with payment dates $T_1, \ldots, T_N$ an constant reset period δ.

Choose the annuity factor

$$A_t = \sum_{i=1}^{N} \delta P_{t,T_i}$$

as the numéraire N_t. The probability measure associated with this numéraire is known as the *forward swap measure*. Let $\mathbb{E}_t^A$ be the expectation with respect to this measure. The price of the swaption is

$$\text{PS}_t = \mathbb{E}_t^A \left(A_t \frac{\text{PS}_T}{A_T} \right)$$

Given notional principal L; swap rate s_T and swaption strike rate s_K, the swaption payoff at T is $LA_T[s_T - s_K]^+$ and the swaption price is

$$\text{PS}_t = \mathbb{E}_t^A \left(A_t \frac{LA_T[s_T - s_K]^+}{A_T} \right) = LA_t \mathbb{E}_t^A [s_T - s_K]^+$$

Assume that s_T, the swap rate at time T, has a lognormal conditional distribution with volatility σ^2. Then from Theorem 13.1 in the appendix, we have

$$\text{PS}_t = LA_t(\mathbb{E}_t^A s_T \, \Phi(d_1) - K\Phi(d_2))$$

where

$$d_1 = \frac{\log(\mathbb{E}_t^A s_T / K) + \frac{1}{2}\sigma^2(T-t)}{\sigma\sqrt{T-t}}, d_2 = d_1 - \sigma$$

It remains to show that $\mathbb{E}_t^A s_T$ is the forward swap rate at time t.

From the definition in Section 12.5.2, the forward swap rate s_{t,T,T_N} at time t is

$$s_{t,T,T_N} = \frac{P_{t,T} - P_{t,T_N}}{A_t}$$

As the numerator is the price of a combination of zero-coupon bonds, the fraction is a martingale under the measure with A_t as numéraire. Hence

$$s_{t,T,T_N} = \mathbb{E}_t^A s_{T,T,T_N} = \mathbb{E}_t^A s_T$$

since the forward swap rate becomes the realized swap rate when we reach time T.

13.6 Inconsistency of Black's model

The models presented in the previous section provide a theoretical basis for the traditional market practice of pricing caps and swaptions with Blacks's formula. Assuming lognormal forward interest rates allowed caplets to be priced consistently with Black's model while lognormal forward swap rates were assumed in order to price swaptions consistently with Black's model. However, these two assumptions are mutually incompatible as both forward interest rate and forward swap rates cannot be lognormal. This is a problem if pricing exotic derivatives with cap and swaption features and also risk managing portfolios of interest rate derivatives.

To see this, write the forward swap rate in terms of forward interest rates. Consider a swap paying s_K and receiving a floating rate beginning T_0 and payment dates $T_1, \ldots, T_N$ with constant reset period δ.

Writing each P_{t,T_i} as a product of forward rate terms using the definition in (Equation 12.8), we get

$$P_{t,T_i} = P_{t,T_0} \prod_{j=1}^{i} \frac{P_{t,T_j}}{P_{t,T_{j-1}}} = P_{t,T_0} \prod_{j=1}^{i} \frac{1}{1 + \delta F_{t,T_{j-1},T_j}}$$

then substituting in the forward swap rate formula (Equation 12.11)

$$\begin{aligned} s_{t,T_0,T_N} &= \frac{P_{t,T_0} - P_{t,T_N}}{\sum_{i=1}^{N} \delta P_{t,T_i}} \\ &= \frac{1 - \prod_{j=1}^{N} \frac{1}{1+\delta F_{t,T_{j-1},T_j}}}{\sum_{i=1}^{N} \prod_{j=1}^{i} \delta \frac{1}{1+\delta F_{t,T_{j-1},T_j}}} \end{aligned}$$

Assume we have lognormal forward rates under the T^*-forward measure. Let's change to the forward swap measure thus changing the statistical properties of the forward rates. Then determine the probability distribution of the forward swap rate using the complex relationship above. Without working through the details, it should be clear that forward swap rates are unlikely to be lognormal.[1]

This shortcoming motivated the development of short-rate models which will be considered in the next chapter. Using short-rate models, it is possible to derive pricing formulas that are consistent across interest-rate derivatives.

13.7 Black's model and implied volatilities

Black's model for caplets can be used to generate an implied forward rate volatility in the same way implied stock volatilities were generated by the Black–Scholes model (Section 7.3).

For example, you observe the market price of a caplet on a notional \$100 million starting in one year capping the 6-month rate at strike 3.5% to be \$289,750. The forward rate for the period from one year to 18 months is 4%. The 18 month continuously compounded risk-free rate is 5.1%. Substituting these values into the caplet formula (Equation 13.2)

$$\text{Cap}_t = L\delta P_{t,T^*}\left(F\Phi(d_1) - R_K\Phi(d_2)\right)$$

$$\Longrightarrow\ 289750 = 100m \times 0.5e^{-0.051\times 1.5}\left(0.04\Phi(d_1) - 0.035\Phi(d_2)\right)$$

where

$$d_1 = \frac{\log(F/K) + \frac{1}{2}\sigma^2(T-t)}{\sigma\sqrt{T-t}}$$

$$= \frac{\log(0.04/0.035) + \frac{1}{2}\sigma^2 \times 1}{\sigma \times 1} \quad \text{and } d_2 = d_1 - \sigma\sqrt{T-t}$$

Set up Black's model for pricing a caplet in Excel with the input values in separate cells. Then using the Goal Seek feature, the volatility that calibrates to this price is found to be 21.1%.

It should not matter which options are used to calibrate the model but in reality caplets/floorlets with different strikes and maturities imply different volatilities to match their market prices. Once again implied volatilities generate a smile or skew. This would indicate that Black's model with constant volatility is incorrect leading Rebonato (2004) to famously describe an implied volatility as the 'wrong number to plug into the wrong formula to get the right price'.

Jarrow and Chatterjea (2024) argue that, when considered as a statistical model, Black's model calibrated to market prices can be used legitimately to price other caplets/floorlets. Just as an econometric model incorporates all relevant parameters with coefficients estimated from empirical data, the model used in this way provides the analyst with a method to extrapolate beyond the market data. However, its use should be confined to only pricing other caplets/floorlets.

For other purposes such as portfolio risk management, Black's model is inappropriate. Calibration to market prices indicates that forward rate volatility is dependent on not only strike and maturity but also the current forward rate (McInerney and Zastawniak, 2015). Delta hedging for caplets will be inaccurate if these relationships are not explicitly modeled. Delta hedging for swaptions has the added complication that correlations between forward rates must also be modeled since the forward swap rate is a function of multiple forward rates. This motivated the search for improvements on Black's approach capable of generating realistic implied volatility surfaces. Models that achieve this are known as *stochastic volatility models*.

13.8 Summary

This chapter introduced the use of Black's model for pricing interest rate derivatives such as bond options, caps, floors and swaptions. The theoretical foundation is then clarified: Black's model, though originally heuristic, can be justified rigorously by using risk-neutral valuation and changes of numéraire. In these models, zero-coupon bonds serve as tradable assets and contingent claims are priced as expectations under appropriately chosen numéraires with lognormality assumptions. The result are prices identical to Black's model.

There is an inherent incompatibility between the assumption of lognormal forward interest rates (needed for caps) and the assumption of lognormal forward swap rates (needed for swaptions). Theory dictates that they cannot both be lognormal. This shortcoming motivated the development of short-rate models considered next.

Market data indicates that this incompatibility is not as major an issue as first appears. Forward swap rates tend to be approximately lognormal when lognormal forward interest rates are assumed. The issue is then more an academic challenge than an obstacle to application by practitioners.

A more problematic feature of the model is the nature of forward volatilities implied by caplet prices. Forward volatilities display dependencies and correlations not captured in Black's approach. The final chapter on the Heath–Jarrow–Morton model describe how Black's model can be extended to allow for joint dynamics.

Despite its flaws, Black's model is simple, fast and widely understood. It is still the standard for quoting implied volatilities. But for trading and risk management, more sophisticated models are often preferred.

Key things to remember

- Application of Black's model to pricing bonds, caps and swaptions (Sections 13.2, 13.3 and 13.4).
- Limitations of Black's model (Sections 13.6 and 13.7).

Further reading

Veronesi (2010) offers a practitioner-friendly introduction to fixed income pricing, including Black's model for caps and swaptions. The model is linked to practical hedging and risk management strategies. Brigo et al. (2001) is a comprehensive reference that situates Black's model within the wider family of interest rate models. McInerney and Zastawniak (2015) builds up from Black's model to more advanced arbitrage-free term structure models showing both theoretical foundations and computational techniques. Brace et al. (2001) and Jarrow et al. (2007) address extensions and market shortcomings.

Note

1. However, Brace et al. (2001) argue that the forward swap rate distribution will be approximately lognormal in practice. The incompatibility is to an extent more of a problem for academic theoreticians.

Questions

1 Find the price of a one-year option on a bond with face value \$100. The strike price is \$101, and the volatility of the one-year forward bond price is 10% per annum. The bond matures in 3 years. The bond pays annual coupons at a coupon rate of 5%. Coupons have just been paid. The yield curve is flat at 4% continuously compounded.

> **Solution**
>
> The present value of the coupon paid in one-year is
>
> $$I = 5e^{-0.04} = 4.80$$
>
> The current bond price is
>
> $$\mathrm{BP}_t = 5e^{-0.04\times 1} + 5e^{-0.04\times 2} + 105e^{-0.04\times 3} = 102.55$$
>
> The forward bond price for a contract maturing in 1 year is
>
> $$F = (102.55 - 4.80)e^{0.04} = 101.73$$
>
> Now apply Black's formula.
>
> $$d_1 = \frac{\log(101.73/101) + \frac{1}{2}0.1^2}{0.1} = 0.122 \text{ and } d_2 = 0.122 - 0.1 = 0.022$$
>
> And the bond option price BO_t is
>
> $$\begin{aligned}\mathrm{BO}_t &= e^{-0.04}\Big(101.73\,\Phi(0.122) - 101\,\Phi(0.022)\Big)\\ &= \$4.25\end{aligned}$$

2 Calculate the value of a one-year European call option on a zero-coupon bond with face value \$1 that will mature five years from today using Black's model. The current bond price is \$0.65. The strike price of the option is \$0.70, the one-year risk-free interest rate is 3% per annum with continuous compounding, and the forward bond price volatility for the bond is 20% per annum.

Solution The forward zero coupon bond price for a contract maturing in 1 year is

$$F = \frac{P_{0,5}}{P_{0,1}} = \frac{0.65}{e^{-0.03}} = 0.67$$

Now apply Black's formula.

$$d_1 = \frac{\log(0.67/0.70) + \frac{1}{2}0.2^2}{0.2} = -0.121 \text{ and } d_2 = -0.121 - 0.2 = -0.321$$

And

$$\begin{aligned} \mathrm{BO}_t &= e^{-0.03}\left(0.67\,\Phi(-0.121) - 0.70\,\Phi(-0.321)\right) \\ &= \$0.04 \end{aligned}$$

3 Find the price of an option that caps the three-month rate, starting in 15 months' time, at 4.5% (quarterly compounded) on a notional of \$100,000. The forward interest rate for this period is 3.7% per annum (quarterly compounded) and the volatility of this forward rate is 15% per annum. The 18-month risk-free interest rate (continuously compounded) is 4% per annum.

Solution Applying Black's formula.

$$d_1 = \frac{\log(0.037/0.045) + \frac{1}{2}0.15^2 \times 1.25}{0.15\sqrt{1.25}} = -1.083 \text{ and}$$

$$d_2 = -1.083 - 0.15\sqrt{1.25} = -1.251$$

And

$$\begin{aligned} \mathrm{Cap}_t &= 100\mathrm{k} \times 0.25 \times e^{-0.04 \times 1.5}\left(0.37\Phi(-1.083) - 0.045\Phi(-1.251)\right) \\ &= \$9.64 \end{aligned}$$

4 You observe the following for a 1-year cap struck at 3%, which resets every 3 months and makes a payment at the end of each period based on the rates at the beginning of that period. The initial yield curve is flat at 3%.

Tenor (yrs)	*Cap price per 10,000*
0.5	2.946
0.75	6.874
1	11.396

(a) Calculate the time $T = 1$ caplet price.
(b) Calculate the implied volatility for the $T = 1$ caplet for Black's model.

[FETE - Fall 2010 Q9]

Solution

(a) The cap price for any period is the sum of the caplet prices over the period. Therefore, the time $T = 1$ caplet price is $11.396 - 6.874 = 4.522$.
(b) Use the Goal Seek function in Microsoft Excel to find the σ that gives a $T = 1$ caplet price of 4.522. The solution is $\sigma = 18\%$.

5 On January 1, your company, the Bank of Mom and Dad (BMD), offers a promotional loan to prospective college students. You are given the following:

- Loan amount is $50,000.
- The term of the loan is two-years.
- The annual interest rate is 5%.
- Interest on the loan is payable annually.

BMD will fund the loan by borrowing at the 12-month LIBOR rate for two years. BMD would like to use interest rate derivatives to limit its interest rate risk. Two derivatives being considered are a cap and a swaption.

LIBOR/swap zero curve: 4.0% flat
Volatility of all forward rates: 30%

(a) Calculate the price on January 1 of a derivative that caps the 12-month LIBOR rate, starting in 3 months, at 4.1% for 2 years, using the standard market model.
(b) Calculate the price on January 1 of a 3-month swaption giving BMD the right to enter a 2-year interest rate swap, paying 4.1% fixed interest in exchange for 12-month risk-free rate, using the standard market model.

[FETE - Spring 2013 Q9]

Solution

(a) There are two caplets in this cap. The first caplet begins at $T = 0.25$ lasting to the first reset at $T^* = 1.25$. The second caplet begins at $T = 1.25$ lasting to the next reset at $T^* = 2.25$.

A flat yield curve where spot rates are constant implies forward rates are constant from Section 12.2. Therefore $F = 4\%$.

Pricing the first caplet:

$$d_1 = \frac{\log(0.04/0.041) + \frac{1}{2}0.3^2 \times 0.25}{0.3\sqrt{0.25}} = -0.090 \text{ and}$$
$$d_2 = -0.090 - 0.3\sqrt{0.25} = -0.240$$

$$\text{Cap}_t = 50\text{k} \times 1 \times e^{-0.04\times 1.25}\Big(0.04\,\Phi(-0.090) - 0.041\,\Phi(-0.240)\Big)$$
$$= \$92.93$$

Pricing the second caplet:

$$d_1 = \frac{\log(0.04/0.041) + \frac{1}{2}0.3^2 \times 1.25}{0.3\sqrt{1.25}} = 0.094 \text{ and}$$
$$d_2 = 0.094 - 0.3\sqrt{1.25} = -0.241$$

$$\text{Cap}_t = 50\text{k} \times 1 \times e^{-0.04\times 2.25}\Big(0.04\,\Phi(0.094) - 0.041\,\Phi(-0.241)\Big)$$
$$= \$224.30$$

The price of the cap is \$92.93 + \$224.30 = \$317.23.

(b) Calculating the annuity term

$$A_t = e^{-0.04\times 1.25} + e^{-0.04\times 2.25} = 1.87$$

As was shown in Section 12.5.2, a flat yield curve implies forward swap rates are constant therefore $F = 4\%$.

Applying Black's formula

$$d_1 = \frac{\log(0.04/0.041) + \frac{1}{2}0.3^2 \times 0.25}{0.3\sqrt{0.25}} = -0.090 \text{ and}$$
$$d_2 = -0.090 - 0.3\sqrt{0.25} = -0.240$$

And the payer swaption price is

$$\text{PS}_t = 50\text{k} \times 1.87\Big(0.04\,\Phi(-0.090) - 0.041\,\Phi(-0.240)\Big)$$
$$= \$182.22$$

6 At time 0, the prices of three zero-coupon bonds, which mature at time t, $P_{0,t}$ are as follows:
$P_{0,1}=0.9608$
$P_{0,2}=0.9241$
$P_{0,3}=0.8896$

Assume annual payments for all interest rate swaps.

(a) Demonstrate that the swap rate on a 2-year interest rate swap which starts in 1 year is 3.926%.

You are given the following information about a European swaption:

- Notional of \$10,000,000
- Exercisable after 1 year into a 2-year interest rate swap paying 4% fixed
- The volatility for the forward swap is 15% per annum

(b) Calculate the value of the swaption.

[FETE - Spring 2013 Q15(c)(d)]

Solution

(a) The forward swap rate is

$$\frac{P_{0,1} - P_{0,3}}{P_{0,2} + P_{0,3}} = \frac{0.9608 - 0.8896}{0.9241 + 0.8896} = 0.03926$$

(b) Calculating the annuity term

$$A_t = P_{0,2} + P_{0,3} = 0.9241 + 0.8896 = 1.814$$

Applying Black's formula

$$d_1 = \frac{\log(0.03926/0.04) + \frac{1}{2}0.15^2}{0.15} = -0.050 \text{ and}$$
$$d_2 = -0.050 - 0.15 = -0.200$$

And the payer swaption price is

$$\text{PS}_t = 100\text{m} \times 1.814\left(0.04\,\Phi(-0.050) - 0.03926\,\Phi(-0.200)\right)$$
$$= \$36,565$$

Appendix

A.1 Lognormal result

Theorem 13.1. *Assuming X lognormal with $\log X \sim N(\mu, \sigma^2)$ then*

$$\mathbb{E}[X - K]^+ = \mathbb{E}(X)\Phi(d_1) - K\Phi(d_2)$$

where $d_1 = \frac{\log(\mathbb{E}X/K)+\frac{\sigma^2}{2}}{\sigma}$, $d_2 = d_1 - \sigma$.

Proof. This a generalization of the derivation of the Black–Scholes formula for the price of a European call option (Question 7 in Chapter 10).

Let $Y = \log X \sim N(\mu, \sigma^2)$.

$$\begin{aligned}
\mathbb{E}[X - K]^+ &= \frac{1}{\sqrt{2\pi}} \int_{z^*}^{\infty} (e^Y - K)e^{-\frac{1}{2}z^2}\,dz \qquad \text{where } X = e^Y \geq K \text{ when } z \geq z^* \\
&= \frac{1}{\sqrt{2\pi}} \int_{z^*}^{\infty} (e^{\mu+\sigma z} - K)e^{-\frac{1}{2}z^2}\,dz \\
&= \frac{1}{\sqrt{2\pi}} \int_{z^*}^{\infty} e^{\mu+\sigma z}e^{-\frac{1}{2}z^2}\,dz - \frac{1}{\sqrt{2\pi}} \int_{z^*}^{\infty} Ke^{-\frac{1}{2}z^2}\,dz \\
&= e^{\mu+\frac{\sigma^2}{2}} \int_{y^*}^{\infty} \frac{1}{\sqrt{2\pi}} e^{-\frac{1}{2}y^2}\,dy - K \int_{z^*}^{\infty} \frac{1}{\sqrt{2\pi}} e^{-\frac{1}{2}z^2}\,dz \text{ where } y = z - \sigma \\
&= e^{\mu+\frac{\sigma^2}{2}} \int_{-\infty}^{-y^*} \frac{1}{\sqrt{2\pi}} e^{-\frac{1}{2}y^2}\,dy - K \int_{-\infty}^{-z^*} \frac{1}{\sqrt{2\pi}} e^{-\frac{1}{2}z^2}\,dz \\
&= \mathbb{E}X\Phi(d_1) - K\Phi(d_2)
\end{aligned}$$

The last line follows from Equation (2.9).

Also, z^* is the value for which $X = K$.

$$z^* = \frac{\log K - \mu}{\sigma} = \frac{\log K - \log \mathbb{E}X + \sigma^2/2}{\sigma} = \frac{\log(K/\mathbb{E}X) + \sigma^2/2}{\sigma}$$

Therefore

$$d_1 = -y^* = -z^* + \sigma = \frac{\log(\mathbb{E}X/K) + \sigma^2/2}{\sigma} \text{ and}$$

$$d_2 = -z^* = d_1 - \sigma$$

□

14 Short-rate models

Introduction

The previous chapter highlighted an inherent anomaly in applying Black's model separately to caps and swaptions. Ideally, we would like to use a theoretical framework whereby all derivatives could be priced consistently without running the risk of price contradictions and arbitrage opportunities. This is the goal of short-rate models.

In this chapter, the evolution of interest rates will be described by a model of the short rate, that is, the spot rate over a very short period of time met in Section 12.2. A short-rate model leads to a model of the full yield curve using the relationships developed in that section. In this way, short-rate models allow products with a range of terms and maturities to be priced consistently. Also correlations between movements at different points along the yield curve can be modeled. This may be a significant factor affecting the price of some derivatives.

The exposition mainly focuses on pricing the simplest fixed-income security, zero-coupon bonds, as the mathematics of interest-rate derivatives gets complicated. The approach is similar to pricing equity options. Using a hedging argument, we derive a partial differential equation satisfied by any zero-coupon bond. The solution to this equation is the bond price. As before, we will choose a risk-neutral approach to finding this solution.

The short-rate models considered will have only one source of uncertainty and are known as *one-factor models*. Their main drawback is that they only capture simple parallel shifts in the yield curve but not flattening or steepening of the yield curve or changes in its curvature. These kinds of changes will typically occur during economic shocks or when monetary policy changes.

14.1 Pricing a zero-coupon bond

In this section, the model is first established. The risk-neutral probability $\mathbb{Q}$ for this model can then be determined along with the representation of the short rate under $\mathbb{Q}$. The process $\frac{P_{t,T}}{B_t}$ is a martingale under this probability measure. A

DOI: 10.4324/9781003477624-18

partial differential equation satisfied by all zero-coupon bonds is then derived using the fact that martingales are driftless processes.

Let the interest rate be stochastic. Assume the short rate follows

$$dr = a(t, r_t)dt + \sigma(t, r_t)dW \tag{14.1}$$

where W_t is standard Brownian motion. The price at time t of the zero-coupon bond $P_{t,T}$ with payoff \$1 at time T is assumed to depend on r_t i.e. $P_{t,T} = P(t, T, r_t)$. Similar to before, an investment of cash evolves according to the process

$$dB = r_t B_t dt$$

but now the interest rate r_t is stochastic. And assuming the security is worth \$1 at time zero

$$dB = r_s B_s ds \implies B_t = e^{\int_0^t r_s ds}$$

Using the time-dependent Itô rule and dropping the dependency of a and σ on (t, r_t) for clarity:

$$\begin{aligned} dP &= \frac{\partial P}{\partial t}dt + \frac{\partial P}{\partial r}dr + \frac{1}{2}\frac{\partial^2 P}{\partial r^2}(dr)^2 \\ &= \frac{\partial P}{\partial t}dt + \frac{\partial P}{\partial r}(adt + \sigma dW) + \frac{1}{2}\frac{\partial^2 P}{\partial r^2}\sigma^2 dt \\ &= \left(\frac{\partial P}{\partial t} + \frac{\partial P}{\partial r}a + \frac{1}{2}\frac{\partial^2 P}{\partial r^2}\sigma^2\right)dt + \sigma\frac{\partial P}{\partial r}dW \end{aligned} \tag{14.2}$$

By letting

$$\alpha P_t = \frac{\partial P}{\partial t} + \frac{\partial P}{\partial r}a + \frac{1}{2}\frac{\partial^2 P}{\partial r^2}\sigma^2$$

$$\text{and } \beta P_t = \sigma\frac{\partial P}{\partial r}$$

for functions α and β of t and r_t we can rewrite Equation (14.2) in geometric Brownian motion form as

$$dP = \alpha(t, r_t)P_t dt + \beta(t, r_t)P_t dW \tag{14.3}$$

As in Example 10.1, the market price of risk defines the Radon-Nikodým derivative to change from the real-world measure $\mathbb{P}$ to the risk-neutral measure

$\mathbb{Q}$ under which $\frac{P_{t,T}}{B_t}$ is a martingale. So

$$\gamma_t = \frac{\alpha - r_t}{\beta}$$

and the stochastic process $\widetilde{W}_t$ defined by

$$d\widetilde{W} = \gamma_t dt + dW$$

is a standard Brownian motion under $\mathbb{Q}$.

Under $\mathbb{Q}$, the short-rate process then becomes

$$dr = adt + \sigma dW = adt + \sigma(d\widetilde{W} - \gamma_t dt) = (a - \sigma\gamma_t)dt + \sigma d\widetilde{W}$$

Letting

$$\mu(t, r_t) :- a - \sigma\gamma_t \tag{14.4}$$

for function μ of t and r_t we proceed using the following general representation of the short rate under $\mathbb{Q}$

$$dr = \mu(t, r_t)dt + \sigma(t, r_t)d\widetilde{W} \tag{14.5}$$

Repeating the application of the time-dependent Itô rule, Equation (14.2) is now

$$dP = \left(\frac{\partial P}{\partial t} + \frac{\partial P}{\partial r}\mu + \frac{1}{2}\frac{\partial^2 P}{\partial r^2}\sigma^2\right)dt + \sigma\frac{\partial P}{\partial r}d\widetilde{W}$$

Applying the stochastic quotient rule,

$$\begin{aligned} d\frac{P}{B} &= \frac{1}{B_t}dP - \frac{P}{B_t^2}dB \\ &= \frac{1}{B_t}\left((\frac{\partial P}{\partial t} + \frac{\partial P}{\partial r}\mu + \frac{1}{2}\frac{\partial^2 P}{\partial r^2}\sigma^2)dt + \sigma\frac{\partial P}{\partial r}d\tilde{W}\right) - \frac{P}{B_t^2}(r_t B_t dt) \\ &= \frac{1}{B_t}\left(\frac{\partial P}{\partial t} + \frac{\partial P}{\partial r}\mu + \frac{1}{2}\frac{\partial^2 P}{\partial r^2}\sigma^2 - r_t P\right)dt + \frac{\sigma}{B_t}\frac{\partial P}{\partial r}d\tilde{W} \end{aligned} \tag{14.6}$$

If $\frac{P_{t,T}}{B_t}$ is a martingale then

$$\frac{\partial P}{\partial t} + \frac{\partial P}{\partial r}\mu(t, r_t) + \frac{1}{2}\frac{\partial^2 P}{\partial r^2}\sigma(t, r_t) - r_t P = 0 \tag{14.7}$$

since martingales are driftless processes.

In models where the short rate is the only source of uncertainty, this partial differential equation must be satisfied by any zero-coupon bond pricing formula. The price can be found from this PDE along with the terminal condition $P_{T,T} = 1$. This then is the analogue of the Black-Scholes equation which describes equity derivative prices.

Instead of attempting to solve the PDE we will use risk-neutral pricing to price zero-coupon bonds just as we did in the case of equity options.

Given that $\frac{P_{t,T}}{B_t}$ is a martingale then

$$\frac{P_{t,T}}{B_t} = \mathbb{E}_t^{\mathbb{Q}} \frac{P_{T,T}}{B_T}$$

Since the payoff at time T is \$1, the zero-coupon bond price $P_{t,T}$ is given by

$$P_{t,T} = B_t \mathbb{E}_t^{\mathbb{Q}} \frac{1}{B_T} = \mathbb{E}_t^{\mathbb{Q}} \frac{B_t}{B_T} = \mathbb{E}_t^{\mathbb{Q}} \frac{e^{\int_0^t r_s ds}}{e^{\int_0^T r_s ds}} = \mathbb{E}_t^{\mathbb{Q}} e^{-\int_t^T r_s ds}$$

Once $P_{t,T}$ is calculated then the spot rate $R_{t,T}$ is found from rearranging the relationship in Equation (12.2)

$$R_{t,T} = -\frac{\log P_{t,T}}{T - t}$$

In this way, the short rate determines the price of zero-coupon bonds at different maturities and hence the entire yield curve.

The price $P_{t,T}$ calculated from the risk-neutral expectation is dependent on the model used for the short-rate r_t. In the next section, we will derive the solution for the *Vasicek model*

$$dr = \alpha(\mu - r_t)du + \sigma d\tilde{W} \text{ for } \alpha, \mu, \sigma \text{ non-negative constants}$$

Other short-rate models considered are the *Cox–Ingersoll–Ross model* (often abbreviated as CIR)

$$dr = \alpha(\mu - r_t)dt + \sigma\sqrt{r_t}d\tilde{W} \text{ for } \alpha, \mu, \sigma \text{ non-negative constants}$$

and an extension of the Vasicek model known as the *Hull-White model*

$$dr = \alpha(\mu(t) - r_t)dt + \sigma d\tilde{W} \text{ for } \alpha, \sigma \text{ constants}$$

There are many other one-factor models in the academic literature.

The Vasicek and CIR models are known as 'equilibrium models' as they are based on an economic understanding of how interest rates behave. These models are typically built from macroeconomic or utility-maximization frameworks and the resulting equilibrium determines the dynamics of the short rate. The parameters of the model are chosen according to this economic understanding.

The Vasicek model has the same representation as the Ornstein-Uhlenbeck model described in Section 6.7.3. This process was seen to move to a long-run mean level. The economic intuition in the Vasicek model is that interest rates do not tend to grow exponentially like equity prices but instead revert to a long-term mean. This is because high interest rates slow down spending and investment leading to a macroeconomic state where central banks adjust interest rates downwards. While low interest rates lead to economic expansion and a macroeconomic state where monetary policy favors higher interest rates. The long-run equilibrium level for interest rates is then determined by long-run monetary policy, technological innovation and overall economic strength.

On the other hand, the Hull-While model is a 'no arbitrage model'. Unlike equilibrium models, these types of models do not attempt to explain rate behavior – instead they are designed to match the term structure of interest rates as indicated by current market prices. Financial markets are presumed to be efficient and free of arbitrage opportunities. The extra flexibility of having a time-dependent μ_t term allows parameter values to be chosen in the Hull-White model so that the current term structure can be replicated.

Once zero-coupon bonds have been priced for each model coupon-bearing bonds are straightforward to price. Simply apply the zero-coupon bond pricing formula to each coupon separately and multiply by the coupon amount to determine the coupon bond price.

14.2 Vasicek model

The previous section established the risk-neutral pricing formula for zero-coupon bonds

$$P_{t,T} = \mathbb{E}_t^{\mathbb{Q}} e^{-\int_t^T r_s ds}$$

In the Vasicek model the short rate follows the process

$$dr = \alpha(\mu - r_t)dt + \sigma d\tilde{W}$$

where α, μ, σ are non-negative constants. This is an Ornstein-Uhlenbeck process for which the solution was derived in Exercise 6.5.

$$r_t = \mu + (r_0 - \mu)e^{-\alpha t} + \sigma \int_0^t e^{\alpha(s-t)} d\tilde{W} \tag{14.8}$$

And the statistical properties of r_t were established in Equation (6.7):

$$r_t \sim N\left(\mu + (r_0 - \mu)e^{-\alpha t}, \frac{\sigma^2}{2\alpha}(1 - e^{-2\alpha t})\right)$$

14.2.1 *Zero-coupon bond price*

The next step in the risk-neutral pricing formula is to integrate the solution. Then,

$$
\begin{aligned}
r_s &= \mu + (r_0 - \mu)e^{-\alpha s} + \sigma \int_0^s e^{\alpha(u-s)} d\tilde{W} \\
\Longrightarrow \int_0^T r_s ds &= \mu T + \frac{(r_0 - \mu)}{\alpha}(1 - e^{-\alpha T}) + \sigma \int_0^T \int_0^s e^{\alpha(u-s)} d\tilde{W} ds \\
&= \mu T + \frac{(r_0 - \mu)}{\alpha}(1 - e^{-\alpha T}) + \sigma \int_0^T \int_u^T e^{\alpha(u-s)} ds\, d\tilde{W} \\
&= \mu T + \frac{(r_0 - \mu)}{\alpha}(1 - e^{-\alpha T}) + \frac{\sigma}{\alpha} \int_0^T (1 - e^{\alpha(u-T)}) d\tilde{W} \quad (14.9)
\end{aligned}
$$

In the third line, we interchange the order of the integrals according to the result given in Example 1.3

$$\int_a^b \int_a^x f(x, y)\, dy\, dx = \int_a^b \int_y^b f(x, y)\, dx\, dy$$

This result had been established only when integrating non-stochastic variables x and y. However, Fubini's theorem in stochastic calculus provides a generalization of the result as long as $f(x, y)$ meets a number of quite general regularity and integrability conditions (see Exercise 12.3 in Klebaner (2012)).

We have from Theorem 6.3 that

$$\int_0^T (1 - e^{\alpha(u-T)}) d\tilde{W} \sim N\left(0, \int_0^T (1 - e^{\alpha(u-T)})^2 du\right)$$

So, from Equation (14.9), we can conclude that $\int_0^T r_s ds$ is also normally distributed with properties

$$
\begin{aligned}
\mathbb{E}^{\mathbb{Q}}\left(\int_0^T r_s ds\right) &= \mu T + \frac{(r_0 - \mu)}{\alpha}(1 - e^{-\alpha T}) \\
\text{and } var^{\mathbb{Q}}\left(\int_0^T r_s ds\right) &= \frac{\sigma^2}{\alpha^2} \int_0^T (1 - e^{\alpha(u-T)})^2 du \\
&= \frac{\sigma^2}{\alpha^2} \int_0^T (1 - 2e^{\alpha(u-T)} + e^{2\alpha(u-T)}) du \\
&= \frac{\sigma^2}{\alpha^2}\left(T - \frac{2}{\alpha}(1 - e^{-\alpha T}) + \frac{1}{2\alpha}(1 - e^{-2\alpha T})\right)
\end{aligned}
$$

Putting this information into the risk-neutral pricing formula for zero-coupon bonds and using the result that $\mathbb{E}e^X = e^{\mu + \frac{1}{2}\sigma^2}$ when $X \sim N(\mu, \sigma^2)$

(Equation 2.9)

$$P_{0,T} = \mathbb{E}^{\mathbb{Q}} e^{-\int_0^T r_s ds} = \exp\left(-\left\{ \mu T + \frac{(r_0 - \mu)}{\alpha}(1 - e^{-\alpha T}) \right\} \right.$$

$$\left. + \left\{ \frac{\sigma^2}{2\alpha^2} \left(T - \frac{2}{\alpha}(1 - e^{-\alpha T}) + \frac{1}{2\alpha}(1 - e^{-2\alpha T}) \right) \right\} \right) \tag{14.10}$$

This can be written more succinctly as

$$P_{0,T} = e^{A_{0,T} - B_{0,T} r_0}$$

where

$$A_{0,T} = -(T - B_{0,T})\left(\mu - \frac{\sigma^2}{2\alpha^2} \right) - \frac{\sigma^2 B_{0,T}^2}{4\alpha}$$

$$\text{and } B_{0,T} = \frac{(1 - e^{-\alpha T})}{\alpha}$$

14.2.2 Yield curve

Given the parameter values for A_T and B_T, the price of zero-coupon bonds at all maturities are easily determined. Then using Equation (12.2), we can generate the yield curve.

$$R_{0,T} = -\frac{\log P_{0,T}}{T} = \frac{-A_{0,T} + B_{0,T} r_0}{T} \tag{14.11}$$

There are only three possible shapes of yield curve in the Vasicek model – upward sloping, downward sloping or slightly hump-shaped (Figure 14.1). One limitation of this model is therefore that many yield curve shapes observed in the market cannot be replicated.

In addition, any shocks to the yield curve are perfectly correlated so that the entire yield curve will move in an almost parallel direction when the initial rate abruptly changes.

To see this let r_0 vary. There is no other random variation in Equation (14.11) hence correlation between yields at different maturities T_1 and T_2 is given by

$$corr(R_{0,T_1}, R_{0,T_2}) = corr(r_0, r_0) = 1$$

Shocks to the yield curve are therefore perfectly correlated across maturities. This is a problem as historical data indicates that changes in yields for bonds with different maturities are not perfectly correlated.

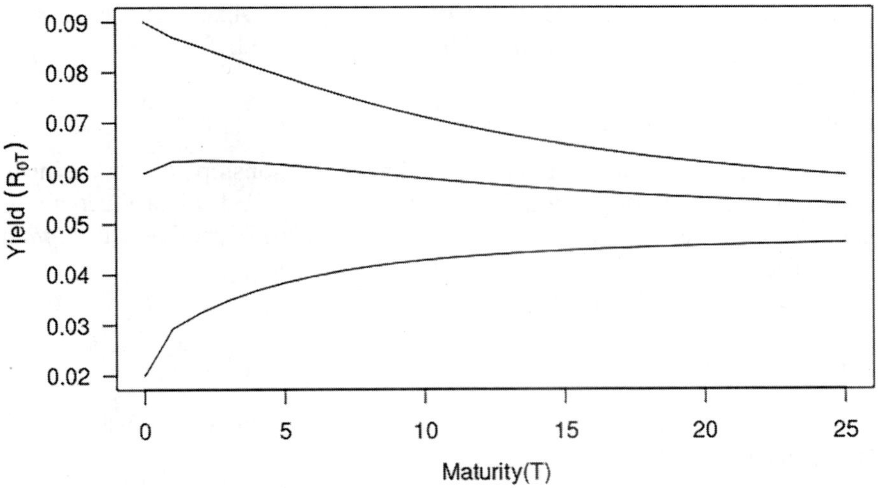

Figure 14.1 Possible yield curve shapes in the Vasicek model

The response of the yield to a unit shock in the initial short rate is

$$\frac{dR_{0,T}}{dr_0} = \frac{B_{0,T}}{T}$$

For typical parameters, changes at all maturities T are of similar size and the yield curve moves almost parallel. More complicated changes in the yield curve such as changes in curvature or changes limited to certain maturities are not possible. This is a major disadvantage in the risk management of portfolios of financial instruments dependent on the relationship between interest rates across the yield curve or in pricing products dependent on rates at different maturities on the yield curve such as swaptions.

14.2.3 Volatility

Volatility is a key determinant of optimal strategies to hedge interest-rate risk and is crucial to the accurate pricing of interest-rate options. The specification of the diffusion term in short-rate models is therefore fundamental to their usefulness.

Another limitation of the Vasicek model is the simplicity of the diffusion term with constant volatility σ. Chan et al. (1992) compare the fit of a range of models to short-term interest rate dynamics and find that the Vasicek model performs poorly due to the specification of volatility in the model.

Calibrating caplet prices as determined by Black's model to market prices for different maturities results in a term structure of forward-rate volatilities that is typically hump-shaped Brigo et al. (2001). Forward rate volatilities derived from the diffusion term do not produce these typical shapes in the Vasicek model. Caps and floors will then potentially be mispriced.

14.3 Affine term-structure models

The yield in the Vasicek model has a linear relationship with r_0. Linear functions are a one-dimensional case of what are called *affine functions* in mathematics. The Vasicek model then belongs to a class of models called *affine term structure* models.

Filipovic (2009) shows that a short-rate model

$$dr = \mu(t, r_t)dt + \sigma(t, r_t)d\tilde{W}$$

will produce an affine term structure if and only if the drift and diffusion terms are of the form

$$\mu(t, r_t) = b(t) + c(t)r_t \text{ and } \sigma^2(t, r_t) = u(t) + v(t)r_t$$

In the Vasicek model $\mu(t, r_t) = \alpha(\mu - r_t) = \alpha\mu - \alpha r_t$ and $\sigma^2(t, r_t) = \sigma^2$ so this model will produce an affine term structure.

If we know in advance that the short-rate model will produce an affine term structure then pricing a zero-coupon bond becomes a lot simpler. An affine term structure implies the zero-coupon bond price at time t is of the form

$$P_{t,T} = e^{A_{t,T} - B_{t,T} r_t} \tag{14.12}$$

then we can apply the partial differential equation that must be satisfied by any zero-coupon bond pricing formula (Equation 14.7) to find the value of $A_{t,T}$ and $B_{t,T}$. Question 11 at the end of chapter demonstrates the approach for the Vasicek model.

14.4 Cox–Ingersoll–Ross model

In the Vasicek model, r_t is not always strictly positive. Negative interest rates are avoided in the CIR model.

Although negative interest rates have historically been considered a very unconventional policy instrument and therefore inappropriate for a short-rate model, more recent experience in many countries with negative Central Bank Policy Rates have made avoiding negative short-rates a less essential feature of models (Figure 14.2). A negative policy rate means commercial banks are paying the central bank to hold their reserves rather than the central bank paying them interest on their reserves. During the previous decade many Central Banks – including Denmark, Sweden, Japan, Switzerland and the European Central Bank – used negative rates to force banks to lend or in some cases to deter foreign capital inflows.

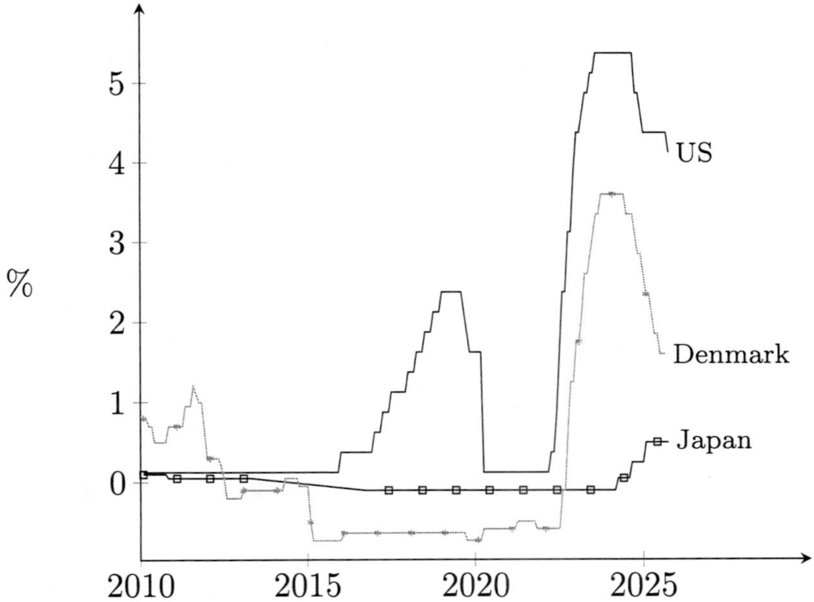

Figure 14.2 Central Bank Policy Rates
Source: Bank of International Settlements

The CIR model of the short rate is

$$dr = \alpha(\mu - r_t)dt + \sigma\sqrt{r_t}d\tilde{W}$$

where α, μ, σ are non-negative constants. For $\sigma^2 \leq 2\alpha\mu$, if the process approaches zero then volatility disappears. With a positive drift, the process then moves away from zero back toward the long-run mean. In this way, negative short rates are not possible in the CIR model.[1]

We have met this process in Question 6.8 in Chapter 6 where the solution was found to be

$$r_t = \mu + (r_0 - \mu)e^{-\alpha t} + \sigma \int_0^s e^{\alpha(s-t)}\sqrt{r_s}d\tilde{W} \tag{14.13}$$

Unlike in the Vasicek model, the integrand here is not deterministic and hence in this case r_t does not have a normal distribution. It can be shown that instead it has a 'non-central' χ^2 distribution with mean and variance as given in Question 6.8 in Chapter 6. Whereas a χ^2 distribution with n degrees of freedom is the sum of the square of n independent standard normal variables, a non-central χ^2 distribution with n degrees of freedom and non-centrality parameter λ is the sum of the square of n normal variables with mean μ_i; $i = 1, ..., n$ and variance 1 such that $\lambda = \sum_1^n \mu_i^2$.

The partial differential equation in Equation (14.7) becomes:

$$\frac{\partial P}{\partial t} + \frac{\partial P}{\partial r}(\alpha\mu - \alpha r_t) + \frac{1}{2}\frac{\partial^2 P}{\partial r^2}(\sigma^2 r_t) - r_t P = 0$$

Here the drift term $\mu(t, r_t) = \alpha\mu - \alpha r_t$ and the diffusion term squared $\sigma^2(t, r_t) = \sigma^2 r_t$ so this model will produce an affine term structure. Let $P_{t,T} = e^{A_{t,T} - B_{t,T} r_t}$ then

$$\left(\frac{\partial A}{\partial t} - \frac{\partial B}{\partial t} r_t\right) P - BP(\alpha\mu - \alpha r_t) + \frac{1}{2} B^2 P(\sigma^2 r_t) - r_t P = 0$$

$$\Longrightarrow \left(\frac{\partial A}{\partial t} - B\alpha\mu\right) + \left(-\frac{\partial B}{\partial t} + \alpha B + \frac{\sigma^2}{2} B^2 - 1\right) r_t = 0$$

In order for the left-hand side to be zero, both terms in brackets need to equate to zero leading to two ordinary differential equations:

$$\frac{\partial A}{\partial t} - B\alpha\mu = 0 \tag{14.14}$$

$$-\frac{\partial B}{\partial t} + \alpha B + \frac{\sigma^2}{2} B^2 - 1 = 0 \tag{14.15}$$

Also since the price of a zero-coupon bond at maturity $P_{T,T}$ is \$1 then $A_{T,T} = 0$ and $B_{T,T} = 0$.

These equations are easily solved using standard integration techniques. Details are given in Appendix A.2.

Solving Equation (14.15)

$$B_{0,T} = \frac{2(e^{\gamma T} - 1)}{(\alpha + \gamma)(e^{\gamma T} - 1) + 2\gamma}$$

where $\gamma = \sqrt{\alpha^2 + 2\sigma^2}$. To obtain $B_{t,T}$, time to maturity T in the above expression is replaced by $T - t$. Substitute for $B_{t,T}$ in Equation (14.14) and integrate to get

$$A_{0.T} = \frac{2\alpha\mu}{\sigma^2} \log\left(\frac{2\gamma e^{(\alpha+\gamma)(T/2)}}{(\gamma + \alpha)(e^{\gamma T} - 1) + 2\gamma}\right)$$

Then the zero-coupon bond price at time $t = 0$ is $P_{0,T} = e^{A_{0,T} - B_{0,T} r_0}$ with $A_{0.T}$ and $B_{0.T}$ as given above.

Having determined the zero-coupon bond price, the yield curve is then easily generated. However, the CIR model has the same limitations as the Vasicek model in that only the yield curve shapes in Figure 14.1 are obtained. Therefore, many observed yield curves cannot be replicated by the model. And as before, shocks to the curve will lead to almost parallel movements due to the affine term structure.

Although, the dependency of volatility on the level of the interest rates improves the fit to one-month Treasury bills compared to using Vasicek, Chan et al. (1992) find the CIR model also performs poorly relatively to other short-rate models. The fit to the term structure of forward-rate volatilities as indicated by caplet market prices is also inadequate.

14.5 Estimating parameters

Although both the Vasicek and Cox–Ingersoll–Ross models generally provide a poor fit to the yield curve, simple one-factor models may be useful in other applications such as valuing interest rate sensitive cashflows or calculating regulatory capital. Reasonable parameter values must be chosen for such purposes. They may be provided by regulatory criteria or estimated by the financial analyst from suitable data.

The first step is to decide whether the data leads to estimates for the parameters of a real-world short-rate process (14.1) or a risk-neutral short-rate process (14.5). Once the market price of risk is estimated , the drift can be adjusted to translate between the real-world short-rate model and the risk-neutral short-rate model.

Using interest rate data for a short period of time such as overnight interbank rates or yields on the shortest maturity Treasury bills will give parameter estimates for a real-world short-rate process. The difficulty with using a historical time series to determine the model parameters is that there is no guarantee that the parameters of the short-rate model are time-invariant. Monetary policy changes, shifts in the credibility of government economic policy or geopolitical events will lead to a new set of parameters.

Using zero-coupon bond prices or a yield curve derived from prices leads to parameter estimates for the risk-neutral short-rate process since all the analysis in the preceding sections used the parameters of a risk-neutral interest rate process as a starting point to derive the zero-coupon bond price. The advantage of this approach is that multiple data points corresponding to different maturities are available at every time step.

14.5.1 Real-world estimates: Euler–Maruyama discretization

The simplest approach is to use the Euler–Maruyama discretization.

Given

$$dr = a(t, r_t)dt + \sigma(t, r_t)dW$$

then

$$
\begin{aligned}
r_t - r_{t-1} &= \int_{t-1}^{t} a(s, r_s)\, dt + \int_{t-1}^{t} \sigma(s, r_s) dW \\
\Longrightarrow \; r_t &= r_{t-1} + \int_{t-1}^{t} a(s, r_s) dt + \int_{t-1}^{t} \sigma(s, r_s) dW
\end{aligned}
$$

In the Euler–Maruyama discretization, the integrals are approximated by a product of the lower limit value and the discrete change

$$
r_t = r_{t-1} + a(t-1, r_{t-1})\Delta t + \sigma(t-1, r_{t-1})\Delta W
$$

We will use asterisks to highlight that we are estimating the parameters of a real-world short-rate model. Remember the drift is adjusted when changing probability measure while the diffusion coefficient is unaltered (Section 14.1). In the Vasicek model

$$
\begin{aligned}
dr &= \alpha^*(\mu^* - r_t)dt + \sigma dW \\
\Longrightarrow \; r_t &= r_{t-1} + \alpha^*(\mu^* - r_{t-1})\Delta t + \sigma \Delta W \\
\Longrightarrow \; r_t &= \alpha^* \mu^* \Delta t + (1 - \alpha^* \Delta t)\, r_{t-1} + \sigma \Delta W
\end{aligned}
$$

Then use ordinary least squares to estimate parameters in this basic model

$$
r_t = \beta_0 + \beta_1 r_{t-1} + \epsilon_t, \quad \epsilon_t \sim N(0, s^2)
$$

And comparing the discretized model with the regression estimates

$$
\hat{\beta}_0 = \alpha^* \mu^* \Delta t, \quad \hat{\beta}_1 = 1 - \alpha^* \Delta t, \quad \hat{s} = \sigma \sqrt{\Delta t}
$$

In the CIR model

$$
\begin{aligned}
dr &= \alpha^*(\mu^* - r_t)dt + \sigma\sqrt{r_t} dW \\
\Longrightarrow \; r_t &= r_{t-1} + \alpha^*(\mu^* - r_{t-1})\Delta t + \sigma\sqrt{r_{t-1}} \Delta W \\
\Longrightarrow \; \frac{r_t}{\sqrt{r_{t-1}}} &= \alpha^* \mu^* \Delta t \left(\frac{1}{\sqrt{r_{t-1}}} \right) + (1 - \alpha^* \Delta t)\sqrt{r_{t-1}} + \sigma \Delta W
\end{aligned}
$$

Estimate parameters in this model using ordinary least squares

$$
\frac{r_t}{\sqrt{r_{t-1}}} = \beta_1 \frac{1}{\sqrt{r_{t-1}}} + \beta_2 \sqrt{r_{t-1}} + \epsilon_t, \quad \epsilon_t \sim N(0, s^2)
$$

And comparing the discretized model with the regression estimates

$$
\hat{\beta}_1 = \alpha^* \mu^* \Delta t, \quad \hat{\beta}_2 = 1 - \alpha^* \Delta t, \quad \hat{s} = \sigma \sqrt{\Delta t}
$$

14.5.2 Real-world estimates: Maximum Likelihood

Instead of approximating the stochastic differential equation (SDE) with the discrete representation above we can use the solutions to the Vasicek and CIR models to exactly represent the process.

From (14.8) and (14.13), we have the following generalized framework for the solution

$$\begin{aligned} r_t &= \mu^* + (r_{t-1} - \mu^*)e^{-\alpha^*\Delta t} + \eta_t \\ &= \underbrace{\mu^*(1 - e^{-\alpha^*\Delta t})}_{a} + \underbrace{e^{-\alpha^*\Delta t}}_{b} r_{t-1} + \eta_t \end{aligned}$$

where $\eta_t \sim N\left(0, \frac{\sigma^2}{2\alpha^*}(1 - e^{-2\alpha^*\Delta t})\right)$ in the Vasicek model and η_t has a non-central ξ^2 distribution in the CIR model.[2] The parameters of these models can then be estimated using econometric techniques where the error distribution is first specified such as Maximum-Likelihood or Generalized Method of Moments.[3]

The model parameters are retrieved from the estimates

$$\begin{aligned} b = e^{-\alpha^*\Delta t} &\implies \alpha^* = -\frac{\log b}{\Delta t} \\ a = \mu^*(1 - e^{-\alpha^*\Delta t}) &\implies \mu^* = \frac{a}{(1 - e^{-\alpha^*\Delta t})} = \frac{a}{1 - b} \end{aligned}$$

For the Vasicek model, specify the error distribution as $\eta_t \sim N(0, c^2)$ then

$$c^2 = \frac{\sigma^2}{2\alpha^*}(1 - e^{-2\alpha^*\Delta t}) \implies \sigma = c\left(\frac{1 - e^{-2\alpha^*\Delta t}}{2\alpha^*}\right)^{-\frac{1}{2}} = c\sqrt{\frac{2\log b}{\Delta t(b^2 - 1)}}$$

The parameters of the more complicated non-central ξ^2 error distribution in the CIR model can also be retrieved.

> **Exercise 14.1.** *Estimate the parameters of the real-world Vasicek model using daily data from January 4rd 2023 to December 31st 2024 for the one-month T-bill rate.*

Proof.

- Download the data from the Federal Reserve web site (http://www.federalreserve.gov/releases/h15/data.htm).

- Remove all observations with blanks and ND.
- Run the following code in R.

Estimate of real-world parameters of Vasicek model

```
library(readxl)
library(stats4)
# Prepare data
data <- read_excel("YOUR LOCATION")
rt =data$rt
Delta = 1/252
N = length(rt)
y = rt[2:N]
x = rt[1:(N-1)] # lag variable
# Define the log-likelihood function
ll <- function(a, b, c) {
        y_pred <- a + b * x
        sum(-dnorm(y, mean = y_pred, sd = c,
log = TRUE))
}
# Provide starting values and run Maximum Likelihood
Estimation
fit <- mle(ll, start = list(a = 1, b = 1, c = 1))
summary(fit)
# Calculate Vasicek parameters
alpha = -log(fit@coef[2])/Delta
mu =  fit@coef[1]/ (1- fit@coef[2])
sigma = fit@coef[3]*(2*alpha/(1-fit@coef[2]^2))^.5
# Print result
cat(''alpha ='', alpha, '', mu ='', mu, '',
sigma ='', sigma, ''\n'')
```

The Maximum-Likelihood Estimates are $\alpha^* = 5.082057, \mu^* = 5.228078, \sigma = 1.376687$. □

Note that for a linear model with normally distributed error terms, as is the case for the Vasicek model, Ordinary Least Squares would give identical regression coefficients (a and b) and very similar c estimate. So for the Vasicek model with a large number of observations, Ordinary Least Squares estimation of a, b, c would suffice. Maximum-Likelihood or Generalized Method of Moments is needed to retrieve the parameters of the error distribution in the CIR model.

14.5.3 Risk neutral estimates

Given a dataset of n zero-coupon prices at different maturities T_i, short-rate model estimates can be found by adjusting the model parameters so as to minimize the squared pricing error between the theoretical prices

$$P_{0,T_i} = e^{A_{0,T_i} - B_{0,T_i} r_0}$$

and the prices given in the data, that is, minimize

$$J(\alpha, \mu, \sigma) = \sum_{1}^{n} \left(P_{0,T_i}^{\text{Theoretical}} - P_{0,T_i}^{\text{Observed}} \right)^2$$

Exercise 14.2. *Estimate the parameters of the risk-neutral Vasicek model using the data for US Treasury Yields at Oct 6th 2025 below.*

Maturity(T)	*Yield*
0.083	*0.0422*
0.25	*0.0402*
0.5	*0.0381*
1	*0.0366*
2	*0.036*
3	*0.0363*
5	*0.0375*
7	*0.0395*
10	*0.0418*
20	*0.0473*
30	*0.0476*

Source: S & P Capital IQ.

Proof. Paste the data into a spreadsheet and rename the columns "T" and "zero". Estimate the parameters by applying a non linear least squares model to the data using the R code below. The estimation model is non linear since the Vasicek model parameters relate to $P_{0,T}$ through the exponential of the functions $A_{0,T}$ and $B_{0,T}$ given in Section 14.2.1.

```
# Vasicek calibration to observed zero curve
library(readxl)
library(numDeriv)

# —— 1. Load data ——
df <- read_excel("YOUR LOCATION")
df <- as.data.frame(df)

# Market discount factors
df$P_market <- exp(- df$zero * df$T)

# —— 2. Vasicek helper functions ——
# B(T)
B_vasicek <- function(alpha, T) {
        (1 - exp(-alpha * T)) / alpha
}

# A(T)
A_vasicek <- function(alpha, mu, sigma, T) {
        B <- B_vasicek(alpha, T)
        term1 <- (mu - (sigma^2) / (2 * alpha^2))
 * (B - T)
        term2 <- (sigma^2 * B^2) / (4 * alpha)
        term1 - term2
}

# Model P(0,T) and R(0,T)
P_model_given_params <- function(par, T) {
        # par: vector with (alpha, mu , sigma, r0)
        alpha <- par[1]; mu <- par[2]
        sigma <- par[3]; r0 <- par[4]
        B <- B_vasicek(alpha, T)
        A <- A_vasicek(alpha, mu , sigma, T)
        P <- exp(A - B * r0)
        return(P)
}

z_model_given_params <- function(par, T) {
        P <- P_model_given_params(par, T)
        z <- -log(P) / T
        return(z)
}
```

```
# —— 3. Objective function for calibration ——
# Minimize sum of squared errors (market vs model).
obj_fun <- function(params, T, z_market) {
        alpha <- params[1] ; mu <- params[2]
        sigma <- params[3] ; r0 <- params[4]
        par <- c(alpha, mu , sigma, r0)
        z_model <- z_model_given_params(par, T)
        res <- z_model - z_market
        sum(res^2)
}

# —— 4. Starting values ——
r0_start <- df$zero[which.min(abs(df$T -
min(df$T)))]
alpha_start <- 0.5 ; mu_start <- 0.03
sigma_start <- 0.01 ; r0_start <- r0_start

params_start  <- c( alpha_start, mu_start,
 sigma_start, r0_start)

# —— 5. Run optimization ——
optim_res <- optim(par = params_start,
        fn = obj_fun,
        T = df$T,
        z_market = df$zero,
        method = "BFGS",
        control = list(maxit = 5000),
        hessian = TRUE)

# —— 6. Extract parameters ——
par_hat <- optim_res$par
alpha_hat <- par_hat[1]
mu_hat <- par_hat[2]
sigma_hat <- par_hat[3]
r0_hat <- par_hat[4]

estimates <- c(alpha = alpha_hat, mu = mu_hat,
sigma = sigma_hat, r0 = r0_hat)
print("Estimated (risk-neutral) Vasicek
parameters:")
print(estimates)
```

The non linear least squares estimates are

$$\alpha = 0.01001653, \mu = 0.1157073, \sigma = 0.00001659116, r_0 = 0.03764623$$

This model provides a poor fit to the US Treasury yield data due to the limited yield curve shapes possible in the Vasicek model.

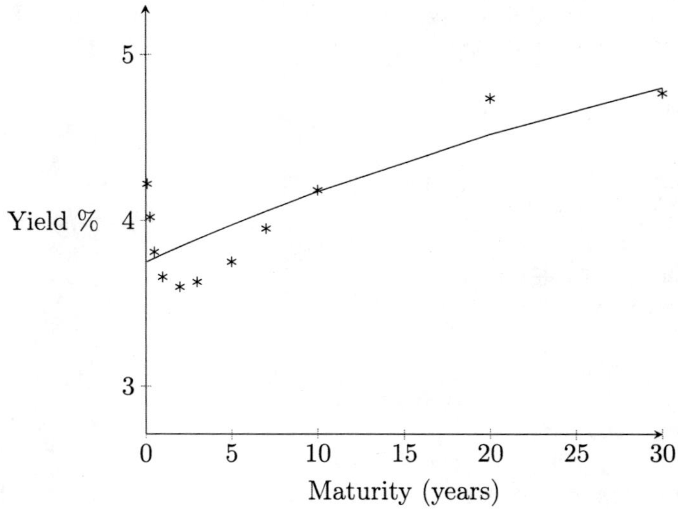

□

The data in the previous question are Constant Maturity Treasury (CMT) yields rather than the yield on STRIPS. They are often used for modeling. The Treasury takes yields on all actively traded Treasury securities each with different maturities and some with coupons in order to fit a yield curve. From that fitted curve, they extract the yield corresponding to specific maturities ("Constant Maturities"). CMT yields are then par yields not zero-coupon yields.

If available market data is limited, σ may be estimated from historical data using the techniques in the previous subsections while using the market data to estimate the rest of the model parameters. This is possible because the σ parameter is unaltered moving from the short-rate SDE under the real-world probability measure to the short-rate SDE under the risk-neutral measure. Or an estimate of σ from historical data could be used as an initial value in the optimization procedure above.

14.6 Hull-White model

The Hull-White model is a 'no-arbitrage model' and is designed to be consistent with the term structure of interest rates. In this model, the short rate follows

$$dr = \alpha(\mu(t) - r_t)dt + \sigma d\tilde{W}$$

This is sometimes called an “Extended Vasicek model" as it modifies the Vasicek model by allowing μ to be time-dependent.[4] By choosing appropriate values of $\mu(t)$, the desired yield curve can be reproduced. This process is described at the end of this section.

The PDE for zero-coupon bond prices (Equation 14.7) becomes:

$$\frac{\partial P}{\partial t} + \frac{\partial P}{\partial r}(\alpha\mu - \alpha r_t) + \frac{1}{2}\frac{\partial^2 P}{\partial r^2}(\sigma^2) - r_t P = 0$$

The drift and diffusion terms here are of the correct form to produce an affine term structure. Let $P_{t,T} = e^{A_{t,T} - B_{t,T} r_t}$ then

$$\left(\frac{\partial A}{\partial t} - \frac{\partial B}{\partial t} r_t\right)P - BP(\alpha\mu - \alpha r_t) + \frac{1}{2}B^2 P(\sigma^2) - r_t P = 0$$

$$\Longrightarrow \left(\frac{\partial A}{\partial t} - B\alpha\mu + \frac{\sigma^2}{2}B^2\right) + \left(-\frac{\partial B}{\partial t} + \alpha B - 1\right)r_t = 0$$

Leading to two ordinary differential equations:

$$\frac{\partial A}{\partial t} - B\alpha\mu + \frac{\sigma^2}{2}B^2 = 0$$

$$-\frac{\partial B}{\partial t} + \alpha B - 1 = 0$$

The solution for $B_{0,T}$ is easily found from the second equation using the technique in Section 1.16 and the terminal condition $B_{T,T} = 0$.

$$-\frac{\partial B}{\partial t} + \alpha B - 1 = 0$$

$$\Longrightarrow e^{\int -\alpha dt} B = -\int e^{\int -\alpha dt} dt$$

$$\Longrightarrow e^{-\alpha t} B_{t,T} = \frac{e^{-\alpha t}}{\alpha} + C$$

$$\Longrightarrow e^{-\alpha t} B_{t,T} = \frac{e^{-\alpha t}}{\alpha} - \frac{e^{-\alpha T}}{\alpha}$$

$$\Longrightarrow B_{t,T} = \frac{1 - e^{-\alpha(T-t)}}{\alpha}$$

Then

$$B_{0,T} = \frac{1 - e^{-\alpha T}}{\alpha}$$

and

$$A_{0,T} = \int_0^T (\alpha\mu(t)B_{t,T} - \frac{\sigma^2}{2}B_{t,T}^2)\,dt$$

from the first equation. The zero-coupon bond price at time $t = 0$ is $P_{0,T} = e^{A_{0,T} - B_{0,T} r_0}$ with these parameter values.

To calibrate the Hull-White model, we need to develop a relationship between $\mu(t)$ and market data related to the current yield curve.

In Appendix A.3, the following relationship is derived

$$\mu(t) = f_{0,t} + \alpha^{-1}\frac{\partial}{\partial t}f_{0,t} + \frac{\sigma^2}{2\alpha^2}(1 - e^{-2\alpha t}) \tag{14.16}$$

where $f_{0,t}$ is the instantaneous forward rate.

The procedure is to first obtain a set of zero-coupon bond prices $\{P_{0,t_1}, P_{0,t_2}, \ldots P_{0,T}\}$ consistent with the observed yield curve. Then construct a mathematical function $P_{0,t}$ that best fits these data. The next step is to calculate the instantaneous forward rate using Equation (12.5)

$$f_{0,t} = -\frac{\partial \log P_{0,t}}{\partial t}$$

Finally, $\mu(t)$ is determined by substituting $f_{0,t}$ in the relationship above. The Hull-White model with this choice of $\mu(t)$ will reproduce the observed yield curve.

However, forward volatility term structures implied by the model generally do not match shapes implied by market data.

Exercise 14.3. *The following function is found to best fit the market price of zero-coupon bonds maturing at time t:*

$$P_{0,t} = e^{-0.05t + 0.2(1 - e^{-0.15t})}$$

(i) Find the instantaneous forward rate $f_{0,t}$ based on this model for the zero-coupon bond price.

(ii) Given $\alpha = 0.1$ and $\sigma = 0.04$, calibrate the function $\mu(t)$ in the Hull-White model to be consistent with the given zero-coupon bond price model.

Proof.

(i)

$$f_{0,t} = -\frac{\partial \log P_{0,t}}{\partial t}$$

$$= 0.05 - 0.03e^{-0.15t}$$

(ii)

$$\mu(t) = f_{0,t} + \alpha^{-1}\frac{\partial}{\partial t}f_{0,t} + \frac{\sigma^2}{2\alpha^2}(1 - e^{-2\alpha t})$$

$$= 0.05 - 0.03e^{-0.15t} + \frac{1}{0.1}0.0045e^{-0.15t}$$

$$+ \frac{0.04^2}{2 \times 0.1^2}(1 - e^{-2\times 0.1t})$$

$$= 0.13 + 0.015e^{-0.15t} - 0.045e^{-0.2t}$$

□

14.7 Duration and convexity

In Section 12.3, the modified duration D^* and convexity C of a bond were defined in terms of the sensitivity of the bond price P to changes in the yield y:

$$D^* := -\frac{1}{P}f'(y) \text{ and } C := \frac{1}{P}f''(y)$$

As the short rate is driving bond prices in the models of this chapter, the relevant variable of interest is the sensitivity of the bond price to the short-rate r. The duration and convexity are therefore defined here as

$$D := -\frac{1}{P}\frac{\partial P}{\partial r} \text{ and } C := \frac{1}{P}\frac{\partial^2 P}{\partial r^2}$$

When the model has an affine term structure (Equation 14.12) these simplify to

$$D = -\frac{1}{P_{t,T}}\left(B_{t,T}P_{t,T}\right) = B_{t,T}$$

and

$$C = \frac{1}{P_{t,T}}\left(B^2_{t,T}P_{t,T}\right) = B^2_{t,T}$$

14.8 Option on a zero-coupon bond

The inconsistency of Black's model was highlighted in the previous chapter. Log-normality is assumed for the forward interest rate and forward swap rate in pricing caps and swaptions respectively. But it is not possible that both are log-normal. The advantage of using short-rate models is that it is possible to derive pricing formulas that are consistent across all interest-rate derivatives. The next sections show how this is done. The derivations will follow the structure of Section 13.5 while incorporating the short-rate model.

Our market consists of zero-coupon bonds for every maturity up to a given time horizon. A risk-neutral measure $\mathbb{Q}$ exists such that $\frac{P_{t,T}}{B_t}$ is a martingale, where $P_{t,T}$ is the price of a zero-coupon bond maturing at T and B_t is a riskless investment of cash. Assume the short rate is given by the Vasicek model.

Consider an option on a zero-coupon bond with expiry at time T. This bond matures at a later time T^*.

For any option ZBO_t based on the zero-coupon bond, the quotient $\frac{\text{ZBO}_t}{B_t}$ is also a martingale. Therefore

$$\frac{\text{ZBO}_t}{B_t} = \mathbb{E}_t^{\mathbb{Q}} \frac{\text{ZBO}_T}{B_T}$$

The option price ZBO_t at time t is

$$\text{ZBO}_t = \mathbb{E}_t^{\mathbb{Q}} B_t \frac{\text{ZBO}_T}{B_T} = \mathbb{E}_t^{\mathbb{Q}} e^{-\int_t^T r_s ds} \text{ZBO}_T \tag{14.17}$$

The difficulty with evaluating this expectation is that unlike in the treatment of option pricing in Chapter 10, interest rates now are stochastic. The option payoff C_T involves the price P_{T,T^*} of the zero-coupon bond at T which is a function of the short rate in the Vasicek model. The payoff will therefore be correlated with the first term $e^{-\int_t^T r_s ds}$, which also involves the short rate.

From the properties of random variables (Equation 2.3), the expectation of the product of two random variables X and Y is

$$\mathbb{E}(XY) = Cov(X, Y) + \mathbb{E}(X)\mathbb{E}(Y)$$

The expectation of the product in the zero-coupon bond option pricing formula is complicated by having to compute this covariance term. However, the formula can be simplified by a change of probability measure.

Change to the T-forward probability measure (see Section 13.5.1) from the risk-neutral measure $\mathbb{Q}$.

From Equation (13.4), the zero-coupon bond option pricing formula can be rewritten as:

$$\text{ZBO}_t = P_{t,T} \mathbb{E}_t^T \text{ZBO}_T$$

The expectation term has become more straightforward, and the option price is easily evaluated. For example, consider a call option on a zero-coupon bond price with strike price K.

$$\mathrm{ZBC}_t = P_{t,T}\mathbb{E}_t^T[P_{T,T^*} - K]^+$$

If we assume that the T^*-bond price P_{T,T^*} has a lognormal conditional distribution with volatility σ_F^2 then from Theorem 13.1

$$\mathrm{ZBC}_t = P_{t,T}(\mathbb{E}_t^T P_{T,T^*}\Phi(d_1) - K\Phi(d_2)) \tag{14.18}$$

where

$$d_1 = \frac{\log(\mathbb{E}_t^T P_{T,T^*}/K) + \frac{\sigma_F^2}{2}}{\sigma_F}, d_2 = d_1 - \sigma_F$$

The fraction $\frac{P_{t,T^*}}{P_{t,T}}$ is a martingale under the measure with $P_{t,T}$ as numéraire.

$$\frac{P_{t,T^*}}{P_{t,T}} = \mathbb{E}_t^T \frac{P_{T,T^*}}{P_{T,T}} = \mathbb{E}_t^T P_{T,T^*}$$

Then Equation (14.18) becomes

$$\mathrm{ZBC}_t = P_{t,T^*}\Phi(d_1) - P_{t,T}K\Phi(d_2) \tag{14.19}$$

where

$$d_1 = \frac{\log(P_{t,T^*}/(P_{t,T}K)) + \frac{\sigma_F^2}{2}}{\sigma_F}, d_2 = d_1 - \sigma_F$$

The results above rely on the lognormality of the T^*-bond at time T. In the Vasicek model

$$P_{T,T^*} = e^{A_{T,T^*} - B_{T,T^*} r_T}$$

is lognormally distributed under $\mathbb{Q}$ since r_T is normal. It is shown in Appendix A.4 that changing the numéraire to $\mathbb{Q}_T$ only changes the drift term in r_T therefore P_{T,T^*} is still lognormally distributed under $\mathbb{Q}_T$ with the same volatility.

We can determine this volatility when the short rate is given by the Vasicek model. The variance of the short-rate r_T in Section 14.2 was

$$\frac{\sigma^2}{2\alpha}(1 - e^{-2\alpha(T-t)})$$

It follows that the standard deviation σ_F of $\log P_{T,T^*} = A_{T,T^*} - B_{T,T^*} r_T$ is

$$\sigma_F = B_{T,T^*}\sigma\sqrt{\frac{1 - e^{-2\alpha(T-t)}}{2\alpha}}$$

It can be shown that the expression for σ_F in the Hull-White model is the same.

Zero-coupon bond puts are priced analogously where

$$\text{ZBP}_t = P_{t,T} K \Phi(-d_2) - P_{t,T^*}\Phi(-d_1)$$

14.9 Option on a coupon bond

Consider a European call option on a coupon bond with strike price K and expiry at time T. After the option maturity, the bond provides a series of payments $\{c_1, ..., c_N\}$ at times $\{T_1, ..., T_N\}$. The payoff from this option is then

$$[\text{Bond price} - K]^+$$

where the bond price is the discounted value of the payments:

$$\text{Bond price} = \sum_{i=1}^{N} c_i P_{T,T_i}$$

The logic of the previous section where we appealed to Theorem 13.1 in order to proceed will not work here with multiple discount terms in the payoff. If we could separate out the payoff into the sum of payoffs with only one discount term then we could use the zero-coupon bond call option pricing formula multiple times. This is achieved by the *Jamshidian*-decomposition (Jamshidian, 1989).

Remembering that each P_{T,T_i} is dependent on the short rate at time T, the approach is to find the short rate r^* for which

$$\sum_{i=1}^{N} c_i P_{T,T_i,r^*} = K$$

The payoff from the option then becomes

$$\left[\sum_{i=1}^{N} c_i (P_{T,T_i,r_T} - P_{T,T_i,r^*})\right]^+$$

Since P_{T,T_i} monotonically decreases with the short rate then for all T_i

$$P_{T,T_i,r_T} > P_{T,T_i,r^*} \text{ when } r_T > r^* \text{ and}$$
$$P_{T,T_i,r_T} < P_{T,T_i,r^*} \text{ when } r_T < r^*$$

This allows us to break up the payoff into the sum of payoffs with only one discount term

$$\sum_{i=1}^{N} c_i \left[P_{T,T_i,r_T} - P_{T,T_i,r^*} \right]^+$$

Each term in the sum is the payoff from a zero-coupon bond call option with strike price given by P_{T,T_i,r^*} and can be easily priced using the result from the previous section.

14.10 Caps/floors and swaptions

Given the pricing formulas above, caps/floors can also be priced by reconfiguring the option as a portfolio of zero-coupon bond options.

Given cap rate R_K, reset period δ and principal L, the price of a caplet is

$$\begin{aligned}
\text{Cap}_t &= \mathbb{E}_t^{\mathbb{Q}} e^{-\int_t^{t_{k+1}} r_s ds} (L\delta[R_k - R_K]^+) \\
&= L\mathbb{E}_t^{\mathbb{Q}} e^{-\int_t^{t_k} r_s ds} P_{t_k,t_{k+1}} (\delta[R_k - R_K]^+) \\
&= L\mathbb{E}_t^{\mathbb{Q}} e^{-\int_t^{t_k} r_s ds} P_{t_k,t_{k+1}} \left[\frac{1}{P_{t_k,t_{k+1}}} - 1 - \delta R_K \right]^+
\end{aligned}$$

The last line follows since the definition of the spot rate with discrete compounding is the bond yield $R_{t,T}$ such that $P_{t,T}(1 + R_{t,T}(T - t)) = 1$.

$$\begin{aligned}
\text{Cap}_t &= L\mathbb{E}_t^{\mathbb{Q}} e^{-\int_t^{t_k} r_s ds} \left[1 - P_{t_k,t_{k+1}} (1 + \delta R_K) \right]^+ \\
&= L(1 + \delta_k R_K) \mathbb{E}_t^{\mathbb{Q}} e^{-\int_t^{t_k} r_s ds} \left[\frac{1}{1 + \delta R_K} - P_{t_k,t_{k+1}} \right]^+
\end{aligned}$$

Therefore, the caplet price can be written as $L(1 + \delta_k R_K)$ times the price of a European put with maturity t_k and strike price $\frac{1}{1+\delta_k R_K}$ written on a zero-coupon bond with maturity t_{k+1} and unit nominal amount. The cap can then be priced as a portfolio of these European put options. By the same logic, floors are priced as a portfolio of European calls.

Also, a swaption can be reinterpreted as a coupon bond put option and priced using the approach in Section 14.9. From Equation (12.9), the present value of the floating leg of a swap with payment times $\{T_1, ..., T_N\}$ added to a final payment of the principal L at T_N is

$$L(1 - P_{0,T_N}) + LP_{0,T_N} = L$$

Then the floating leg of the swap becomes a floating-rate bond. Likewise, the fixed leg becomes a fixed-rate bond when a final payment of the principal is added. Therefore the swap is a contract to exchange a fixed-rate bond for an amount given by the principal. And a swaption is the option to do this at some point in time.

Consider a payer swaption with reset period δ where the holder has the right to pay a rate s_K and receive a floating rate starting in T years with payment times $\{T_1, ..., T_N\}$ and notional principal L. This swaption is then a put option with maturity T and strike price L on a bond with maturity T_N and coupons $s_K\delta$. Receiver swaptions are priced as coupon bond call options.

14.11 Multi-factor models

The Vasicek, CIR and Hull-White models discussed in this chapter are one-factor models, that is, they have only one source of uncertainty $\tilde{W}_t$. The strength of these models is that they are relatively simple to implement and interpret.

Compared to more sophisticated approaches, they are easy to work with mathematically. One-factor models usually have closed-form solutions allowing easy derivation of bond prices and yields as well as bond option prices.

Having fewer parameters to estimate, calibration is less computationally intensive but the fit to real-world data can be poor as only a limited range of yield curve shapes can be replicated by these models. Some models such as Hull-White can be calibrated to fit the yield curve exactly through the introduction of a set of time-dependent parameters for each possible maturity. Matching the shape of market volatility curves requires a further extension with extra parameters.

One-factor models will not capture changes to the yield curve such as non-parallel shifts or changes in curvature. They are then of limited use in the risk management of portfolios sensitive to changes across the yield curve or in pricing products dependent on correlations across the yield curve. Also, one-factor models generally hold volatility constant ignoring the stochastic nature of volatility meaning that sustained periods of low or high volatility observed in the market cannot be modeled.

However, one-factor models are often effective at capturing short-term interest rate movements, particularly for liquid and highly traded instruments. Although not used for pricing complex derivatives, they can provide reasonable predictions for short-term interest rate changes and are used in applications such as valuing interest rate sensitive cashflows, calculating regulatory capital and the valuation of short-term straightforward derivative contracts.

The introduction of more sources of uncertainty relieves some of the limitations in one-factor models. For example, the two-factor Vasicek model is given below:

$$r_t = x_t + y_t$$

$$dx = \alpha_x(\mu_x - x_t)dt + \sigma_x d\tilde{W}_1$$

$$dy = \alpha_y(\mu_y - y_t)dt + \sigma_y d\tilde{W}_2$$

where $\alpha_x, \mu_x, \sigma_x, \alpha_y, \mu_y, \sigma_y$ are constants, while the two sources of uncertainty $\tilde{W}_{1,t}$ and $\tilde{W}_{2,t}$ are potentially correlated standard Brownian motions.

In this model, greater yield curvature is possible than in the single factor Vasicek model and yields at different maturities lose the restriction of being perfectly correlated allowing closer calibration to products sensitive to correlations along the yield curve such as swaptions. Despite these improvements, negative interest rates are still possible and further parameters need to be added to match the yield curve as was done when moving from the Vasicek to the Hull-White model.

Why stop at two factors ? Adding more factors improves the fit to market data at the risk of losing ease of use in applications. A compromise must therefore be reached. Empirical analysis of yield curve behavior in different countries indicates that two or three factors are sufficient to model variation in the yield curve (Jamshidian and Zhu, 1996; Litterman et al., 1991; Rebonato, 1996). These factors are typically interpreted as representing the level, slope and curvature of the yield curve. Consideration of such models is left to more advanced texts. Brigo et al. (2001) is excellent on this material.

14.12 Summary

Once the short rate is specified, the full yield curve can be determined and interest-rate derivatives can be priced consistently. This chapter examined some simple one-factor models of the short-rate. They have a number of shortcomings. The Vasicek and CIR models cannot replicate the variety of yield curve shapes observed in the market. In contrast, the Hull-White model is designed to allow calibration to actual yield curves. In all one-factor models, yield curve changes are limited to parallel shifts; volatility is held constant over time and forward volatility surfaces provide a poor fit to market data. They therefore have limited application in pricing interest-rate derivatives and risk management.

Brigo et al. (2001) highlight a number of questions which should be considered when choosing an interest rate model including

- Do the dynamics imply positive rates ?
- What distribution do the dynamics imply for the short rate ?

- Are bond prices (and therefore spot rates, forward rates and swap rates) explicitly computable from the dynamics?
- Are bond-option (and cap, floor, swaption) prices explicitly computable from the dynamics?
- Is the model mean-reverting ?
- Do the volatility structures implied by the model look reasonable ?
- Do the chosen dynamics allow for calibration to market data?

Adding factors or additional parameters may result in a model with the desired characteristics but at the risk of losing mathematical tractability. Some more advanced models are considered in the next chapter.

Key things to remember

- The Zero-Coupon Bond price $P_{t,T} = E_t^{\mathbb{Q}} e^{-\int_t^T r_s ds}$: Section 14.1.
- Derivation of the Zero-Coupon Bond price in the Vasicek model: Section 14.2.
- The limitations of one-factor short-rate models: Section 14.11.

Further reading

Hirsa and Neftci (2013); McDonald (2013); Veronesi (2010) and Mazzoni (2018) provide accessible introductions to derivative pricing and the practical application of short-rate models within the broader derivatives market context. Brigo et al. (2001) remains a standard reference for practitioners and researchers, presenting both one-factor and multi-factor extensions. The mathematical framework underlying these models can be found in Shreve (2004) and Klebaner (2012). For a rigorous treatment of the theoretical structure, implementation, and calibration of affine and short-rate models, Bolder (2001), Dai and Singleton (2000) and Filipovic (2009) offer comprehensive discussions. Empirical insights into the behavior and suitability of alternative short-rate specifications are provided by Chan et al. (1992). Finally, Ambagaspitiya and Ford (2023) offers a modern perspective on the calibration of interest rate models, emphasizing current market practices and numerical methods.

Notes

1. For mathematical proof, see Klebaner (2012) Exercise 6.18.
2. Details of this distribution are given in Brigo et al. (2001) p.65.
3. Chan et al. (1992) also argue that GMM estimators are consistent where errors are conditionally heteroskedastic which may be an issue here.
4. It is possible also to have α a function of time.

Questions

1 The short rate is being modeled by the Vasicek model with parameters $\mu = 0.05, \alpha = 0.5, \sigma = 0.10$. The current short rate is 4%.

(a) Find the probability that the short rate will be negative in 1 year's time.

(b) Find the probability that the short rate will be negative in 1 year's time if the short rate is being modeled by the CIR model with the same parameters.

Solution

(a) In the Vasicek model

$$r_t \sim N\left(\mu + (r_0 - \mu)e^{-\alpha t}, \frac{\sigma^2}{2\alpha}(1 - e^{-2\alpha t})\right)$$

With the parameters given

$$r_1 \sim N\left(0.05 + (0.04 - 0.05)e^{-0.5}, \frac{0.1^2}{2 \times 0.5}(1 - e^{-2\times 0.5})\right)$$

$$= N(0.0439, 0.0063)$$

Therefore,

$$P(r_1 < 0 = 0.2903)$$

(b) Here the condition $\sigma^2 \leq 2\alpha\mu$ is satisfied so negative short rates are not possible.

2 Assume the short rate of interest, r_t, follows the Vasicek model:

$$dr = \alpha(\mu - r_t)dt + \sigma d\widetilde{W}_t$$

where $\alpha = 0.6, \mu = 5\%, \sigma = 15\%, r_0 = 3\%$ p.a. and $\widetilde{Wt}$ is a standard Brownian motion under the risk-neutral probability measure.

(i) Calculate the price of a zero-coupon bond with par value \$100 maturing after 1 year.

(ii) Calculate the price of a zero-coupon bond with par value \$100 maturing after 2 years.

(iii) Find the spot rates implied by the short-rate model for year 1 and year 2.

(iv) Find the forward rate from year 1 to year 2 implied by the model.

Solution

(i)

$$B_{0,T} = \frac{(1-e^{-\alpha T})}{\alpha} \implies B_{0,1} = \frac{(1-e^{-0.6\times 1})}{0.6} = 0.7520$$

$$A_{0,T} = -(T - B_T)\left(\mu - \frac{\sigma^2}{2\alpha^2}\right) - \frac{\sigma^2 B_T^2}{4\alpha}$$

$$\implies A_{0,1} = -(1-0.7520)\left(0.05 - \frac{0.15^2}{2\times 0.6^2}\right) - \frac{0.15^2 0.7520^2}{4\times 0.6}$$

$$= -0.0100$$

$$P_{0,T} = e^{A_T - B_T r_0} \implies \text{Price} = 100e^{-0.0100-0.7520\times 0.03} = 96.80$$

(ii) $\text{Price} = 100e^{-0.0284-1.165\times 0.03} = 93.86$

(iii) $R_{0,T} = \dfrac{-A_{0,T} + B_{0,T}\, r_0}{T} \implies R_{0,1} = 0.0100 + 0.7520\times 0.03 =$ 3.25%

$$R_{0,2} = \frac{0.0284 + 1.165\times 0.03}{2} = 3.17\%$$

(iv) $e^{R_{0,1}} e^{F_{0,1,2}\times 1} = e^{R_{0,2}\times 2} \implies F_{0,1,2} = R_{0,2}\times 2 - R_{0,1}$

$$= 0.0317\times 2 - 0.0325 = 3.08\%$$

3 Consider the Cox–Ingersoll–Ross (CIR) model:

$$dr = 0.1(0.03 - r_t)dt + \sqrt{0.004\, r_t}\, dW$$

where W is standard Brownian motion and $r_0 = 0.023$.

(a) Verify the technical condition that guarantees the interest rate is always positive is satisfied.

(b) Calculate the bond price $P_{0,10}$.

[QFIQF – Fall 2022 Q8]

Solution

(a) From the main text, given the CIR model in this form

$$dr = \alpha(\mu - r_t)dt + \sigma\sqrt{r_t}\, d\tilde{W}$$

negative rates are prevented if $\sigma^2 \le 2\alpha\mu$.

Using the question parameter values $\sigma^2 = 0.004$ and $2\alpha\mu = 2 \times 0.1 \times 0.03 = 0.006$ so the condition is satisfied.

(b) First calculate $\gamma = \sqrt{\alpha^2 + 2\sigma^2} = \sqrt{0.1^2 + 2 \times 0.004} = 0.1342$

$$B_{0,T} = \frac{2(e^{\gamma T} - 1)}{(\alpha + \gamma)(e^{\gamma T} - 1) + 2\gamma}$$

$$\implies B_{0,10} = \frac{2(e^{0.1342 \times 10} - 1)}{(0.1 + \gamma)(e^{0.1342 \times 10} - 1) + 2 \times 0.1342} = 6.076$$

$$A_{0,T} = \frac{2\alpha\mu}{\sigma^2} \log\left(\frac{2\gamma e^{(\alpha+\gamma)(T/2)}}{(\gamma + \alpha)(e^{\gamma T} - 1) + 2\gamma}\right)$$

$$\implies A_{0,10} = \frac{2 \times 0.1 \times 0.03}{0.004} \log$$

$$\left(\frac{2 \times 0.1342 \times e^{(0.1+0.1342)(10/2)}}{(0.1342 + 0.1)(e^{0.1342 \times 10} - 1) + 2 \times 0.1342}\right)$$

$$= -0.108$$

$$P_{0,T} = e^{A_T - B_T r_0} \implies P_{0,10} = e^{-0.108 - 6.076 \times 0.023} = 0.7805$$

4 You have been asked to consider the following interest rate model for valuing a zero-coupon bond.

The CIR model for the short-term interest rate r_t:

$$dr = \alpha(\mu - r_t)dt + \sigma\sqrt{r_t}d\tilde{W}$$

where α, μ and σ are positive constant with $\sigma^2 < 2\alpha\mu$, and $\tilde{W}t$ is a standard Brownian motion.

The price of a zero-coupon bond is given by $P_{t,T} = e^{A_{t,T} - B_{t,T} r_t}$, where $A_{t,T}$ and $B_{t,T}$ are functions of t and T

You are given: $B_{0,10} = 4.256073, \gamma = \sqrt{\alpha^2 + 2\sigma^2} = 0.212132, r_0 = 0.04, \mu = 0.041$

a. Calculate α, σ^2.

b. Calculate $A_{0,10}, P_{0,10}$.

[QFIQF – Spring 2023 Q7]

Solution

$$B_{0,T} = \frac{2(e^{\gamma T}-1)}{(\alpha+\gamma)(e^{\gamma T}-1)+2\gamma}$$

$$\Longrightarrow 4.256073 = \frac{2(e^{0.212132\times 10}-1)}{(\alpha+0.212132)(e^{0.212132\times 10}-1)+2\times 0.212132}$$

$$\Longrightarrow \alpha = 0.2$$

$$\gamma = \sqrt{\alpha^2+2\sigma^2} = \sqrt{0.2^2+2\sigma^2} = 0.212132 \Longrightarrow \sigma^2 = 0.0025$$

$$A_{0,T} = \frac{2\alpha\mu}{\sigma^2}\log\left(\frac{2\gamma e^{(\alpha+\gamma)(T/2)}}{(\gamma+\alpha)(e^{\gamma T}-1)+2\gamma}\right)$$

$$\Longrightarrow A_{0,10} = \frac{2\times 0.2\times 0.041}{0.0025}\log\left(\frac{2\times 0.212132\times e^{(0.2+0.212132)(10/2)}}{(0.1342+0.2)(e^{0.212132\times 10}-1)+2\times 0.212132}\right) = -0.231$$

$$P_{0,T} = e^{A_T - B_T r_0} \Longrightarrow P_{0,10} = e^{-0.231-4.256073\times 0.04} = 0.6697$$

5 Suppose that $\alpha = 0.3, r_0 = 4\%$ and $\mu = 10\%$ in both the Vasicek and the CIR models. Let σ in the Vasicek model be 0.02. To make the models comparable initially set $\sigma_{CIR}\sqrt{r_0} = \sigma_{Vasicek}$.

(a) Compare the prices given by the models for a zero-coupon bond that pays \$1 in year 10.

(b) Find the spot rates implied by the zero-coupon bond prices. Hence, draw the yield curves generated by both models from year one to year ten.

(c) Change $\sigma_{Vasicek}$ to 0.15 and draw the new yield curves. What do you notice ?

Solution

(a) Using the zero-coupon bond price formula $P_{0,T} = e^{A_{0,T}-B_{0,T}r_0}$
Vasicek: $P_{0,10} = e^{-0.6714-3.1674\times 0.04} = 0.4502.$
CIR: $P_{0,10} = e^{-0.6673-3.0459\times 0.04} = 0.4542$

(b) Using the yield formula $R_{0,T} = \frac{-A_{0,T}+B_{0,T}r_0}{T}$

Vasicek: $R_{0,10} = \frac{0.6714+3.1674\times 0.04}{10} = 7.98\%$

$$\text{CIR: } R_{0,10} = \frac{0.6673 + 3.0459 \times 0.04}{10} = 7.89\%$$

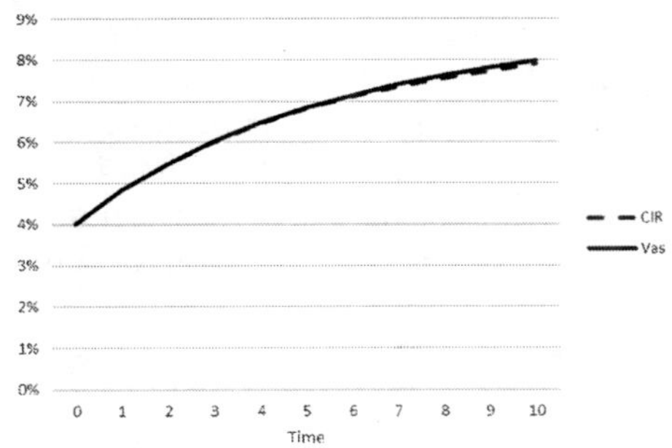

Time	0	1	2	3	4	5	6	7	8	9	10
Vasicek (%)	4.00	4.81	5.47	6.01	6.46	6.83	7.14	7.40	7.63	7.82	7.98
CIR (%)	4.00	4.81	5.47	6.00	6.44	6.80	7.10	7.35	7.56	7.74	7.89

(c)

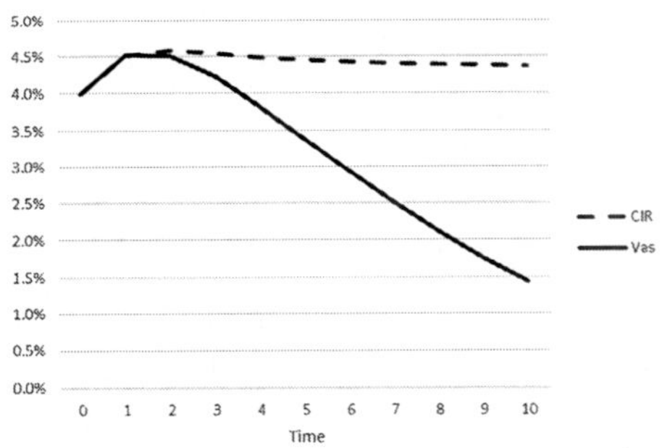

Time	0	1	2	3	4	5	6	7	8	9	10
Vasicek (%)	4.00	4.52	4.51	4.23	3.83	3.38	2.93	2.51	2.12	1.76	1.44
CIR (%)	4.00	4.51	4.58	4.54	4.49	4.45	4.42	4.40	4.39	4.38	4.37

When σ is low both models give similar prices and yields but when σ is high outputs from the models diverge.

The risk-neutral pricing formula for zero-coupon bonds takes account of the variation in the short-rate realizations. This is seen clearly in Equation (14.10).

When σ is low (and μ high r_0 high) negative short rates do not occur in many realizations of the Vasicek model. Negative short rates are not allowed at all in the CIR model. Therefore there is little difference between the models for low σ.

When σ is high, differences between the models in prices and yields are more pronounced.

6 You are an ALM (Asset and Liability Management) analyst for a public pension fund, and the fund has just completed an ALM study. During the study, the main focus was on risk quantification, with a particular emphasis on funding status. In efforts to mitigate interest rate risk, the investment division has begun using derivatives for hedging purposes.

a. Recommend whether it would be more appropriate to use a real-world model or a risk-neutral model for the following purposes.
   i. Measuring a 5% worst-case interest rate level for funding status.
   ii. Valuation of derivatives.

The Vasicek model of interest rates is

$$dr = \alpha(\mu - r_t)dt + \sigma dW$$

where α and σ are positive constants, and W is a standard Brownian motion.

In order to determine the interest rate sensitivity of the current funding ratio, you use the Vasicek model with the following parameter values.

Real World		Risk-Neutral	
Parameter	*Value*	*Parameter*	*Value*
α	0.3261	α	0.4653
r_0	3.02%	r_0	3.02%
μ	5.09%	μ	6.34%
σ	2.21%	σ	2.21%

The investment division is currently hedging with bond options. You have been requested to compare the Vasicek model with the CIR model given by:

$$dr = \alpha^*(\mu^* - r_t)dt + \sigma^*\sqrt{r_t}dW$$

where W is standard Brownian motion. The parameters of the CIR model are given below:

Parameter	*Value*
α^*	0.25
μ^*	6.59%
σ^{*2}	0.0134

(b) i. Verify that the calibrated CIR model does not allow negative rates.
   ii. Plot a chart displaying the zero-coupon bond yield for both the Vasicek and CIR models for each maturity from 1-year to 10-years with an increment of 1 year.

[QFIQF – Spring 2024 Q4]

Solution

(a) To determine whether to use a real-world model or a risk-neutral model, we must consider the goal of the analysis – whether it relates to risk measurement (real-world probabilities) or pricing (risk-neutral probabilities).

i The object here is to understand the likelihood of adverse outcomes in a real-world scenario analysis. Real-world models reflect actual probabilities making them suitable for simulating worst-case interest rate levels.

ii Pricing derivatives relies on the use of risk-neutral probabilities.

(b) i From the main text, negative rates are prevented if $\sigma^{*2} \leq 2\alpha^*\mu^*$.
Using the question parameter values $\sigma^{*2} = 0.0134$ and $2\alpha^*\mu^* = 2 \times 0.25 \times 0.0659 = 0.03295$ so the condition $\sigma^{*2} \leq 2\alpha^*\mu^*$ is satisfied.

ii Using the yield formula $R_{0,T} = \frac{-A_{0,T}+B_{0,T}\, r_0}{T}$

Time	0	1	2	3	4	5	6	7	8	9	10
Vasicek (%)	3.02	3.68	4.16	4.52	4.79	5.00	5.17	5.30	5.40	5.49	5.56
CIR (%)	3.02	3.43	3.76	4.04	4.27	4.46	4.62	4.76	4.87	4.97	5.06

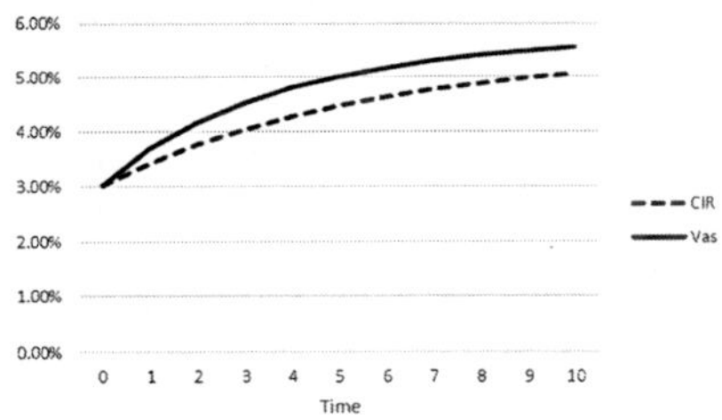

The risk-neutral parameters are used to estimate the yield in the Vasicek model.

7 You are given the following $B_{0,T}$ and $A_{0,T}$ values associated with the Vasicek model.

Risk neutral

T in years	$B_{0,T}$	$A_{0,T}$
0.5	0.4477	–0.00166
1 0	0.8053	–0.00615

Real world

T in years	$B_{0,T}$	$A_{0,T}$
0.5	0.4392	–0.00185
1 0	0.7762	–0.00785

From the market you observed, the following prices for two short-term coupon bonds. The current short-term interest rate is $r_0 = 2.10\%$.

(a) Calculate the fitted bond prices for the two bonds using the Vasicek model.

Bond reference	*Time to maturity*	*Coupon rate*	*Bid price*	*Ask price*
1	0.5 years	5%	101.32	101.38
2	1 year	6%	103.73	103.87

(b) Identify any mispricing.

[QFIQF – Spring 2019 Q14]

Solution In the United States, coupon payments on bonds are typically made semiannually. This will be assumed in this solution.

The nominal amount is not given. Judging by the bid and ask prices, it would appear that the nominal amount is \$100.

The notation in the question is slightly different to the notation used in the text. $\tau = T - t$ is the time to maturity. So $B(\tau)$ corresponds to $B_{t,T}$ in this chapter or $B_{0,T}$ if t is set to zero.

(a) Using the risk-neutral table values, the prices of a zero-coupon bond with payoff \$1 at maturities 0.5 years and 1 year are

$$P_{0,0.5} = e^{-0.00166-0.4477\times 0.021} = 0.9890$$

$$P_{0,1} = e^{-0.00615-0.8053\times 0.021} = 0.9772$$

Therefore,

$$\begin{aligned}\text{Price Bond 1} &= (100 \times 0.05/2 + 100) \times 0.9890 = 101.37 \\ \text{Price Bond 2} &= 100 \times 0.06/2 \times 0.9890 \\ &\quad + (100 \times 0.06/2 + 100) \times 0.9772 = 103.62\end{aligned}$$

(b) Taking the average of the bid and ask prices we compare

Bond	*Theoretical price*	*Market price*
1	101.37	101.35
2	103.62	103.80

Bond 2 is overpriced by the market.

8 The Vasicek model of interest rates is

$$dr = \gamma(\mu - r_t)dt + \sigma dW$$

where γ and σ are positive constants, and W is a standard Brownian motion.

Suppose that under the Vasicek model:
The short rate r is 4% and its real-world process is

$$dr = 0.1(0.05 - r_t)dt + 0.01dW$$

The risk-neutral process is

$$dr = 0.1(0.09 - r_t)dt + 0.01dW$$

(a) Determine the market price of interest risk.

The price of a zero-coupon bond with \$1 principal at time t with maturity date T is given by

$$P_{t,T} = e^{A_{t,T} - B_{t,T} r}$$

where $B_{t,T} = \frac{1-e^{-\gamma(T-t)}}{\gamma}$.

Let P be the price process with T=10.

(b) Compute the drift and the diffusion of $\frac{dP}{P}$ for the risk-neutral process.

(c) Compute the drift and the diffusion of $\frac{dP}{P}$ for the real-world process.

[QFIQF – Spring 2024 Q5]

Solution

(a) From Equation (14.4), the relationship between the drift in the risk-neutral model $\mu(t, r_t)$ and the drift in the risk-neutral model $a(t, r_t)$ is

$$\mu(t, r_t) = a(t, r_t) - \sigma(t, r_t)\gamma_t$$

Therefore

$$\gamma_t = \frac{a(t, r_t) - \mu(t, r_t)}{\sigma(t, r_t)} = \frac{(0.1(0.05 - r_t)) - (0.1(0.09 - r_t))}{0.01}$$
$$= \frac{-0.004}{0.01} = -0.4$$

(b) Repeating the logic leading to Equation (10.3), substitute $dW = d\widetilde{W} - \gamma_t dt$ into Equation (14.3) where $\gamma_t = \frac{\alpha(t,r_t)-r_t}{\beta(t,r_t)}$

$$dP = \alpha P dt + \beta P dW$$
$$= \alpha P dt + \beta P(d\widetilde{W} - \gamma_t dt)$$

$$= \alpha P dt - P(\alpha - r_t)dt + \beta P d\widetilde{W}$$
$$= r_t P dt + \beta P d\widetilde{W}$$
$$\Longrightarrow \frac{dP}{P} = r_t dt + \beta d\widetilde{W}$$

where $\beta P = \sigma \frac{\partial P}{\partial r_t}$.

In the question, the drift of $\frac{dZ}{Z}$ is $r_t = 0.04$ and the diffusion is $\beta = \sigma \times (-B_{t,T}) = \sigma \times \frac{e^{-\gamma(T-t)}-1}{\gamma} = 0.01 \times \frac{e^{-0.1\times 10}-1}{0.1} = -0.0632$

a From above

$$\frac{dP}{P} = \alpha dt + \beta dW$$

and $\gamma_t = \frac{\alpha(t,r_t)-r_t}{\beta(t,r_t)}$, which implies that the drift term

$$\alpha = \gamma_t \beta + r_t = (-0.4) \times (-0.0632) + 0.04 = 0.06528$$

The diffusion term is unchanged from the risk-neutral process.

9 The Vasicek term structure model is described by the following SDE for the short-rate r_t:

$$dr = \alpha(\mu - r_t)dt + \sigma dW$$

where α, μ, σ are constants, and W is standard Brownian motion

Next consider the corresponding CIR model for the short rate $\mathfrak{r}_t$:

$$d\mathfrak{r} = \alpha(\mu - \mathfrak{r}_t)dt + \sigma\sqrt{\mathfrak{r}_t}dW$$

It can be shown that the solution to the SDE satisfies the following:

$$\mathfrak{r}_t = \mu + (\mathfrak{r}_0 - \mu)e^{-\alpha t} + e^{-\alpha t}\int_0^t e^{\alpha s}\sigma\sqrt{\mathfrak{r}_s}dW$$

Now you are to compare the solutions r_t and $\mathfrak{r}_t$ in terms of their expectations and variances.

(a) i Determine the expectations of r_t and $\mathfrak{r}_t$ respectively.
ii Determine the variances of r_t and $\mathfrak{r}_t$ respectively.

(b) Show that the expectations and the variances in (a) have the following limits as $t \to \infty$

i $\lim_{t\to\infty} \mathbb{E}(r_t) = \lim_{t\to\infty} \mathbb{E}(\mathfrak{r}_t) = \mu$

ii $\lim_{t\to\infty} var(r_t) = \dfrac{\sigma^2}{2\alpha}$ and $\lim_{t\to\infty} var(\mathfrak{r}_t) = \dfrac{\sigma^2\mu}{2\alpha}$

[QFIQF – Fall 2018 Q8]

Solution

The solution to the Vasicek model is

$$r_t = \mu + (r_0 - \mu)e^{-\alpha t} + \sigma \int_0^t e^{\alpha(s-t)} dW$$

(a) Vasicek (i) and (ii)
From Example 6.5

$$\mathbb{E}\, r_t = \mu + (r_0 - \mu)e^{-\alpha t}$$

$$\text{and } var\, r_t = \frac{\sigma^2}{2\alpha}(1 - e^{-2\alpha t})$$

CIR (i) and (ii)

$$\mathfrak{r}_t = \mu + (\mathfrak{r}_0 - \mu)e^{-\alpha t} + e^{-\alpha t}\int_0^t e^{\alpha s}\sigma\sqrt{\mathfrak{r}_s}dW$$

$$\implies \mathbb{E}\,\mathfrak{r}_t = \mu + (\mathfrak{r}_0 - \mu)e^{-\alpha t}$$

$$\begin{aligned}
\text{and } var\,\mathfrak{r}_t &= var\left(e^{-\alpha t}\int_0^t e^{\alpha s}\sigma\sqrt{\mathfrak{r}_s}dW\right)\\
&= \mathbb{E}\left(e^{-\alpha t}\int_0^t e^{\alpha s}\sigma\sqrt{\mathfrak{r}_s}dW\right)^2\\
&= e^{-2\alpha t}\int_0^t e^{2\alpha s}\sigma^2\mathbb{E}\,\mathfrak{r}_s ds\\
&= \sigma^2 e^{-2\alpha t}\int_0^t e^{2\alpha s}(\mu + (\mathfrak{r}_0 - \mu)e^{-\alpha s})ds\\
&= \frac{\sigma^2}{2\alpha}(\mu(1 - e^{-2\alpha t}) + \frac{\sigma^2}{\alpha}(\mathfrak{r}_0 - \mu)e^{-\alpha t}(1 - e^{-\alpha t}))\\
&= \frac{\sigma^2\mu}{2\alpha}(1 - e^{-2\alpha t})^2 + \frac{\sigma^2\mathfrak{r}_0}{\alpha}e^{-\alpha t}(1 - e^{-\alpha t})
\end{aligned}$$

(b) Since

$$\lim_{t \to \infty} e^{-\alpha t} = 0$$

then

$$\lim_{t \to \infty} \mathbb{E}(r_t) = \lim_{t \to \infty} \mathbb{E}(z_t) = \mu$$

and

$$\lim_{t \to \infty} var(r_t) = \frac{\sigma^2}{2\alpha}$$

$$\lim_{t \to \infty} var(z_t) = \frac{\sigma^2 \mu}{2\alpha}$$

10 Which of the following short-rate models have an affine term structure?

Authors	*Model*
Ho and Lee(1986)	$dr = \mu(t)dt + \sigma dW$
Dothan (1978)	$dr = \theta r_t dt + \sigma r_t dW$
Black and Karasinski (1991)	$dr = \phi(t)(\log \mu(t) - \log r_t)dt + \sigma(t)dW$

Solution A short-rate model

$$dr = \mu(t, r_t)dt + \sigma(t, r_t)d\tilde{W}$$

will produce an affine term structure if and only if the drift and diffusion terms are of the form

$$\mu(t, r_t) = b(t) + c(t)r_t \text{ and } \sigma^2(t, r_t) = u(t) + v(t)r_t$$

Hence Ho and Lee(1986) is the only model in this table that has an affine term structure.

11 Under the Vasicek interest rate model, the fundamental pricing equation is as follows:

$$\frac{\partial P}{\partial t} + \frac{\partial P}{\partial r}\alpha(\mu - r_t) + \frac{1}{2}\frac{\partial^2 P}{\partial r^2}\sigma^2 - r_t P = 0$$

where α, μ, σ are constants, r_t is the short rate, and P is an interest rate security. Let $P_{t,T}$ represent the price at time t of a zero-coupon bond maturing at time T. You are given that $P_{t,T}$ has the form:

$$P_{t,T} = e^{A_{t,T} - B_{t,T} r_t}$$

where $A_{t,T}$ and $B_{t,T}$ satisfy the following ordinary differential equations (ODEs).

$$\frac{dA}{dt} = B\alpha\mu - \frac{\sigma^2}{2}B^2$$

$$\frac{dB}{dt} = \alpha B - 1$$

Derive the expression for $P_{t,T}$ by solving the above ODEs.
(Hint: The solutions of an ODE $\frac{df}{dx} = f(x)\alpha - 1$ are $f(x) = \frac{1}{\alpha} - Ce^{\alpha x}$, where C is a constant.)

[QFIQF – Fall 2023 Q6]

Solution
Here we are using the affine term structure property of the Vasicek model and the pricing partial differential equation to find the solution for the zero-coupon bond price. This is an alternative to the derivation given in the chapter.
From the hint

$$\frac{dB}{dt} = \alpha B - 1$$

$$\implies B = \frac{1}{\alpha} - Ce^{\alpha t}$$

$$\implies B_{t,T} = \frac{1 - e^{-\alpha(T-t)}}{\alpha} \qquad \text{(using the terminal condition } B_{T,T} = 0\text{)}$$

Substituting for $B_{t,T}$

$$
\begin{aligned}
\frac{dA}{dt} &= (1 - e^{-\alpha(T-t)})\mu - \frac{\sigma^2}{2\alpha^2}(1 - 2e^{-\alpha(T-t)} + e^{-2\alpha(T-t)}) \\
\implies A &= (t - \frac{1}{\alpha}e^{-\alpha(T-t)})\mu \\
&\quad - \frac{\sigma^2}{2\alpha^2}\left(t - \frac{2}{\alpha}e^{-\alpha(T-t)} + \frac{1}{2\alpha}e^{-2\alpha(T-t)}\right) + C \\
\implies A_{t,T} &= (t - \frac{1}{\alpha}e^{-\alpha(T-t)})\mu \\
&\quad - \frac{\sigma^2}{2\alpha^2}\left(t - \frac{2}{\alpha}e^{-\alpha(T-t)} + \frac{1}{2\alpha}e^{-2\alpha(T-t)}\right) \\
&\quad - T\mu + \frac{\mu}{\alpha} + \frac{\sigma^2}{2\alpha^2}\left(T - \frac{3}{2\alpha}\right) \qquad (\text{ since } A_{T,T} = 0) \\
&= -(T-t)\mu + \frac{\sigma^2}{2\alpha}(T-t) + \mu\left(\frac{1 - e^{\alpha(T-t)}}{\alpha}\right) \\
&\quad - \frac{\sigma^2}{4\alpha^2}\left(-\frac{4}{\alpha}e^{-\alpha(T-t)} + \frac{1}{\alpha}e^{-2\alpha(T-t)}\right) \\
&\quad + \frac{\sigma^2}{2\alpha^2}\left(-\frac{1}{\alpha} - \frac{1}{2\alpha}\right) \\
&= -(T-t)\mu + \frac{\sigma^2}{2\alpha}(T-t) + \mu\left(\frac{1 - e^{\alpha(T-t)}}{\alpha}\right) \\
&\quad - \frac{\sigma^2}{4\alpha^2}\left(\frac{1}{\alpha} - \frac{2}{\alpha}e^{-\alpha(T-t)} + \frac{1}{\alpha}e^{-2\alpha(T-t)}\right) \\
&\quad - \frac{\sigma^2}{2\alpha^2}\left(\frac{1}{\alpha} - \frac{1}{\alpha}e^{-\alpha(T-t)}\right) \\
&= -(T-t)\mu + \frac{\sigma^2}{2\alpha}(T-t) + \mu B_{t,T} - \frac{\sigma^2}{4\alpha}B_{t,T}^2 - \frac{\sigma^2}{2\alpha^2}B_{t,T} \\
&= -((T-t) - B_{t,T})\left(\mu - \frac{\sigma^2}{2\alpha^2}\right) - \frac{\sigma^2 B_{t,T}^2}{4\alpha}
\end{aligned}
$$

Therefore

$$
P_{t,T} = e^{A_{t,T} - B_{t,T}\, r_t}
$$

where

$$A_{t,T} = -((T-t) - B_{t,T})\left(\mu - \frac{\sigma^2}{2\alpha^2}\right) - \frac{\sigma^2 B_{t,T}^2}{4\alpha} \text{ and}$$

$$B_{t,T} = \frac{(1 - e^{-\alpha(T-t)})}{\alpha}$$

12 Given that $\sigma = 2\%$, $r_0 = 3\%$ in the Vasicek model and the following information:

(a) Observed zero-coupon bond prices in the market are

$$P_{0,1} = 0.97, P_{0,2} = 0.94, P_{0,3} = 0.87, P_{0,4} = 0.83 \text{ and } P_{0,5} = 0.79$$

(b) Average yield of 1-month T-bill over the past 50 years is 5.1%.

Three potential choices for parameters of α and μ are given below:

	Choice 1	*Choice 2*	*Choice 3*
α	0.9	0.45	0.45
μ	5%	5%	6%

Choose the best parametrization α and μ.

[QFIQF – Fall 2023 Q6]

Solution Following the approach in Section 14.5.3 minimize

$$J(\alpha, \mu, \sigma) = \sum_{1}^{n} \left(P_{0,T_i}^{\text{Theoretical}} - P_{0,T_i}^{\text{Observed}}\right)^2$$

Observed and theoretical zero-coupon bond prices:

T	*Observed*	*Choice 1*	*Choice 2*	*Choice 3*
1	0.97	0.964	0.967	0.965
2	0.94	0.922	0.929	0.923
3	0.87	0.879	0.890	0.878
4	0.83	0.837	0.851	0.833
5	0.79	0.797	0.812	0.788
J		0.00054	0.00145	0.00040

Choice 3 is the best parametrization.

13 You have one-month daily treasury bill yields (annualized) over 500 consecutive trading days and 20 treasury bond yields with various maturities. There are 252 trading days per year. You would like to fit the Vasicek model for the datasets.

Outputs are below.

Dependent variable: One-month daily treasury bill yields (annualized) over 500 trading days

Maximum-likelihood estimates

	Estimate	*Std. error*	*t value*	$P(>\|t\|)$
Intercept	0.0001815	0.0001620	1.121	0.263
r_{t-1}	0.9964045	0.0038202	260.822	0.000

Residual standard error: 0.001300116

We use sigma value that was estimated in the real world situation

Dependent variable: Bond yields

	Std. error	*t value*	$P(>\|t\|)$
α	0.031458254	15.49071	0.000
μ	0.001384681	50.90106	0.000

You are using the real-world Vasicek model,

$$dr = \alpha^*(\mu^* - r_t)dt + \sigma dW$$

where W is standard Brownian motion.

(a) Estimate the parameters of your model.

You are using the risk-neutral Vasicek model,

$$dr = \alpha(\mu - r_t)dt + \sigma d\tilde{W}$$

where $\tilde{W}$ is standard Brownian motion.

(b) Estimate the parameters of your new model.

[QFIQF – Spring 2024 Q6]

Solution

(a) Using data on yields on the shortest maturity Treasury bills will give parameter estimates for a real-world short-rate process.

The model parameters are related to the maximum-likelihood estimates by the following equations

$$\alpha^* = -\frac{\log b}{\Delta t}$$
$$\mu^* = \frac{a}{1-b}$$
$$\sigma = c\sqrt{\frac{2\log b}{\Delta t\left(b^2-1\right)}}$$

From the maximum-likelihood estimates:

$$a = 0.0001815, b = 0.9964045, c = 0.001300116$$

and $\Delta t = 1/252$. The model parameters are therefore:

$$\alpha^* = 0.9076988, \mu^* = 0.0504798, \sigma = 0.0206742$$

(b) Using data on bond yields will give parameter estimates for a risk-neutral short-rate process.

The estimates are not given but are easily determined from t-values and standard errors since $t\text{-value} = \frac{\text{estimate}}{\text{standard error}}$.

The model parameters are therefore:

$$\alpha = 0.4873107, \mu = 0.0704817$$

14 You have been asked to compare the prices of zero-coupon bonds to the Treasury STRIPS data for various maturities at time 0 under the single factor Vasicek model. You are given:

- The SDE of the short rate for the single factor Vasicek model is given by

$$dr = \alpha(\mu - r_t)dt + \sigma dW$$

  where Wt is a standard Brownian motion.
- Each STRIPS has face value of 100.
- $P_{0,T_i}^{\text{Vasicek}}$ is the Vasicek model price at time 0 of STRIPS with maturity $T_i, i = 1, 2, ...$ given the initial rate r_0.

You obtain estimates $\alpha = 0.029, \mu = 2.09\%$ and $\sigma = 1.78\%$.
In addition, you are given r_0 =1.8% along with the following information:

Time to maturity (τ)	$A_{0,\tau}$	$B_{0,\tau}$	$P_{0,\tau}^{Vasicek}$
1.139	–0.00031	1.12039	
3.139			94.6128

(a) i. Calculate $P_{0,3.139}$

ii. Calculate $A_{0,3.139}$ and $B_{0,3.139}$ based on the estimates given.

Consider a portfolio with two zero-coupon bonds, long one unit of a bond with maturity date T_1 and short Δ units of another bond with maturity date $T_2 > T_1$.

Let P_{t,T_1,r_t} and P_{t,T_2,r_t} be the two prices with the Vasicek model for the T_1 and T_2 maturity dates, respectively, at time t when the current interest rate is r_t.

Denote the value of this portfolio at time t by

$$\Pi_{t,r_t} = P_{t,T_1,r_t} - \Delta P_{t,T_2,r_t}$$

(b) Prove that if Δ makes the portfolio insensitive to r_t at time t, then

$$\Delta = \frac{B_{t,T_1} P_{t,T_1}}{B_{t,T_2} P_{t,T_2}}$$

(c) Calculate the Δ in part (b) at time 0 when T_1 =1.139 and T_2 = 3.139, using the data given and results derived earlier.

Note: STRIPS is an acronym for Separate Trading of Registered Interest and Principal of Securities. These are zero-coupon bonds formed from Treasury securities.

[QFIQF – Fall 2019 Q9]

Solution

(a) i $P_{0,3.139} = 100e^{A(1.139)-B(1.139)r_0} = 100e^{-0.00031-1.12039\times 1.8\%} =$ 97.9728

ii

$$A_{0,3.139} = -(3.139 - 3.000365)\left(2.09\% - \frac{(1.78\%)^2}{2(0.029)^2}\right) - \frac{(1.78\%)^2(3.000365)^2}{4\times 0.029} - \frac{(1.78\%)^2(3.000365)^2}{4\times 0.029} = -0.00137$$

$$\text{and } B_{0,3.139} = \frac{(1 - e^{-0.029\times 3.139})}{0.029} = 3.000365$$

(b) Since $P_{t,T,r_t} = e^{A_{t,T} - B_{t,T} r_t}$

$$\frac{d\Pi}{dr} = -B_{t,T_1} P_{t,T_1,r_t} + \Delta B_{t,T_2} P_{t,T_2,r_t} = 0$$

$$\Longrightarrow \Delta = \frac{B_{t,T_1} P_{t,T_1,r_t}}{B_{t,T_2} P_{t,T_2,r_t}}$$

(c)

$$\Delta = \frac{1.12039 \times 97.9728}{3.000365 \times 94.6128} = 0.3867$$

15 Using the current yield curve and cap prices, you have to calibrate the Hull-White model below

$$dr = \alpha(\mu(t) - r_t)dt + \sigma d\tilde{W}$$

The dataset contains 20 treasury bond prices and 5 cap prices. You fitted the following curve for $R_{0,t}$ (continuously compounded yield between 0 and t)

$$R_{0,t} = 0.01091858598 + 0.01251008594t - 0.000140114635t^2 + 0.005654825t^3$$

and estimated values of α, σ to be 0.19, 0.0196 respectively.

(a) Derive an expression for the instantaneous forward rate at time 0 $f_{0,t}$.

(b) Derive an expression for $\mu(t)$.

[QFIQF – Spring 2023 Q5]

Solution

(a)

$$\begin{aligned} f_{0,t} &= -\frac{\partial \log P_{0,t}}{\partial t} = \frac{\partial(tR_{0,t})}{\partial t} \\ &= 0.01091858598 + 0.01251008594 \\ &\quad \times 2t - 0.000140114635 \times 3t^2 + 0.005654825 \times 4t^3 \\ &= 0.01091858598 + 0.02502017188t \\ &\quad - 0.00042034391t^2 + 0.02261930t^3 \end{aligned}$$

(b)

$$\frac{\partial}{\partial t} f_{0,t} = 0.02502017188 - 0.000840688t + 0.0678579t^2 \quad \text{and}$$

$$\frac{\sigma^2}{2\alpha^2} = 0.005320776$$

$$\mu(t) = f_{0,t} + \alpha^{-1}\frac{\partial}{\partial t} f_{0,t} + \frac{\sigma^2}{2\alpha^2}(1 - e^{-2\alpha t})$$
$$= 0.14792447676 + 0.020595499t + 0.356726498t^2 + 0.02261930t^3 - 0.005320776e^{-0.38t}$$

16 You are given the following Hull-White model, which is calibrated to the current market bond prices, to value at time 0, 1 year and 2 months European call option on a zero-coupon bond that will mature in 2.5 years.

$$dr = \alpha(\mu(t) - r_t)dt + \sigma d\tilde{W}$$

where α and σ are positive constants, and Wt is a standard Brownian motion.

The price of a zero-coupon bond with \$1 principal at time t with maturity date T is given by

$$P_{t,T} = e^{A_{t,T} - B_{t,T} r_t}$$

Let $f_{t,T} = -\frac{\partial \log P_{t,T}}{\partial T}$ be the continuously compounded instantaneous forward rate.

You are given the following:

- $f_{0,t} = c + mt$ for $t \geq 0$, where c and m are positive constants.
- $B_{t,T} = \frac{1 - e^{-\alpha(T-t)}}{\alpha}$, where $t \leq T$.

(a) Demonstrate that $\alpha\mu(t) = m + \alpha c + \alpha mt + \frac{\sigma^2}{2\alpha}(1 - e^{-2\alpha t})$

(b) Show that:

(i) $A_{0,T} = \frac{c}{\alpha}(1 - e^{-\alpha T}) - cT - \frac{mT^2}{2}$

(ii) $P_{0,T} = e^{\frac{-mT^2}{2} - r_0 T}$

In addition, you are given the following:

- $\alpha\mu_{\frac{1}{2\alpha}} - \alpha\mu_0 = 0.084$
- $m = 0.0862$
- $\sigma = 0.2$
- r_0= Initial short-term interest rate =4%

(c) Calculate α.

[QFIQF – Fall 2023 Q4]

Solution

(a)

$$\alpha\mu(t) = \alpha f_{0,t} + \frac{\partial}{\partial t} f_{0,t} + \frac{\sigma^2}{2\alpha}(1 - e^{-2\alpha t})$$
$$= \alpha(c + mt) + m + \frac{\sigma^2}{2\alpha}(1 - e^{-2\alpha t})$$
$$= m + \alpha c + \alpha m t + \frac{\sigma^2}{2\alpha}(1 - e^{-2\alpha t})$$

(b) (i)

$$f_{0,T} = -\frac{\partial \log P_{0,T}}{\partial T} = \frac{\partial}{\partial T}(-A_{0,T} + B_{0,T}\, r_0) = c + mT$$
$$\Longrightarrow cT + \frac{1}{2} m T^2 = -A_{0,T} + B_{0,T}\, r_0$$
$$\Longrightarrow A_{0,T} = \frac{c}{\alpha}(1 - e^{-\alpha T}) - cT - \frac{mT^2}{2}$$

since $r_0 = f_{0,0} = c$

(ii)

$$P_{0,T} = e^{A_{0,T} - B_{0,T}\, r_0}$$
$$= e^{\frac{c}{\alpha}(1 - e^{-\alpha T}) - cT - \frac{mT^2}{2} - \frac{1 - e^{-\alpha T}}{\alpha} c}$$
$$= e^{\frac{-mT^2}{2} - r_0 T}$$

(c)

$$\alpha\mu_{\frac{1}{2\alpha}} - \alpha\mu_0 = \left(m + \alpha c + \alpha m \frac{1}{2\alpha} + \frac{\sigma^2}{2\alpha}(1 - e^{-2\alpha \frac{1}{2\alpha}})\right)$$
$$- (m + \alpha c) = \frac{m}{2} + \frac{\sigma^2}{2\alpha}(1 - e^{-1})$$

Therefore $\frac{0.0862}{2} + \frac{0.2^2}{2\alpha}(1 - e^{-1}) = 0.084 \Longrightarrow \alpha = 0.309105$

17 Using the information from the question above, calculate the price at $t = 0$ of the 1 year and 2 months European call with strike price of \$90, written on a zero-coupon bond with face value \$100 and maturity of 2.5 years.

[QFIQF – Fall 2023 Q4]

Solution The volatility is:

$$\sigma_F(1\tfrac{1}{6}, 2.5) = \frac{1-e^{-0.309105(2.5-1\frac{1}{6})}}{0.309105} \times 0.2 \times \sqrt{\frac{1-e^{-2\times 0.309105\times 1\frac{1}{6}}}{2\times 0.309105}} = 0.1992$$

And.

$$P_{0,1\frac{1}{6}} = e^{-0.0662-0.9795\times 0.04} = 0.900, P_{0,2.5} = e^{-0.2997-1.7414\times 0.04} = 0.6912$$

Then the ZCB option price is:

$$100 \times (P_{0,2.5} \times \Phi(-0.6968) - 0.9P_{0,1\frac{1}{6}} \times \Phi(-0.8960)) = 1.798$$

18 You would like to price interest rate derivatives, assuming that the interest rate follows the Vasicek model:

$$dr = \alpha(\mu - r_t)dt + \sigma dW$$

where α, μ, σ are constants, r_t is the short rate, and W is standard Brownian motion under the risk-neutral measure.

Let $P_{t,T}$ be the price at time t of a zero-coupon bond maturing at T under the Vasicek model.

(a) Derive formulas for the spot-rate duration and convexity of $P_{t,T}$.
(b) Explain the impact of increasing α on the duration and convexity.

[QFIQF – Spring 2019 q15]

Solution

(a) $D = B_{t,T} = \dfrac{1 - e^{-\alpha(T-t)}}{\alpha}$

and $C = B_{t,T}^2 = \left(\dfrac{1 - e^{-\alpha(T-t)}}{\alpha}\right)^2$

(b) As $\alpha \to \infty$, $\dfrac{1 - e^{-\alpha(T-t)}}{\alpha} \to 0$
therefore duration and convexity decrease. The greater the speed adjustment to the mean, the lower the sensitivity of the bond to fluctuations in the short rate.

19 Consider the SDE for the spot rate r_t

$$dr = \alpha(\mu - r_t)dt + \sigma dW$$

where α, μ, σ are constant, and W_t is a standard Wiener process.

The risk-neutral parameters are $\mu = 0.0634, \alpha = 0.4653$, with $r_0 = 0.0162, \sigma = 0.0221$. You are also given the following:

	Call option
Notional (of underlying bond)	100
Coupon (of underlying bond)	5% per annum paid semiannually
Maturity (of underlying bond)	2 years
Exercise date (T)	1 year
Strike price	100

The interest rate r^* that makes the value at T of the underlying bond of the call option equal to the strike price is 4.60%.

Calculate the bond option price at time 0.

[QFIQF – Spring 2023 Q6]

Solution The short rate follows the Vasicek model.

Break the coupon bond option into two options on zero-coupon bonds.

Strike prices are:

$P_{1,1.5,0.046} = 0.9764$ and $P_{1,2,0.046} = 0.9518$

The volatilities are:

$\sigma_F(1, 1.5) = 0.080$ and $\sigma_F(1, 2) = 0.0143$

And

$P_{0,1} = 0.9747, P_{0,1.5} = 0.9570$ and $P_{0,2} = 0.9371$

Then the ZCB option prices are:

For $T_1 = 1.5$, $P_{0,1.5} \times \Phi(0.7032) - P_{0,1} \times 0.9764 \times \Phi(0.6952) = 0.0064$

For $T_2 = 2$, $P_{0,2} \times \Phi(0.7095) - P_{0,1} \times \times\Phi(0.6952) = 0.0112$

The coupon bond option price is then:

$$2.5 \times 0.0064 + 102.5 \times 0.0112 = 1.167$$

20 You are given the following Hull-White model, which is calibrated to the current market bond prices, to value at time 0, 2.1-year European call option on a semi-annual coupon bond that will mature in 3 years.

$$dr = (\mu(t) - \alpha r_t)dt + \sigma d\tilde{W}$$

where α and σ are positive constants, and Wt is standard Brownian motion.

The price at time t of a zero-coupon bond with \$1 principal with maturity date $T \geq t$ is given by

$$P_{t,T} = e^{A_{t,T} - B_{t,T}\, r_t}$$

where $B_{t,T} = \frac{1-e^{-\alpha(T-t)}}{\alpha}$ and $A_{t,T} = \frac{a}{\alpha}(1 - e^{-\alpha(T-t)}) - a(T-t) - \frac{b(T-t)^2}{2}$.

You are given the following:

- $\sigma = 0.2$, $\alpha = 0.309105$
- $a = 0.04$, $b = 0.0862$
- r_0 = Initial short-term interest rate = 4%

(a) Calculate
  (i) $B_{2.1,2.5}$ and $B_{2.1,3}$
  (ii) $A_{2.1,2.5}$ and $A_{2.1,3}$

Let X_{t,r_t} be the price at time $t \leq 3$ of a semi-annual coupon bond with

- Annual coupon rate =2%
- Principal = \$100
- Maturity T=3

You are given r^*=4.08541%.

(b) (i) Calculate $P_{2.1,2.5,r^*}$
  (ii) Calculate $P_{2.1,3,r^*}$
  (iii) Show that $X_{2.1,r^*} = 95$

Consider a European Call option on the coupon bond with strike 95 to be matured in 2.1 years.

You are given the following:

- The price at t=0 of the 2.1-year European call with strike price of $P_{2.1,3,r^*}$, written on zero-coupon bond with face value \$1 and maturity of 3 years is 0.010116858.

(c) Compute the value at $t = 0$ of the above European Call option on the coupon bond.

[QFIQF – Fall 2024 Q6]

Solution

(a) (i) $B_{2.1,2.5} = 0.3763$ and $B_{2.1,3} = 0.7857$
  (ii) $A_{2.1,2.5} = -0.0078$ and $A_{2.1,3} = -0.0395$

(b) (i) $P_{2.1,2.5,r^*} = 0.9771$
  (ii) $P_{2.1,3,r^*} = 0.9309$
  (iii) $X_{2.1,r^*} = 1 \times 0.9771 + 101 \times 0.9309 = 95$

(c) The ZCB option prices are:
For $T_1 = 2.5, 0.6912\times\Phi(-0.843)-0.7603\times0.9771\times\Phi(-0.9242) = 0.0061$
For $T_2 = 3$, 0.010116858 (given in question)
The coupon bond option price is then:

$$1 \times 0.0061 + 101 \times 0.010116858 = 1.028$$

21 Show that the instantaneous forward rate for the Vasicek model can be expressed as:

$$f_{t,T} = r_t e^{-\alpha\tau} + \left(\mu - \frac{\sigma^2}{2\alpha^2}\right)(1 - e^{-\alpha\tau}) + \frac{\sigma^2}{2\alpha^2}(1 - e^{-\alpha\tau})e^{-\alpha\tau}$$

where $\tau = T - t$

Solution

$$P_{t,T} = \exp\Bigg(- \{\mu\tau + \frac{(r_t - \mu)}{\alpha}(1 - e^{-\alpha\tau})\} + \left\{\frac{\sigma^2}{2\alpha^2}(\tau - \frac{2}{\alpha}(1 - e^{-\alpha\tau}) + \frac{1}{2\alpha}(1 - e^{-2\alpha\tau}))\right\}\Bigg)$$

Noting that $\frac{\partial\tau}{\partial T} = 1$

$$\begin{aligned} f_{t,T} &= -\frac{\partial \log P_{t,T}}{\partial T} \\ &= \mu + \frac{(r_t - \mu)}{\alpha}(\alpha e^{-\alpha\tau}) - \{\frac{\sigma^2}{2\alpha^2}(1 - \frac{2}{\alpha}(\alpha e^{-\alpha\tau}) + \frac{1}{2\alpha}(2\alpha e^{-2\alpha\tau}))\} \\ &= r_t e^{-\alpha\tau} + \left(\mu - \frac{\sigma^2}{2\alpha^2}\right)(1 - e^{-\alpha\tau}) + \frac{\sigma^2}{2\alpha^2}(1 - e^{-\alpha\tau})e^{-\alpha\tau} \end{aligned}$$

Note:
when $\tau = 0, f_{t,T} = r_T$, the 'short' rate at T
when $\tau \to \infty, f_{t,T} \to \left(\mu - \frac{\sigma^2}{2\alpha^2}\right) = L$, the 'long' rate

Appendix

A.1 Calculus result

$$
\begin{aligned}
ax^2 + bx + c &= a\left(x^2 + \frac{b}{a}x + \frac{b^2}{4a^2}\right) + \frac{4ac - b^2}{4a} \\
&= a\left(x + \frac{b}{2a}\right)^2 + \frac{4ac - b^2}{4a}
\end{aligned}
$$

If $4ac - b^2 < 0$, then

$$
\begin{aligned}
\int \frac{dx}{ax^2 + bx + c} &= \frac{1}{a}\int \frac{dx}{(x + \frac{b}{2a})^2 - (\frac{b^2-4ac}{4a^2})} \\
&= \frac{1}{a}\int \frac{dx}{(x + \frac{b}{2a})^2 - (\frac{\sqrt{b^2-4ac}}{2a})^2} \\
&= \frac{1}{\sqrt{b^2 - 4ac}}\int \left(\frac{1}{x + \frac{b}{2a} - \frac{\sqrt{b^2-4ac}}{2a}} - \frac{1}{x + \frac{b}{2a} + \frac{\sqrt{b^2-4ac}}{2a}}\right)dx \\
&= \frac{1}{\sqrt{b^2 - 4ac}} \log\left|\frac{x + \frac{b}{2a} - \frac{\sqrt{b^2-4ac}}{2a}}{x + \frac{b}{2a} + \frac{\sqrt{b^2-4ac}}{2a}}\right| + C \\
&= \frac{1}{\sqrt{b^2 - 4ac}} \log\left|\frac{2ax + b - \sqrt{b^2 - 4ac}}{2ax + b + \sqrt{b^2 - 4ac}}\right| + C
\end{aligned}
$$

A.2 Derivation of CIR zero-coupon bond price

Solving 14.15

$$
\begin{aligned}
&\frac{\partial B}{\partial t} = \frac{\sigma^2}{2}B^2 + \alpha B - 1 \\
\implies &\int_0^T \frac{dB}{\frac{1}{2}\sigma^2 B^2 + \alpha B - 1} = T
\end{aligned}
$$

Using the integration result from the Appendix A.1 and letting $\gamma = \sqrt{\alpha^2 + 2\sigma^2}$ then

$$\frac{1}{\gamma}\log\left|\frac{\alpha-\gamma}{\alpha+\gamma}\right| - \frac{1}{\gamma}\log\left|\frac{\sigma^2 B_{0,T}+\alpha-\gamma}{\sigma^2 B_{0,T}+\alpha+\gamma}\right| = T$$

$$\Longrightarrow \log\frac{(\alpha-\gamma)(\sigma^2 B_{0,T}+\alpha+\gamma)}{(\alpha+\gamma)(\sigma^2 B_{0,T}+\alpha-\gamma)} = \gamma T$$

The absolute value symbol is dropped by assuming $\sigma^2 B_{t,T} + \alpha - \gamma$ does not change sign from $t = 0$ to $t = T$.

$$\Longrightarrow \frac{\sigma^2(\alpha-\gamma)B_{0,T}+(\alpha^2-\gamma^2)}{\sigma^2(\alpha+\gamma)B_{0,T}+(\alpha^2-\gamma^2)} = e^{\gamma T}$$

$$\Longrightarrow \frac{(\alpha-\gamma)B_{0,T}-2}{(\alpha+\gamma)B_{0,T}-2} = e^{\gamma T}$$

$$\Longrightarrow B_{0,T} = \frac{2(e^{\gamma T}-1)}{e^{\gamma T}(\alpha+\gamma)-(\alpha-\gamma)} = \frac{2(e^{\gamma T}-1)}{(\alpha+\gamma)(e^{\gamma T}-1)+2\gamma}$$

$B_{t,T}$ is the above expression with T replaced by $T - t$. Solving 14.14 by first substituting for $B_{t,T}$ and then integrating both sides between the limits 0 and T

$$\frac{\partial A}{\partial t} = B\alpha\mu$$

$$\Longrightarrow -A_{0.T} = 2\alpha\mu\int_0^T \frac{e^{\gamma(T-t)}-1}{e^{\gamma(T-t)}(\alpha+\gamma)-(\alpha-\gamma)}\,dt$$

$$\Longrightarrow A_{0.T} = -2\alpha\mu\left(\int_0^T \frac{e^{\gamma(T-t)}dt}{e^{\gamma(T-t)}(\alpha+\gamma)-(\alpha-\gamma)} - \int_0^T \frac{e^{-\gamma(T-t)}dt}{(\alpha+\gamma)-e^{-\gamma(T-t)}(\alpha-\gamma)}\right)$$

Following the integration by substitution approach, let the denominator in the first integral be u and integrate (see Example 1.16 (iii)). Repeat for the second integral.

$$A_{0.T} = -2\alpha\mu\left(\frac{-1}{\gamma(\alpha+\gamma)}\log\left(e^{\gamma(T-t)}(\alpha+\gamma)-(\alpha-\gamma)\right)\Big|_0^T + \frac{1}{\gamma(\alpha-\gamma)}\log\left((\alpha+\gamma)-e^{-\gamma(T-t)}(\alpha-\gamma)\right)\Big|_0^T\right)$$

$$= \frac{-2\alpha\mu}{\gamma}\left(\left(\frac{-1}{\alpha+\gamma}+\frac{1}{\alpha-\gamma}\right)\log 2\gamma + \frac{1}{(\alpha+\gamma)}\log\right.$$

$$\left(e^{\gamma T}(\alpha+\gamma)-(\alpha-\gamma)\right)$$

$$-\frac{1}{(\alpha-\gamma)}\log\left(\left((\alpha+\gamma)-e^{-\gamma T}(\alpha-\gamma)\right)\right)$$

Remembering the relationship $\gamma = \sqrt{\alpha^2+2\sigma^2}$ and using the logarithm property $\log(ab)=\log a+\log b$

$$\begin{aligned}
A_{0.T} &= \frac{-2\alpha\mu}{\gamma}\left(\frac{-\gamma}{\sigma^2}\log 2\gamma+\frac{1}{\alpha+\gamma}\log\left(e^{\gamma T}(\alpha+\gamma)-(\alpha-\gamma)\right)\right.\\
&\quad\left.-\frac{1}{\alpha-\gamma}\log\left(e^{-\gamma T}\left(e^{\gamma T}(\alpha+\gamma)-(\alpha-\gamma)\right)\right)\right)\\
&= \frac{-2\alpha\mu}{\gamma}\left(\frac{-\gamma}{\sigma^2}\log 2\gamma+\left(\frac{1}{\alpha+\gamma}-\frac{1}{\alpha-\gamma}\right)\log\right.\\
&\quad\left.\left(e^{\gamma T}(\alpha+\gamma)-(\alpha-\gamma)\right)+\frac{\gamma T}{\alpha-\gamma}\right)\\
&= \frac{-2\alpha\mu}{\gamma}\left(\frac{-\gamma}{\sigma^2}\log 2\gamma+\frac{\gamma}{\sigma^2}\log\left(e^{\gamma T}(\alpha+\gamma)-(\alpha-\gamma)\right)\right.\\
&\quad\left.+\frac{\gamma}{\sigma^2}\log e^{\frac{\sigma^2 T}{\alpha-\gamma}}\right)\\
&= \frac{2\alpha\mu}{\sigma^2}\log\left(\frac{2\gamma e^{\frac{\sigma^2 T}{\gamma-\alpha}}}{e^{\gamma T}(\alpha+\gamma)-(\gamma-\alpha)}\right)\\
&= \frac{2\alpha\mu}{\sigma^2}\log\left(\frac{2\gamma e^{(\alpha+\gamma)(T/2)}}{(\gamma+\alpha)(e^{\gamma T}-1)+2\gamma}\right)
\end{aligned}$$

since $\frac{\sigma^2 T}{\gamma-\alpha}=\frac{\sigma^2 T}{\gamma-\alpha}\times\frac{\gamma+\alpha}{\gamma+\alpha}=\frac{(\alpha+\gamma)\sigma^2 T}{\gamma^2-\alpha^2}=\frac{(\alpha+\gamma)\sigma^2 T}{2\sigma^2}=(\alpha+\gamma)\frac{T}{2}$

A.3 Calibration in Hull-White model

The instantaneous forward rate was defined in Equation (12.5) as

$$f_{0,t}=-\frac{\partial\log P_{0,t}}{\partial t}$$

In the Hull-White model

$$P_{0,t}=e^{A_{0,t}-B_{0,t}r_0}$$

with $B_{0,t} = \frac{1-e^{-\alpha t}}{\alpha}$ and

$$A_{0,t} = \int_0^t (-\alpha\mu(s)B_{s,t} + \frac{\sigma^2}{2}B_{s,t}^2)\, ds$$
$$= \int_0^t \left(-\mu(s)(1 - e^{-\alpha(t-s)}) + \frac{\sigma^2}{2\alpha^2}(1 - 2e^{-\alpha(t-s)} + e^{-2\alpha(t-s)}) \right) ds$$

Therefore,

$$f_{0,t} = -\frac{\partial(A_{0,t} - B_{0,t}r_0)}{\partial t} = -\frac{\partial A_{0,t}}{\partial t} + \frac{\partial B_{0,t}}{\partial t} r_0$$

Evaluating the partial derivatives

$$\frac{\partial B_{0,t}}{\partial t} = e^{-\alpha t}$$

The partial derivative of $A_{0,t}$ requires an additional result. The first fundamental theorem of calculus (Equation 1.1) states that

$$F(t) = \int_0^t f(s)ds \quad \text{then} \quad \frac{dF}{dt} = f(t)$$

When the limits of the integral are also functions of t and $f(s,t)$ a function of two variables, we use a generalization of this theorem known as the *Leibniz integral rule*

$$F(t) = \int_{a(t)}^{b(t)} f(s,t)ds \quad \text{then} \quad \frac{dF}{dt} = f(b(t),t)\frac{db(t)}{dt}$$
$$- f(a(t),t)\frac{da(t)}{dt} + \int_{a(t)}^{b(t)} \frac{\partial f(s,t)}{\partial t} ds \tag{20}$$

With $b(t) = t$ and $a(t) = 0$ and $f(t,t) = 0$ we have

$$\frac{\partial A_{0,t}}{\partial t} = \int_0^t \left(-\alpha\mu(s)e^{-\alpha(t-s)} + \frac{\sigma^2}{2\alpha^2}(2\alpha e^{-\alpha(t-s)} - 2\alpha e^{-2\alpha(t-s)}) \right) ds$$
$$= (\int_0^t \left(-\alpha\mu(s)e^{-\alpha(t-s)})ds + \frac{\sigma^2}{2\alpha^2}(1 - 2e^{-\alpha t} + e^{-2\alpha t}) \right) ds$$

Substituting the two partial derivatives into the formula for the instantaneous forward rate

$$f_{0,t} = \int_0^t \alpha\mu(s)e^{-\alpha(t-s)}ds - \frac{\sigma^2}{2\alpha^2}(1 - 2e^{-\alpha t} + e^{-2\alpha t}) + e^{-\alpha t}r_0$$

$$\Longrightarrow f_{0,t}\, e^{\alpha t} = \int_0^t \alpha\mu(s)e^{\alpha s}ds - \frac{\sigma^2}{2\alpha^2}(e^{\alpha t} - 2 + e^{-\alpha t}) + r_0$$

Differentiating both sides with respect to t and applying the first fundamental theorem of calculus

$$\alpha f_{0,t}\, e^{\alpha t} + \frac{\partial f_{0,t}}{\partial t}\, e^{\alpha t} = \alpha\mu(t)e^{\alpha t} - \frac{\sigma^2}{2\alpha^2}(\alpha e^{\alpha t} - \alpha e^{-\alpha t})$$

Rearranging gives the result

$$\mu(t) = f_{0,t} + \alpha^{-1}\frac{\partial f_{0,t}}{\partial t} + \frac{\sigma^2}{2\alpha^2}(1 - e^{-2\alpha t})$$

A.4 $P_{t,T}$ lognormally distributed under the T-forward measure

$$\Theta = \frac{dQ^T}{dQ} = \frac{P_{t,T}}{P_{0,T}B_t} = P_{0,T}^{-1}\frac{P_{t,T}}{B_t}$$

and from Equation (14.6)

$$d\Theta = P_{0,T}^{-1}d\frac{P_t}{B_t} = P_{0,T}^{-1}\frac{\sigma}{B_t}\frac{\partial P}{\partial r}d\tilde{W} = -P_{0,T}^{-1}\sigma B_{t,T}\frac{P_{t,T}}{B_t}d\tilde{W} = -\sigma B_{t,T}\Theta d\tilde{W}$$

Comparing this to the Radon-Nikodým process in Girsanov's Theorem 10.3 then

$$\gamma_t = \sigma B_{t,T} \text{ and } dW^T = \gamma_t dt + d\tilde{W}$$

where W^T is standard Brownian motion under $\mathbb{Q}_T$. Substitute this into the Vasicek model

$$dr = (\alpha(\mu - r_t) - \sigma^2 B_{t,T})dt + \sigma dW^T$$

Solving for r_t gives a normally distributed variable as before.

15 Heath–Jarrow–Morton model

Introduction

The Heath-Jarrow-Morton (HJM) and LIBOR Market Model (LMM) frameworks represent generalizations of the interest rate models met in previous chapters, extending the ideas behind Black's model and traditional short-rate models.

While Black's model provides a lognormal description of a single forward or swap rate, and short-rate models describe the stochastic evolution of only the instantaneous interest rate, the HJM framework models the entire forward rate curve directly in a continuous-time, arbitrage-free setting. This allows it to capture the joint dynamics of interest rates across all maturities ensuring consistency in pricing multiple securities in a portfolio or pricing securities with multiple payoffs.

The LMM or Brace-Gatarek-Musiela (BGM) model further adapts the HJM framework to the discrete, market-observable forward rates used in practice, ensuring consistency with Black's formula.

Together, the HJM and LMM models form the theoretical and practical foundations of modern term-structure modeling. Short-rate models still have a place either for modeling interest rate risk or applications, where the ease of use or analytical tractability are concerns. The choice of model often depends on the specific use case such as pricing complex derivatives, risk management or asset-liability management and a balance between tractability, computational efficiency and consistency with market data. In practice, financial institutions often further adapt these models by modeling multiple curves simultaneously, allowing for stochastic volatility or incorporating random jumps to better capture complex market dynamics and specific product features.

15.1 Heath, Jarrow and Morton model

The HJM model (Heath et al., 1992) is a framework for modeling the evolution of interest rates over time. Unlike the short-rate models discussed in the previous chapter that specify dynamics for the instantaneous short rate and

DOI: 10.4324/9781003477624-19

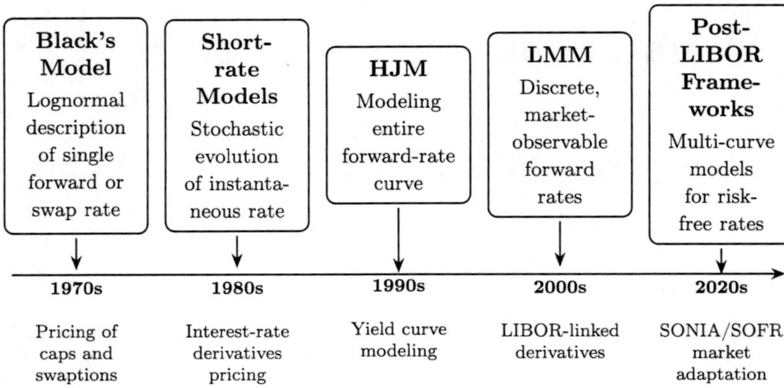

Figure 15.1 Evolution of interest-rate models

derive the entire term structure from it, the HJM framework starts by modeling the entire forward rate curve directly.

A key insight of the HJM framework is that the absence of arbitrage imposes a precise relationship between the drift and volatility of forward rates. Once the volatility structure is specified, the drift is automatically determined making the HJM model internally consistent.

The HJM framework encompasses many well-known short-rate models as special cases (such as the Vasicek, Hull–White and Cox–Ingersoll–Ross models) when specific functional forms for the volatility are chosen. However, some specifications of the volatility lead to path-dependent short-rates making pricing complicated and resource intensive.

A bigger issue is that the HJM framework is incompatible with lognormal forward rates. This is a major problem as lognormality ensures positive forward rates, makes the model mathematically tractable and is consistent with Black's model.

Instead of modeling the short-rate and deriving the zero-coupon bond price as in the previous chapter, the HJM model begins with a specification of the instantaneous forward curve. We can then determine bond prices from the forward rates.

Assuming just one-factor, the forward rate at t for investing at T is described by

$$df = \alpha(t, T)dt + \sigma(t, T)d\tilde{W} \tag{15.1}$$

where $\tilde{W}$ is standard Brownian motion under the risk-neutral probability measure $\mathbb{Q}$.

The zero-coupon bond price is related to the forward rate by Equation (12.5)

$$f_{t,\tau} = -\frac{\partial \log P_{t,\tau}}{\partial \tau}$$

Then integrating with respect to maturity τ

$$\begin{aligned}
\Longrightarrow \int_t^T f_{t,\tau}\, d\tau &= -(\log P_{t,T} - \log P_{t,t}) \\
&= -\log P_{t,T} \\
\Longrightarrow P_{t,T} &= e^{-\int_t^T f_{t,\tau}\, d\tau} \qquad (15.2)
\end{aligned}$$

Theorem 15.1. *(No-arbitrage condition) Assuming that there exists a measure* $\mathbb{Q}$ *such* $\frac{P_{t,T}}{B_t}$ *is a martingale and forward rates satisfy Equation (15.1) then*

$$\alpha(t,T) = \sigma(t,T)\int_t^T \sigma(t,\tau)d\tau$$

Proof. To prove the result, write an expression for $d\frac{P}{B}$ and set the coefficient of dt to zero.

First let $e^{X_t} = P_{t,T}$ and apply Itô's formula

$$\begin{aligned}
d(e^{X_t}) &= e^{X_t} dX + \frac{1}{2}e^{X_t}(dX)^2 \\
\Longrightarrow dP &= P_{t,T}\, dX + \frac{1}{2}P_{t,T}(dX)^2 \qquad (15.3)
\end{aligned}$$

Now applying the stochastic quotient rule and letting $dB = r_t B_t dt$,

$$\begin{aligned}
d\frac{P}{B} &= \frac{1}{B_t}dP - \frac{P_{t,T}}{B_t^2}dB \\
&= \frac{P_{t,T}}{B_t}\left(dX + \frac{1}{2}(dX)^2\right) - \frac{P_{t,T}}{B_t^2}(r_t B_t dt) \\
&= \frac{P_{t,T}}{B_t}\left(dX + \frac{1}{2}(dX)^2 - r_t dt\right) \qquad (15.4)
\end{aligned}$$

From Equation (15.2)

$$dX = d\left(-\int_t^T f_{t,\tau}\, d\tau\right)$$

Then using the Leibniz integral rule (Equation 20 in Appendix A.3 of Chapter 14)

$$dX = f_{t,t}\,dt - \int_t^T df\,d\tau$$

And since $f_{t,t}$ is the short-rate r_t

$$dX = r_t\,dt - \int_t^T (\alpha dt + \sigma d\tilde{W})\,d\tau$$

where we have temporarily dropped the dependency of α and σ on (t, T). Interchanging the integrals using the Stochastic Fubini Theorem

$$dX = \left(r_t - \int_t^T \alpha\,d\tau\right)dt - \left(\int_t^T \sigma\,d\tau\right)d\tilde{W}$$

Letting $\alpha^*(t,T) = \int_t^T \alpha(t,\tau)\,d\tau$ and $\sigma^*(t,T) = \int_t^T \sigma(t,\tau)\,d\tau$

$$dX = (r_t - \alpha^*)dt - \sigma^* d\tilde{W} \tag{15.5}$$

Using this in Equation (15.4)

$$\begin{aligned} d\frac{P}{B} &= \frac{P_{t,T}}{B_t}\left((-\alpha^* + \frac{1}{2}\sigma^{*2})dt - \sigma^* d\tilde{W}\right) \\ &= \frac{P_{t,T}}{B_t}\sigma^*\left((-\frac{\alpha^*}{\sigma^*} + \frac{1}{2}\sigma^*)dt - d\tilde{W}\right) \end{aligned}$$

$\frac{P_{t,T}}{B_t}$ is a martingale if

$$\begin{aligned} -\frac{\alpha^*}{\sigma^*} + \frac{1}{2}\sigma^* &= 0 \\ \Longrightarrow \alpha^* &= \frac{1}{2}\sigma^{*2} \end{aligned}$$

since $\frac{\partial \alpha^*}{\partial T} = \alpha$ and $\frac{\partial \sigma^*}{\partial T} = \sigma$, then differentiating both sides with respect to T leads to

$$\alpha(t,T) = \sigma(t,T)\sigma^*(t,T)$$

This is then the *no-arbitrage condition* since the existence of a risk-neutral probability measure implies no arbitrage according to the first fundamental theorem of asset pricing. □

In order to be arbitrage free, the instantaneous forward rate process must be described by

$$df = \sigma(t,T)\sigma^*(t,T)dt + \sigma(t,T)d\tilde{W} \tag{15.6}$$

That is, the specification of the forward rate volatility $\sigma(t,T)$ completely determines the model of the instantaneous forward rate. The short-rate and zero-coupon bond prices can then be found from this forward rate process.

To find the short rate put the instantaneous forward rate process in integral form giving $f_{t,T}$ and then the spot rate is $r_t = f_{t,t}$.

$$f_{t,T} = f_{0,T} + \int_0^t \sigma(\tau,T)\sigma^*(\tau,T)d\tau + \int_0^t \sigma(\tau,T)d\tilde{W}$$

Now set T to t

$$r_t = f_{t,t} = f_{0,t} + \int_0^t \sigma(\tau,t)\sigma^*(\tau,t)d\tau + \int_0^t \sigma(\tau,t)d\tilde{W}$$

To find the zero-coupon bond price substitute the expression for dX (15.5) in the expression for dP (15.3) and solve the resulting SDE:

$$\begin{aligned} dP &= P_{t,T}\,dX + \frac{1}{2}P_{t,T}(dX)^2 \\ &= P_{t,T}\left((r_t - \alpha^*)dt - \sigma^* d\tilde{W}\right) + \frac{1}{2}P_{t,T}(\sigma^{*2}dt) \end{aligned}$$

And since $\alpha^* = \frac{1}{2}\sigma^{*2}$

$$dP = P_{t,T}(r_t\,dt - \sigma^*(t,T)\,d\tilde{W}) \tag{15.7}$$

The key insight of HJM is that specifying the volatility structure determines the drift term to ensure no arbitrage. This guarantees that all bond prices implied by the model are consistent with each other and arbitrage free.

The next example shows how the Vasicek model falls under the umbrella of the HJM framework.

Exercise 15.1. *Show that the Vasicek short-rate model*

$$dr = \alpha(\mu - r_t)dt + \sigma d\tilde{W}$$

implies the HJM forward-rate volatility

$$\sigma(t,T) = \sigma e^{-\alpha(T-t)}$$

Proof. From Section 14.2.1

$$P_{t,T} = e^{A_{t,T} - B_{t,T} r_t}$$

Using the definition of the instantaneous forward rate (Equation 12.5)

$$\begin{aligned} f_{t,T} &= -\frac{\partial \log P_{t,T}}{\partial T} \\ &= -\frac{\partial A_{t,T}}{\partial T} + r_t \frac{\partial B_{t,T}}{\partial T} \\ \Longrightarrow df &= -\frac{\partial}{\partial t}\left(\frac{\partial A_{t,T}}{\partial T}\right)dt + r_t \frac{\partial}{\partial t}\left(\frac{\partial B_{t,T}}{\partial T}\right)dt + \frac{\partial B_{t,T}}{\partial T} dr \end{aligned}$$

Now substitute for the Vasicek model

$$dr = \alpha(\mu - r_t)dt + \sigma d\tilde{W}$$

using

$$B_{t,T} = \frac{(1 - e^{-\alpha(T-t)})}{\alpha} \Longrightarrow \frac{\partial B_{t,T}}{\partial T} = e^{-\alpha(T-t)}$$

This gives

$$df = [\ldots]dt + \sigma e^{-\alpha(T-t)} d\tilde{W}$$

Then compare with the general representation in Equation (15.6) above focusing on the $d\tilde{W}$ terms.

$$\sigma(t, T) = \sigma e^{-\alpha(T-t)}$$

□

The HJM specification of the forward rate curve is very general. The previous example shows how to specify the model to make it compatible with the Vasicek model. The same is possible for any arbitrage-free model. Table 15.1 shows the particular functional form of $\sigma(t, T)$ corresponding to a number of short-rate models including those considered in the previous chapter.

In practical derivative pricing with the HJM model, the procedure is then to choose the form of $\sigma(\tau, T)$ with parameters estimated from historical market-observed prices and simulate the evolution of forward rates. Then assuming the

Table 15.1 Implied HJM volatility specifications

Model	*SDE*	*HJM volatility* $\sigma(t,T)$
Vasicek	$dr = \alpha(\mu - r_t)dt + \sigma d\tilde{W}$	$\sigma e^{-\alpha(T-t)}$
CIR	$dr = \alpha(\mu - r_t)dt + \sigma\sqrt{r_t}d\tilde{W}$	$\sigma e^{-\alpha(T-t)}\sqrt{r_t}$
Hull-White	$dr = \alpha(\mu(t) - r_t)dt + \sigma d\tilde{W}$	$\sigma e^{-\alpha(T-t)}$
Ho-Lee	$dr = \mu(t)dt + \sigma d\tilde{W}$	σ

payoff can be expressed as a function of forward rates the remaining challenge is to discount the payoff back to the present using the short-rate.

However, the short rate is path dependent or *non-Markov* in some specifications of the HJM volatility. What this means is that evaluating the short rate at the next time-step requires the current value and past values. When this is the case, a tree diagram model becomes non-recombining and Monte Carlo simulation must be used instead. The HJM model becomes Markov only when its volatility structure takes a particular form (Filipović, 1999). The short-rate models studied so far as well as many other models actually used in practice have this form.

Although the HJM framework ensures models are free of arbitrage, it is incompatible with lognormal forward rates. Lognormality is attractive for a number of reasons. It ensures forward rates are always positive. Lognormality is easy to work with mathematically. And lognormal forward rates are consistent with how practitioners price interest rate derivatives in Black's model.

To see why we cannot have lognormal forward rates in the HJM framework set $\sigma(t,T)$ to $\sigma f_{t,T}$ in Equation (15.6) so that the SDE has the correct form to yield a lognormal solution (Section 7.1). Heath et al. (1992) find that an SDE with this structure does not have a finite-valued solution. The forward rates become infinitely big generating zero bond prices.

Although continuously compounded forward rates cannot be lognormal in the HJM framework, Brace et al. (1997) found that it is possible with the discretely compounded forward rates observed in the market. Their model is the subject of the next section.

15.2 Brace, Gatarek and Musiela (BGM) model

Black's model relied on an estimate of forward rate volatility for pricing caplets. The model does not provide structure to these estimates and forward rates may or may not be completely independent of each other. The BGM model or *LIBOR Market Model* or *forward LIBOR model* generalizes Black's model to many forward rates evolving together. The approach has the attraction of modeling discretely compounded forward rates similar to practitioners instead of instantaneous rates not observed in the market as well as the attraction of lognormality. The derivation of the model follows directly from the HJM framework.

15.2.1 From Heath – Jarrow—Morton to the Brace—Gatarek—Musiela model

First we establish the dynamics of the discrete forward rate under the risk-neutral measure $\mathbb{Q}$ with B_t as numéraire.

Let $F_i(t)$ be shorthand for the discrete forward rate $F_{t,T_i,T_{i+1}}$ that applies from time T_i to T_{i+1} observed at time t. From the definition of the discrete forward rate in Equation (12.8)

$$F_i = \frac{1}{\delta}\left(\frac{P_{t,T_i}}{P_{t,T_{i+1}}} - 1\right) \tag{15.8}$$

where $\delta = T_{i+1} - T_i$.

As the discrete forward rate is expressed in terms of a ratio of zero-coupon bonds $\frac{P_{t,T_i}}{P_{t,T_{i+1}}}$, the dynamics of this process must first be determined.

From Theorem 15.1, we have that under $\mathbb{Q}$ the discounted bond process $\frac{P_{t,T}}{B_t}$ is a martingale

$$d\frac{P}{B} = -\frac{P_{t,T}}{B_t}\sigma^*(t,T)d\tilde{W}$$

Following the structure of Exercise 6.4 with non-constant $\sigma^*(t,T)$, the integral form is

$$\begin{aligned} d\log\frac{P_{t,T}}{B_t} &= -\frac{1}{2}\sigma^{*2}(t,T)dt - \sigma^*(t,T)d\tilde{W} \\ \Longrightarrow \log\frac{P_{t,T}}{B_t} - \log\frac{P_{0,T}}{B_0} &= -\frac{1}{2}\int_0^t \sigma^{*2}(s,T)ds - \int_0^t \sigma^*(s,T)d\tilde{W} \\ \Longrightarrow \frac{P_{t,T}}{B_t} &= \frac{P_{0,T}}{B_0}e^{-\frac{1}{2}\int_0^t \sigma^{*2}(s,T)ds - \int_0^t \sigma^*(s,T)d\tilde{W}} \end{aligned} \tag{15.9}$$

Write out expressions for $\frac{P_{t,T_i}}{B_t}$ and $\frac{P_{t,T_{i+1}}}{B_t}$ and then divide to get

$$\begin{aligned} \frac{P_{t,T_i}}{P_{t,T_{i+1}}} &= \frac{P_{0,T_i}}{P_{0,T_{i+1}}}e^{-\frac{1}{2}\int_0^t(\sigma^{*2}(s,T_i)-\sigma^{*2}(s,T_{i+1}))ds - \int_0^t(\sigma^*(s,T_i)-\sigma^*(s,T_{i+1}))d\tilde{W}} \\ &= \frac{P_{0,T_i}}{P_{0,T_{i+1}}}e^{-\frac{1}{2}\int_0^t(\sigma_i^{*2}-\sigma_{i+1}^{*2})ds - \int_0^t(\sigma_i^*-\sigma_{i+1}^*)d\tilde{W}} \end{aligned} \tag{15.10}$$

using the shorthand σ_i^* for $\sigma^*(s,T_i)$, etc.

The second step is to change the probability measure to the T_{i+1}-forward measure $\mathbb{Q}^{T_{i+1}}$ with $P_{t,T_{i+1}}$ as numéraire. Just as in Section 13.5.2 where we priced caplets using a forward measure the discrete forward rate is a martingale under this measure.

Theorem 15.2. *Under the T_{i+1}-forward measure $\mathbb{Q}^{T_{i+1}}$, the stochastic process W_{i+1} defined by*

$$dW_{i+1} = \sigma^*_{i+1} dt + d\tilde{W}$$

is standard Brownian motion.

Proof. By Theorem 10.4, the Radon-Nikodým derivative to change measure from the risk-neutral measure $\mathbb{Q}$ to the T_{i+1}-forward measure $\mathbb{Q}^{T_{i+1}}$ is given by

$$\Theta_t = \frac{d\mathbb{Q}^{i+1}}{d\mathbb{Q}} = \frac{P_{t,T_{i+1}}/P_{0,T_{i+1}}}{B_t/B_0} = \frac{P_{t,T_{i+1}}}{P_{0,T_{i+1}} B_t}$$

From Equation (15.9), this is

$$\Theta_t = e^{-\frac{1}{2}\int_0^t \sigma^{*2}_{i+1} ds - \int_0^t \sigma^*_{i+1} d\tilde{W}}$$

and this time going from integral form to differential form (Exercise 6.4 with non-constant integrand in reverse)

$$d\Theta = -\sigma^*_{i+1} d\tilde{W}$$

From Girsanov's theorem, the stochastic process W_{i+1} defined by

$$dW_{i+1} = \sigma^*_{i+1} dt + d\tilde{W}$$

is standard Brownian motion under $\mathbb{Q}^{T_{i+1}}$. □

Now write the ratio of zero-coupon bonds in terms of W_{i+1} the standard Brownian motion under $\mathbb{Q}^{T_{i+1}}$. Substituting $d\tilde{W} = dW_{i+1} - \sigma^*_{i+1} ds$ into Equation (15.10)

$$\begin{aligned}
\frac{P_{t,T_i}}{P_{t,T_{i+1}}} &= \frac{P_{0,T_i}}{P_{0,T_{i+1}}} e^{-\frac{1}{2}\int_0^t (\sigma_i^{*2} - 2\sigma^*_{i+1}(\sigma^*_i - \sigma^*_{i+1}) - \sigma^{*2}_{i+1}) ds - \int_0^t (\sigma^*_i - \sigma^*_{i+1}) dW_{i+1}} \\
&= \frac{P_{0,T_i}}{P_{0,T_{i+1}}} e^{-\frac{1}{2}\int_0^t (\sigma^*_i - \sigma^*_{i+1})^2 ds - \int_0^t (\sigma^*_i - \sigma^*_{i+1}) dW_{i+1}}
\end{aligned}$$

and going from integral form to differential form

$$d\left(\frac{P_{t,T_i}}{P_{t,T_{i+1}}}\right) = \frac{P_{t,T_i}}{P_{t,T_{i+1}}}(\sigma^*_{i+1} - \sigma^*_i)dW_{i+1}$$

Using the definition of the discrete forward rate from Equation (15.8)

$$\begin{aligned} dF_i &= \frac{1}{\delta}(1 + F_i\delta)(\sigma^*_{i+1} - \sigma^*_i)dW_{i+1} \\ &= F_i\frac{(1 + F_i\delta)}{F_i\delta}(\sigma^*_{i+1} - \sigma^*_i)dW_{i+1} \end{aligned} \tag{15.11}$$

Define

$$\zeta_i := \frac{(1 + F_i\delta)}{F_i\delta}(\sigma^*_{i+1} - \sigma^*_i) \tag{15.12}$$

If we choose $\sigma(t, T)$ in such a way[1] that ζ_i is a deterministic function $\zeta_i(t)$, then Equation (15.11) reduces to

$$dF_i = F_i\,\zeta_i\,dW_{i+1} \tag{15.13}$$

As there is no trend term $F_i(t) := F_{t,T_i,T_{i+1}}$ the discrete forward rate from time T_i to T_{i+1} observed at t is then a martingale under the T_{i+1}-forward measure $\mathbb{Q}^{i+1}$. And furthermore the solution for F_i will be lognormal. Therefore we have achieved our objective of obtaining lognormally distributed discretely compounded forward rates.

In the integral form, we have

$$\log\frac{F_{t,T_i,T_{i+1}}}{F_{0,T_i,T_{i+1}}} = -\frac{1}{2}\int_0^t \zeta_i^2(s)ds + \int_0^t \zeta_i(s)dW_{i+1} \tag{15.14}$$

which from Section 6.6.1 is normally distributed with expectation

$$-\frac{1}{2}\int_0^t \zeta_i^2(s)ds$$

and variance given by the isometry property of Itô integrals

$$\int_0^t \zeta_i^2(s)ds$$

Given a sequence of dates $0 = T_0, T_1, T_2, \ldots, T_k, T_{k+1}, \ldots, T_{i-1}, T_i, T_{i+1}$ we can similarly express all discrete forward rates in terms of standard Brownian motion (SBM) under the appropriate T-forward measure

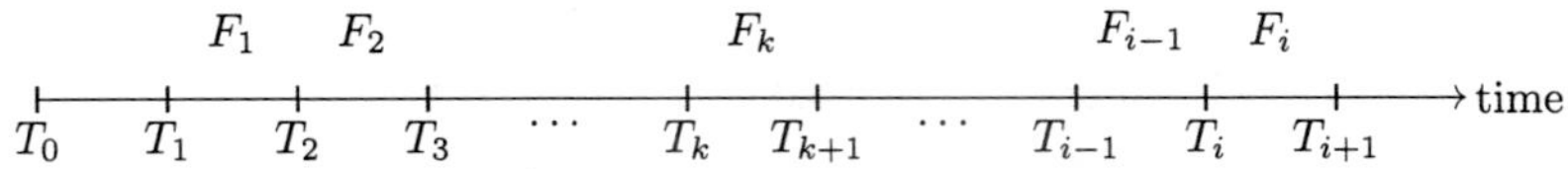

Figure 15.2 Rolling risk-neutral measure

$$
\begin{aligned}
dF_i &= F_i\,\zeta_i\,dW_{i+1} && \text{where } W_{i+1} \text{ is SBM under } \mathbb{Q}^{T_{i+1}} \\
dF_{i-1} &= F_{i-1}\,\zeta_{i-1}\,dW_i && \text{where } W_i \text{ is SBM under } \mathbb{Q}^{T_i} \\
dF_{i-2} &= F_{i-2}\,\zeta_{i-2}\,dW_{i-1} && \text{where } W_{i-1} \text{ is SBM under } \mathbb{Q}^{T_{i-1}} \\
&\ \vdots && \qquad\vdots \\
dF_k &= F_k\,\zeta_k\,dW_{k+1} && \text{where } W_{k+1} \text{ is SBM under } \mathbb{Q}^{T_{k+1}} \\
dF_{k-1} &= F_{k-1}\,\zeta_{k-1}\,dW_k && \text{where } W_k \text{ is SBM under } \mathbb{Q}^{T_k}
\end{aligned}
$$

Each forward rate is a martingale under its own T-forward measure and the measure rolls along the tenor structure changing from one forward measure to the next. This sequence of forward measures is called a 'rolling risk-neutral measure'.

The discrete forward rate for all time intervals can be written under a common measure. This is essential for Monte Carlo simulation. From Theorem 15.2,

$$
\begin{aligned}
dW_{i+1} &= \sigma^*_{i+1}dt + d\tilde{W} \\
\text{and } dW_i &= \sigma^*_i dt + d\tilde{W}
\end{aligned}
$$

eliminating the $d\tilde{W}$ term

$$
\begin{aligned}
dW_{i+1} &= \sigma^*_{i+1}dt + (dW_i - \sigma^*_i dt) \\
&= (\sigma^*_{i+1} - \sigma^*_i)dt + dW_i
\end{aligned}
$$

Using the definition of $\zeta_i(t)$ in (15.12)

$$
dW_{i+1} = \frac{\zeta_i F_i \delta}{1 + F_i \delta} dt + dW_i
$$

This process is then repeated iteratively choosing some reference W_k so that

$$
dW_{i+1} = \sum_{j=k}^{i} \frac{\zeta_j F_j \delta}{1 + F_j \delta} dt + dW_k
$$

Substituting this into the discrete forward rate process under $\mathbb{Q}^{T_{i+1}}$ (15.13) leads to the discrete forward rate process under $\mathbb{Q}^{T_k}$

$$dF_i = F_i \zeta_i \left(\sum_{j=k}^{i} \frac{\zeta_j F_j \delta}{1 + F_j \delta} dt + dW_k \right) \tag{15.15}$$

15.2.2 Pricing power call options

A power call option is a type of non-standard derivative, where the payoff depends on a power of the underlying asset price rather than the price itself. The section describes how to price a power call option where the underlying is the spot rate.

Since $F_{T_i,T_i,T_{i+1}} = R_{T_i,T_{i+1}}$ we have from Equation (15.14) that under $\mathbb{Q}^{T_{i+1}}$

$$\log \frac{F_{T_i,T_i,T_{i+1}}}{F_{0,T_i,T_{i+1}}} = \log \frac{R_{T_i,T_{i+1}}}{F_{0,T_i,T_{i+1}}} \sim N\left(-\frac{1}{2} \int_0^{T_i} \zeta_i^2(s) ds, \int_0^{T_i} \zeta_i^2(s) ds \right) \tag{15.16}$$

The forward caplet volatility σ^i_{fwd} was given in Section 13.5.2 as the volatility of $R_{T_i,T_{i+1}}$ therefore

$$(\sigma^i_{\text{fwd}})^2 T_i = \int_0^{T_i} \zeta_i^2(s) ds \tag{15.17}$$

Substituting this into the distribution in (15.16) above

$$\log \frac{R_{T_i,T_{i+1}}}{F_{0,T_i,T_{i+1}}} \sim N\left(-\frac{1}{2} (\sigma^i_{\text{fwd}})^2 T_i, (\sigma^i_{\text{fwd}})^2 T_i \right) \tag{15.18}$$

and from Theorem 2.6

$$\begin{aligned} \mathbb{E}^{i+1} R^{\alpha}_{T_i,T_{i+1}} &= F^{\alpha}_{0,T_i,T_{i+1}} e^{-\frac{\alpha}{2}(\sigma^i_{\text{fwd}})^2 T_i + \frac{\alpha^2}{2}(\sigma^i_{\text{fwd}})^2 T_i} \\ &= F^{\alpha}_{0,T_i,T_{i+1}} e^{\frac{\alpha(\alpha-1)}{2}(\sigma^i_{\text{fwd}})^2 T_i} \end{aligned}$$

where $\mathbb{E}^{i+1}$ is the expectation with respect to $\mathbb{Q}^{T_{i+1}}$. The volatility of $R^{\alpha}_{T_i,T_{i+1}}$ becomes $\alpha\, \sigma^i_{\text{fwd}}{}^2$

From Section 13.5.2, the power call option price at $t = 0$ is

$$\begin{aligned} \text{power} &= LP_{0,T_{i+1}}\, \mathbb{E}^{i+1}[R^{\alpha}_{T_i,T_{i+1}} - K]^+ \\ &= LP_{0,T_{i+1}} \left(\mathbb{E}^{i+1} R^{\alpha}_{T_i,T_{i+1}}\, \Phi(d_1) - K\Phi(d_2) \right) \end{aligned}$$

where

$$d_1 = \frac{\log\left(\mathbb{E}^{i+1} R^{\alpha}_{T_i,T_{i+1}} / K\right) + \frac{1}{2}(\alpha\sigma^i_{\text{fwd}})^2 T_i}{(\alpha\sigma^i_{\text{fwd}})\sqrt{T_i}}, d_2 = d_1 - (\alpha\sigma^i_{\text{fwd}})\sqrt{T_i}$$

Exercise 15.2. *A power call option pays at maturity time 1 the excess of the square of the 3-month LIBOR over strike:*

$$L[R^2_{0.75,1} - K]^+$$

where

- $R_{0.75,1}$ *is the quarterly compounded LIBOR at time 0.75 with maturity at time 1.*
- *The strike price* $K = (3.12\%)^2$.
- *The discount factor* $P_{0,1} = 0.9530$.
- *Nominal value L is \$1 million.*
- *The current quarterly compounded forward rate* $F_{0,0.75,1} = 3.42\%$.
- *The current forward caplet volatility* $\sigma^i_{fwd} = 35\%$.

Calculate the price of the power call option using the LMM.

[QFIQF – Fall 2019 Q12]

Proof. Since

$$\mathbb{E}^{T=1} R^2_{0.75,1} = 0.0342^2 e^{\frac{2(2-1)}{2} 0.35^2 \times 1} = 0.001282192$$

where $\mathbb{E}^{T=1}$ is the expectation with respect to the 1-forward measure. Then

$$d_1 = \frac{\log\left(0.001282192 / 0.0312^2\right) + \frac{1}{2}(2 \times 0.35)^2 \times 1}{2 \times 0.35 \times \sqrt{1}} = 0.7575$$

$$d_2 = 0.7575 - 2 \times 0.35 \times \sqrt{1} = 0.1513$$

The power call option price is

$$1\text{m} \times 0.9530 \times \left(0.001282192 \times \Phi(0.7575) - 0.0312^2 \times \Phi(0.1513)\right) = \$428.14$$

□

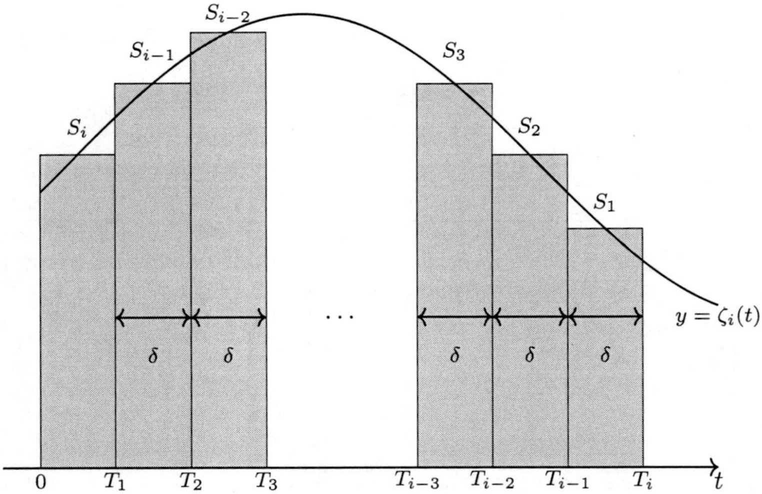

Figure 15.3 The approximation of $\zeta_i(t)$ by a step function

15.2.3 Calibrating forward rate volatilities

Using market data from caps and floors, forward rates in the BGM model can be calibrated in order to price other more general interest rate options.

From Equation (15.17), the forward caplet volatility with maturity T_{i+1} should equal an integral of the forward rate volatility

$$(\sigma^i_{\text{fwd}})^2 T_i = \int_0^{T_i} \zeta_i^2(s)ds$$

To simplify the calibration, it is commonly assumed that $\zeta_i(t)$ is a step function depending on the number of time intervals until T_i.

$$\zeta_i(t) = \begin{cases} S_i & t \le T_1 \\ S_{i-1} & T_1 < t \le T_2 \\ S_{i-2} & T_2 < t \le T_3 \\ & \dots \\ S_1 & T_{i-1} < t \le T_i \end{cases}$$

where $S_1, S_2, \dots$ are positive value constants.

Replacing the integral above by a simple sum, the relationship between the implied caplet forward volatility and the BGM forward volatilities then takes the form

$$(\sigma^i_{\text{fwd}})^2 T_i = S_i^2\, T_1 + S_{i-1}^2\, \delta + \cdots + S_1^2\, \delta \tag{15.19}$$

where S_k is the value of the step function when time is k time intervals from T_i and δ is the length of each time interval.

Exercise 15.3. *You are given the following information about three caplets:*

T_i	*Implied forward volatility* σ^i_{fwd}
$T_1 = 0.25$	0.21
$T_2 = 0.5$	0.24
$T_3 = 0.75$	0.30

where σ^i_{fwd} *is the forward volatility implied by a caplet maturing at* T_{i+1}. *Calculate the corresponding forward rate volatilities* S_i.

[QFIQF – Fall 2022 Q7]

Proof. The relationship between implied caplet forward volatilities and the BGM forward volatilities in Equation (15.19) gives

$$\begin{aligned}
(\sigma^1_{\text{fwd}})^2 T_1 &= S_1^2\, T_1 & &\Longrightarrow 0.21^2 \times 0.25 &&= 0.25\, S_1^2 \\
(\sigma^2_{\text{fwd}})^2 T_2 &= S_2^2\, T_1 + S_1^2\, \delta & &\Longrightarrow 0.24^2 \times 0.5 &&= 0.25\, S_2^2 + 0.25\, S_1^2 \\
(\sigma^3_{\text{fwd}})^2 T_3 &= S_3^2\, T_1 + \delta(S_2^2 + S_1^2) & &\Longrightarrow 0.30^2 \times 0.75 &&= 0.25\, S_3^2 + 0.25(S_2^2 + S_1^2)
\end{aligned}$$

Solving from top to bottom, $S_1 = 0.21, S_2 = 0.268$ and $S_3 = 0.393$. □

15.2.4 Monte Carlo simulation

Monte Carlo simulation is the standard way to price complex interest rate derivatives under the LMM especially for instruments with path-dependent (e.g. Asian options) or multi-period payoffs (e.g. swaptions).

From the process for the forward rate $F_i(t) := F_{t,T_i,T_{i+1}}$ given in Equation (15.15), we have

$$\frac{dF_i}{F_i(t)} = \zeta_i(t) \left(\sum_{j=k}^{i} \frac{\zeta_j(t) F_j(t) \delta}{1 + F_j(t)\delta} dt + dW_k \right) \tag{15.20}$$

where $\zeta_i(t)$ has been calibrated as a step function depending on the number of time intervals until T_i for all i. Freeze the right-hand side drift and diffusion terms at their current values to solve. Therefore, following the structure of Exercise 6.4

$$\log \frac{F_i(t+\Delta t)}{F_i(t)} = \left(\zeta_i(t) \sum_{j=k}^{i} \frac{\zeta_j(t) F_j(t)\delta}{1+F_j(t)\delta} - \frac{1}{2}\zeta_i^2(t) \right) \Delta t + \zeta_i(t)\sqrt{\Delta} Z \qquad (15.21)$$

where Z is a standard normally distributed variable.

If the payoff depends on two forward rates $F_{t,T_i,T_{i+1}}$ and $F_{t,T_m,T_{m+1}}$, then the Monte Carlo procedure is to

- Determine the forward rate at the next time step for the first forward rate $F_{t+\Delta t,T_i,T_{i+1}}$. And repeat for each time step until the forward rate at maturity $F_{T,T_i,T_{i+1}}$ is determined.
- Do the same for the second forward rate $F_{t,T_m,T_{m+1}}$.
- Calculate the payoff.
- Repeat for a large number of forward rate simulations.
- Calculate the expected payoff by taking the average of the simulated payoffs.
- Calculate the discounted expected payoff.

The example below shows the general principle.

Exercise 15.4. *You are given that:*

$$F_{0,0.25,0.5} = 1.1\%$$
$$F_{0,0.5,0.75} = 1.2\%$$
$$F_{0,0.75,1} = 1.3\%$$

and the BGM forward volatilities have been calibrated as in the previous exercise.

Under the T-forward risk-neutral measure, where T = 0.5, the following information is also provided:

Current t	*Forward rate*	*Random simulation from a standard normal*
0.25	$F_{0.25,0.5,0.75}$	$z_1 = 0.1$
0.25	$F_{0.25,0.75,1}$	$z_2 = -0.2$

(a) Simulate one outcome for each of the two forward rates in a single simulation step.

(b) Given a call option maturing at $T = 0.5$ with strike price, K, on a zero-coupon bond with notional value of 1000 and maturity at $T = 1$, describe how the call option on the zero-coupon bond can be priced using Monte Carlo simulations.

[QFIQF – Fall 2022 Q7]

Proof.

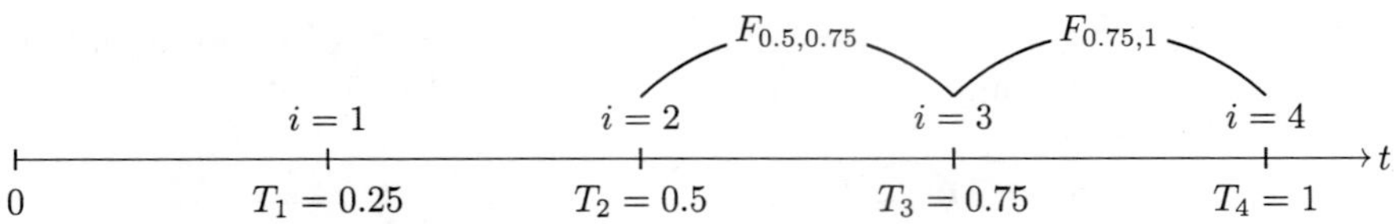

(a) From Exercise 15.3, we have for $T_2 = 0.5$ the step function is

$$\zeta_2(t) = \begin{cases} 0.268 & 0 < t \le 0.25 \\ 0.21 & 0.25 < t \le 0.5 \end{cases}$$

$F_{t,0.5,0.75}$ is a martingale under $\mathbb{Q}^{T_3}$ the forward risk-neutral measure for $T_3 = 0.75$ so we need to roll back one time interval to use $\mathbb{Q}^{T_2}$ the forward risk-neutral measure for $T_2 = 0.5$.

Letting $i = 2, k = 2$ in Equation (15.21)

$$\begin{aligned} \log \frac{F_{0.25,0.5,0.75}}{F_{0,0.5,0.75}} &= \left(\frac{\zeta_2^2(0.25) \times F_{0,0.5,0.75} \times \delta}{1 + F_{0,0.5,0.75} \times \delta} - \frac{1}{2}\zeta_2^2(0.25) \right) 0.25 + \zeta_2(0.25)\sqrt{0.25}Z \\ &= \left(\frac{0.268^2 \times 1.2\% \times 0.25}{1 + 1.2\% \times 0.25} - \frac{1}{2} 0.268^2 \right) 0.25 + 0.268 \times \sqrt{0.25} \times 0.1 \end{aligned}$$

$$\implies F_{0.25,0.5,0.75} = 1.2054\%$$

For $T_3 = 0.75$ the step function is

$$\zeta_3(t) = \begin{cases} 0.393 & 0 < t \le 0.25 \\ 0.268 & 0.25 < t \le 0.5 \\ 0.21 & 0.5 < t \le 0.75 \end{cases}$$

$F_{t,0.75,1}$ is a martingale under $\mathbb{Q}^{T_4}$ the forward risk-neutral measure for $T_4 = 1$ so we need to roll back two time intervals to use $\mathbb{Q}^{T_2}$.

Letting $i = 3, k = 2$ in Equation (15.21)

$$\log \frac{F_{0.25,0.75,1}}{F_{0,0.75,1}} = \left(\zeta_3(0.25)\left(\frac{\zeta_2(0.25) \times F_{0,0.5,0.75} \times \delta}{1 + F_{0,0.5,0.75} \times \delta} + \frac{\zeta_3(0.25) \times F_{0,0.75,1} \times \delta}{1 + F_{0,0.75,1} \times \delta}\right)\right.$$
$$\left. - \frac{1}{2}\zeta_3^2(0.25)\right)0.25 + \zeta_3(0.25)\sqrt{0.25}Z$$
$$= \left(\frac{0.393 \times 0.268 \times 1.2\% \times 0.25}{1 + 1.2\% \times 0.2} + \frac{0.393^2 \times 1.3\% \times 0.25}{1 + 1.3\% \times 0.25}\right.$$
$$\left. - \frac{1}{2} \times 0.393^2\right)0.25 + 0.393 \times \sqrt{0.25} \times -0.2$$
$$\Longrightarrow F_{0.25,0.75,1} = 1.2263\%$$

(b) The payoff from the option can be expressed in terms of forward rates as the notional amount discounted back from $T = 1$ to $T = 0.5$:

$$\text{Payoff} = [\text{BP}_{0.5} - K]^+ = \left[\frac{1000}{(1 + F_{0.5,0.5,0.75} \times 0.25)(1 + F_{0.5,0.75,1} \times 0.25)} - K\right]^+$$

So repeat the simulation above in (a) to determine $F_{0.5,0.5,0.75}$ and $F_{0.5,0.75,1}$. Calculate the payoff. Repeat for a large number of forward rate simulations. Calculate the expected payoff by taking the average of the simulated payoffs. Calculate the discounted expected payoff by multiplying the expected payoff by $\frac{1}{1+R_{0,0.5}\times 0.5}$.

□

15.3 Summary

The HJM framework is a general and internally consistent model for the evolution of the entire forward rate curve. To ensure that the model is free of arbitrage opportunities, the drift term must satisfy the HJM drift condition. Once the volatility structure is specified, this condition determines the drift automatically. The HJM framework therefore provides a very flexible foundation for interest rate modeling.

Some volatility specifications can lead to non-Markov models, which complicates simulation and calibration although many commonly used specifications are Markovian. A more serious problem is the incompatibility of the HJM framework with lognormal continuously compounded forward rates.

In practice, market participants often prefer a more tractable version of the HJM model known as the BGM or LMM. This formulation replaces continuously compounded forward rates with discretely compounded forward rates which are directly observable in the market. Parameters can be chosen to ensure each forward rate is lognormally distributed and behaves as a martingale under its own forward measure. This property ensures consistency with Black's model. The BGM model thus preserves the theoretical rigour of the HJM framework while remaining compatible with market practice and tractable for numerical implementation. It extends Black's model by ensuring pricing across maturities is consistent.

Key things to remember

- No-arbitrage condition in HJM model: Section 15.1.
- Calibration of the BGM model using forward caplet volatilities: Section 15.2.3.

Further reading

Hull (2021); Veronesi (2010) and Mazzoni (2018) offer a clear and intuitive introduction to the term-structure modeling framework, including the relationship between the HJM approach and the market-standard LMM, with a focus on applications. Jarrow and Chatterjea (2024) connects these models to their economic foundations and arbitrage-free pricing theory, reflecting the contributions of Jarrow himself as one of the original architects of the HJM model. For readers interested in rigorous mathematical detail, Shreve (2004) and Klebaner (2012) provide comprehensive derivations of the stochastic processes underlying the HJM framework and the change-of-measure techniques essential to the LMM. Svoboda (2004) presents a detailed discussion of the LMM derivation, implementation and market relevance.

Notes

1. Klebaner (2012), Theorem 12.12 shows how.
2. From Question 6.8 in Chapter 6 if X_t has volatility σ

$$dX = \mu X_t dt + \sigma X_t dW$$

then X_t^n has volatility $n\sigma$ since

$$d(X_t^n) = (n\mu + \frac{1}{2}n(n-1)\sigma^2)X_t^n dt + n\sigma X_t^n dW$$

Questions

1 The HJM volatility $\sigma(t, T)$ in the Ho-Lee model is a constant σ.
(a) Write down an expression for the forward rate process df.
(b) Use this to find the current short-rate r_t.
(c) Then find the zero-coupon bond price $P_{t,T}$ in terms of the current short rate and the initial forward curve.

Solution

(a) From Equation (15.6) the instantaneous forward rate process is

$$\begin{aligned} df &= \sigma(t,T)\sigma^*(t,T)dt + \sigma(t,T)d\tilde{W} \\ &= \sigma\left(\int_t^T \sigma\, d\tau\right)dt + \sigma d\tilde{W} \\ &= \sigma^2(T-t)dt + \sigma d\tilde{W} \end{aligned}$$

(b) Putting the instantaneous forward rate process in integral form

$$\begin{aligned} f_{t,T} &= f(0,T) + \int_0^t \sigma^2(T-\tau)d\tau + \int_0^t \sigma d\tilde{W} \\ &= f_{0,T} + \sigma^2 t(T - \frac{t}{2}) + \sigma\tilde{W}_t \end{aligned}$$

Now set T to t

$$r_t = f_{t,t} = f_{0,t} + \frac{\sigma^2 t^2}{2} + \sigma\tilde{W}_t$$

(c) From Equation (15.2)

$$\begin{aligned} P_{t,T} &= e^{-\int_t^T f_{t,\tau}\, d\tau} \\ &= e^{-\int_t^T f_{0,\tau} d\tau - \int_t^T \sigma^2 t(\tau - \frac{t}{2})d\tau - \int_t^T \sigma\tilde{W}_\tau\, d\tau} \\ &= e^{-\int_t^T f_{0,\tau} d\tau - \sigma^2 tT(T-t)/2 - \sigma\tilde{W}_t(T-t)} \end{aligned}$$

Get rid of $\sigma\tilde{W}_t$ by re-arranging the short-rate expression in (b)

$$P_{t,T} = e^{-\int_t^T f_{0,\tau} d\tau - \sigma^2 t(T-t)^2/2 + (f_{0,t} - r_t)(T-t)}$$

2 You consider a one-factor HJM model with risk-neutral dynamics of the instantaneous forward rates $f_{t,T}$ as follows:

$$df = m(t,T)dt + \sigma e^{-\alpha(T-t)} d\tilde{W}$$

where σ and α are positive constants.

(a) Derive an expression for the drift $m(t,T)$.

(b) Prove that the short-rate r_t is given by

$$r_t = f_{0,t} + \frac{\sigma^2}{2\alpha^2}(1 - e^{-\alpha t})^2 + \sigma \int_0^t e^{-\alpha(t-\tau)} dW$$

(c) Derive the risk neutral dynamics $\frac{dP}{P_{t,T}}$ in terms of r_t and the parameters σ and α.

[QFIQF – Fall 2019 - 12(b)(c)(d)]

Solution

(a)

$$\sigma^*(t,T) = \int_t^T \sigma(t,\tau)\,d\tau = \int_t^T \sigma e^{-\alpha(\tau-t)}\,d\tau$$
$$= \frac{\sigma}{\alpha}(1 - e^{-\alpha(T-t)})$$

Then from Equation (15.6)

$$m(t,T) = \sigma(t,T)\sigma^*(t,T)$$
$$= \sigma e^{-\alpha(T-t)}\left(\frac{\sigma}{\alpha}(1 - e^{-\alpha(T-t)})\right)$$
$$= \frac{\sigma^2}{\alpha}(e^{-\alpha(T-t)} - e^{-2\alpha(T-t)})$$

(b)

$$df = \frac{\sigma^2}{\alpha}\left(e^{-\alpha(T-t)} - e^{-2\alpha(T-t)}\right)dt + \sigma e^{-\alpha(T-t)} d\tilde{W}$$

$$\Longrightarrow f_{t,T} = f_{0,T} + \frac{\sigma^2}{\alpha}\int_0^t \left(e^{-\alpha(T-\tau)} - e^{-2\alpha(T-\tau)}\right)d\tau + \int_0^t \sigma e^{-\alpha(T-\tau)} d\tilde{W}$$
$$= f_{0,T} + \frac{\sigma^2}{\alpha^2}\left(e^{-\alpha(T-t)} - e^{-\alpha T} - \frac{1}{2}e^{-2\alpha(T-t)} + \frac{1}{2}e^{-2\alpha T}\right)$$
$$+ \sigma \int_0^t e^{-\alpha(T-\tau)} d\tilde{W}$$

Now set T to t

$$r_t = f_{t,t} = f_{0,t} + \frac{\sigma^2}{2\alpha^2}\left(2 - 2e^{-\alpha t} - 1 + e^{-2\alpha t}\right) + \sigma \int_0^t e^{-\alpha(t-\tau)} d\tilde{W}$$

$$= f_{0,t} + \frac{\sigma^2}{2\alpha^2}(1 - e^{-\alpha t})^2 + \sigma \int_0^t e^{-\alpha(t-\tau)} dW$$

(c) From Equation (15.7)

$$\frac{dP}{P_{t,T}} = r_t\, dt - \sigma^*(t,T)\, d\tilde{W}$$

$$= r_t\, dt - \frac{\sigma}{\alpha}(1 - e^{-\alpha(T-t)})\, d\tilde{W}$$

3 You are using the HJM model to simulate the behavior of interest rates. Define the instantaneous forward rate process given by the HJM model when the bond price volatility is

$$\frac{\sigma}{\lambda}(e^{-\lambda(T-t)} - 1)$$

where σ and λ are constant.

[FETE - Fall 2008 - Q5(c)]

Solution From 15.7, the bond price volatility is $-\sigma^*(t,T)$

$$\sigma^*(t,T) = -\frac{\sigma}{\lambda}(e^{-\lambda(T-t)} - 1)$$

$$\Longrightarrow \sigma(t,T) = \frac{\partial \sigma^*}{\partial T} = \sigma e^{-\lambda(T-t)}$$

Therefore, from Equation (15.6), the instantaneous forward rate process is

$$df = \sigma(t,T)\sigma^*(t,T)dt + \sigma(t,T)d\tilde{W}$$

$$= -\frac{\sigma^2}{\lambda}(e^{-2\lambda(T-t)} - e^{-\lambda(T-t)})dt + \sigma e^{-\lambda(T-t)}d\tilde{W}$$

4 Your company uses Black's model with caplets' forward volatilities to price caps and is interested in offering more general options. You have chosen the LMM for pricing these options.

- $F_{t,\tau,T}$ is the n-times compounded annual forward rate at t for the period $[\tau, T]$.
- $R_{\tau,T}$ is the n-times compounded annual LIBOR spot rate as at τ for the period $[\tau, T]$.

Consider a call option on the square root of the 6-month LIBOR rate with payoff

$$L\left[\sqrt{R_{\tau,T}} - K\right]^{+}$$

Given that

- L is the notional amount and it is set at \$1 million.
- $\tau = 0.5$.
- $F_{0,0.5,1} = 3\%$.
- The option strike price K is set at 0.2.
- The implied forward volatility σ_{fwd} for a caplet maturing at one-year is 0.2.
- The one-year discount factor $P_{0,1}$ is 0.95.

(a) Calculate the value of the call option.

Suppose that the payoff function in the above is changed to

$$Le^{-|R_{\tau,T}-K|}$$

(b) Outline an algorithm to calculate the option value using the Monte Carlo method.

You plan to set the forward rate volatility $\zeta_i(t)$ of $F_{t,T_i,T_{i+1}}$ as given below:

$$\zeta_i(t) = \begin{cases} S_i & t \le T_1 \\ S_{i-1} & T_1 < t \le T_2 \\ S_{i-2} & T_2 < t \le T_3 \\ & \dots \\ S_1 & T_{i-1} < t \le T_i \end{cases}$$

where $S_1, S_2, \dots$ are positive value constants.

Given the following information:

i	T_i	σ^i_{fwd}
1	0.5	0.03
2	1	0.045
3	1.5	0.05
4	2	0.045

(c) Calculate the corresponding forward rate volatilities S_i in your plan.

[QFIQF – Spring 2020 Q13]

Solution

(a) Since

$$\mathbb{E}^{T=1}\sqrt{R_{0.5,1}} = 0.03^{0.5}e^{\frac{0.5(0.5-1)}{2}0.2^2\times 0.5} = 0.172773$$

then

$$d_1 = \frac{\log\left(0.172773\,/0.2\right) + \frac{1}{2}(0.5\times 0.2)^2\times 0.5}{0.5\times 0.2\times\sqrt{0.5}} = -2.0342$$
$$d_2 = -2.0342 - 0.5\times 0.2\times\sqrt{0.5} = -2.1049$$

The power call option price is

$$1\text{m}\times 0.95\times\left(0.172773\times\Phi(-2.0342) - 0.2\times\Phi(-2.1049)\right) = \$87.77$$

(b) We have from Section 15.2.3 that under $\mathbb{Q}^{T_{i+1}}$

$$\log\frac{R_{T_i,T_{i+1}}}{F_{t,T_i,T_{i+1}}} \sim N\left(-\frac{1}{2}(\sigma^i_{\text{fwd}})^2 T_i, (\sigma^i_{\text{fwd}})^2 T_i\right)$$

For $t = 0, i = 1, T_i = 0.5$

$$\log\frac{R_{0.5,1}}{3\%} \sim N\left(-\frac{1}{2}0.03^2\times 0.5, 0.03^2\times 0.5\right)$$

The Monte Carlo procedure is to

- Randomly generate a sample from this normal distribution and calculate $R_{0.5,1}$.
- Calculate the payoff

$$1\text{m} \times e^{-|R_{0.5,1}-0.2|}$$

- Repeat for a large number of samples.
- Calculate the expected payoff by taking the average of the simulated payoffs.
- Calculate the discounted expected payoff by multiplying the expected payoff by 0.95.

(c)

$$\begin{aligned}
0.03^2 \times 0.5 &= 0.5\, S_1^2 \\
0.045^2 \times 1 &= 0.5\, S_2^2 + 0.5\, S_1^2 \\
0.05^2 \times 1.5 &= 0.5\, S_3^2 + 0.5(S_2^2 + S_1^2) \\
0.045^2 \times 2 &= 0.5\, S_4^2 + 0.5(S_3^2 + S_2^2 + S_1^2)
\end{aligned}$$

Solving from top to bottom, $S_1 = 0.03, S_2 = 0.056, S_3 = 0.059$ and $S_4 = 0.024$.

5 Your company uses Black's model with caplets' forward volatilities to price caps and is interested in offering more general options. You have chosen the LMM for pricing these options.

Given:

- $F_{t,\tau,T}$ is the n-times compounded annual forward rate at t for the period $[\tau, T]$.
- $R_{\tau,T}$ is the n-times compounded annual LIBOR spot rate at τ for the period $[\tau, T]$.
- $\sigma^i_{\text{fwd}}, i = 1, 2, 3, 4$ are the caplet volatilities given in the table below

i	T_i	σ^i_{fwd}
1	0.25	0.2
2	0.50	0.22
3	0.75	0.26
4	1	0.29

- Forward rate volatility ζ_{i+1} of $F_{t,T_i,T_{i+1}}$ is as given below

$$\zeta_i(t) = \begin{cases} S_4 & t \leq T_1 \\ S_3 & T_1 < t \leq T_2 \\ S_2 & T_2 < t \leq T_3 \\ S_1 & T_3 < t \leq T_4 \end{cases}$$

(a) Calculate the corresponding forward rate volatilities $S_i, i = 1, 2, \ldots, 4$.

Consider a security with payoff $G(R_{\tau,T})$ at time T given by

$$G(R_{\tau,T}) = L\delta(2R_{\tau,T} - R_K)$$

where

- L is the notional amount.
- $n = \frac{1}{\delta}$, $\tau = T - \delta$.
- R_K is the strike rate.

(b) (i) Show that the price V at time 0 of the security is given by:

$$V = P_{0,T}\, L\,\delta\,(2F_{0,\tau,T} - R_K)$$

Consider now the case in which the same payoff $L\,\delta\,(2F_{0,\tau,T} - R_K)$ is paid at τ instead of at time T.

(ii) Show that, if the payoff is paid at time τ, the price V at time 0 of the security is given by:

$$V = P_{0,T}\, L\,\delta \left(2F_{0,\tau,T} - R_K - F^n_{0,\tau,T}\, R_K\,\delta + 2\delta\mathbb{E}^T(R_{\tau,T}{}^2)\right)$$

where $\mathbb{E}^T$ is the expectation with respect to T-forward dynamics.

You are provided with the following information:

- Notional amount L is \$100 million.
- $\tau = 0.5$.
- $T = 0.75$.
- $P_{0,0.5} = 0.989$.
- $P_{0,0.75} = 0.982$.
- The implied forward volatility σ_{fwd} for a caplet maturing at T is 0.22.
- $R_K = 0.02$.

(c) Calculate the value of the security using the formula in part (b) (ii) and assuming the LMM.

[QFIQF – Spring 2022 Q6]

Solution

(a)
$$\begin{aligned}
0.2^2 \times 0.25 &= 0.25\, S_1^2 \\
0.22^2 \times 0.5 &= 0.25\, S_2^2 + 0.25\, S_1^2 \\
0.26^2 \times 0.75 &= 0.25\, S_3^2 + 0.25(S_2^2 + S_1^2) \\
0.29^2 \times 1 &= 0.25\, S_4^2 + 0.25(S_3^2 + S_2^2 + S_1^2)
\end{aligned}$$

Solving from top to bottom, $S_1 = 0.20, S_2 = 0.238, S_3 = 0.326$ and $S_4 = 0.366$.

(b) (i) Follow the logic of the derivation of the caplet price (Section 13.5.2).

Using the T-forward measure, the price is

$$V = L\delta P_{0,T}\, \mathbb{E}^T(2R_{\tau,T} - R_K) = L\delta P_{0,T}\,(2F_{0,\tau,T} - R_K)$$

since $\mathbb{E}^T R_{\tau,T} = F_{0,\tau,T}$ under $\mathbb{Q}^T$.

(ii) First discount the payoff back to time τ

$$\frac{2R_{\tau,T} - R_K}{P_{\tau,T}} = (2R_{\tau,T} - R_K)(1 + R_{\tau,T}\delta) = 2R_{\tau,T} - R_K - R_{\tau,T}R_K\delta + 2\delta(R_{\tau,T})^2$$

using the relationship between spot rates and the zero-coupon bond price.

Then the price is

$$\begin{aligned}
V &= L\delta P_{0,T}\, \mathbb{E}^T \frac{2R_{\tau,T} - R_K}{P_{\tau,T}} \\
&= L\delta P_{0,T}\, \mathbb{E}^T(2R_{\tau,T} - R_K - R_{\tau,T}R_K\delta + 2\delta R_{\tau,T}^2) \\
&= L\delta P_{0,T}\left(2F_{0,\tau,T} - R_K - F_{0,\tau,T}R_K\delta + 2\delta\mathbb{E}^T(R_{\tau,T})^2\right)
\end{aligned}$$

(c) $\mathbb{E}^T(R_{0.5,0.75})^2 = (0.0342)^2 e^{\frac{2(2-1)}{2}0.2^2 \times 0.5} = 0.001282192$

From Section 12.2,

$$F_{t,S,T} = \frac{1}{\delta}\left(\frac{P_{t,S}}{P_{t,T}} - 1\right) = \frac{1}{\delta}\left(\frac{P_{t,S}}{P_{t,T}} - 1\right) \Longrightarrow F_{0,0.5,0.75}$$
$$= \frac{1}{0.25}\left(\frac{0.989}{0.98} - 1\right) = 0.0285$$

Then the price is

$$V = 100m \times 0.25 \times 0.98\,\Big(2 \times 0.0285 - 0.02 - 0.0285 \times 0.02 \times 0.25$$
$$+ 2 \times 0.25 \times 0.001282192\Big) = \$915,724$$

References

Abbott, S. (2015), *U*nderstanding analysis, Springer.

Albin, P., Hamza, K. and Klebaner, F. C. (2024), *P*roblems and solutions in stochastic calculus with applications, World Scientific.

Ambagaspitiya, R. and Ford, C. (2023), *Calibrating interest rate models*, SOA Research Institute.

Apostol, T. (1974), *Mathematical Analysis: A Modern Approach to Advanced Calculus*, Pearson.

Atmaz, A. and Basak, S. (2019), 'Option prices and costly short-selling', *J*ournal of Financial Economics **134**(1), 1–28.

Baxter, M. and Rennie, A. (1996), *F*inancial calculus: An introduction to derivative pricing, Cambridge University Press.

BIS (2024), *BIS quarterly review statistical release*, Bank for International Settlements.

Black, F. (1976), 'The pricing of commodity contracts', *J*ournal of financial economics **3**(1-2), 167–179.

Black, F. and Scholes, M. (1973), 'The pricing of options and corporate liabilities', *J*ournal of Political Economy **81**(3), 637–654.

Bolder, D. J. (2001), 'Affine term-structure models: Theory and implementation'.

Brace, A., Dun, T. and Barton, G. (2001), 'Towards a central interest rate model', *O*ption Pricing, Interest Rates and Risk Management, 278–313.

Brace, A., G̦atarek, D. and Musiela, M. (1997), 'The market model of interest rate dynamics', *M*athematical Finance **7**(2), 127–155.

Brace, A., Gellert, K. and Schlögl, E. (2024), 'Sofr term structure dynamicsâŁ"discontinuous short rates and stochastic volatility forward rates', *J*ournal of Futures Markets **44**(6), 936–985.

Brigo, D., Dalessandro, A., Neugebauer, M. and Triki, F. (2009), 'A stochastic processes toolkit for risk management: Geometric Brownian motion, jumps, garch and variance gamma models', *J*ournal of Risk Management in Financial Institutions **2**(4), 365–393.

Brigo, D., Mercurio, F. et al. (2001), *I*nterest rate models: theory and practice, Vol. 2, Springer.

Calin, O. (2015), *A*n informal introduction to stochastic calculus with applications, World Scientific.

Chan, K. C., Karolyi, G. A., Longstaff, F. A. and Sanders, A. B. (1992), 'An empirical comparison of alternative models of the short-term interest rate', *T*he Journal of Finance **47**(3), 1209–1227.

Cieslak, A. and Povala, P. (2016), 'Information in the term structure of yield curve volatility', *T*he Journal of Finance **71**(3), 1393–1436.

Coleman, T. F. and Li, Y. (2010), Optimization & finance, in '*E*ncyclopedia of Quantitative Finance', John Wiley & Sons Ltd. Chichester, UK., pp. 1322–1327.

Constantinides, G. M. and Richard, S. F. (1978), 'Existence of optimal simple policies for discounted-cost inventory and cash management in continuous time', *O*perations research **26**(4), 620–636.

Dahlquist, M. (1996), 'On alternative interest rate processes', *J*ournal of Banking & Finance **20**(6), 1093–1119.

Dai, Q. and Singleton, K. J. (2000), 'Specification analysis of affine term structure models', *T*he Journal of Finance **55**(5), 1943–1978.

Damodaran, A. (2018), *T*he dark side of valuation: Valuing *Y*oung, distressed, and complex businesses, 3rd ed., Pearson FT Press.

Durrett, R. (2012), *E*ssentials of stochastic processes, Vol. 1, Springer.

Fabozzi, F. J. and Fabozzi, F. A. (2021), *B*ond markets, analysis, and strategies, MIT Press.

Fama, E. F. (1965), 'The behavior of stock-market prices', *T*he Journal of Business **38**(1), 34–105.

Fama, E. F. and French, K. R. (1988), 'Permanent and temporary components of stock prices', *J*ournal of Political Economy **96**(2), 246–273.

Filipović, D. (1999), 'A note on the Nelson–Siegel family', *M*athematical Finance **9**(4), 349–359.

Filipovic, D. (2009), *T*erm-structure models: A graduate course, Springer Science & Business Media.

Geman, H., El Karoui, N. and Rochet, J.-C. (1995), 'Changes of numeraire, changes of probability measure and option pricing', *J*ournal of Applied Probability **32**(2), 443–458.

Goncalves, F. and Issler, J. V. (1996), 'Estimating the term structure of volatility and fixed income derivative pricing', *J*ournal of Fixed Income **6**(1).

Heath, D., Jarrow, R. and Morton, A. (1992), 'Bond pricing and the term structure of interest rates: A new methodology for contingent claims valuation', *E*conometrica: Journal of the Econometric Society, 77–105.

Hirsa, A. and Neftci, S. (2013), An introduction to the mathematics of financial derivatives, Academic press.

Hull, J. C. (2021), *O*ptions, futures, and other derivatives, 11th ed., Pearson Education.

Indriawan, I., Jiao, F. and Tse, Y. (2021), 'The sofr and the fedâĿ™s influence over market interest rates', *E*conomics Letters **209**, 110095.

Jamshidian, F. (1989), 'An exact bond option formula', *T*he Journal of Finance **44**(1), 205–209.

Jamshidian, F. and Zhu, Y. (1996), 'Scenario simulation: Theory and methodology', *F*inance and Stochastics **1**, 43–67.

Jarrow, R. A. and Chatterjea, A. (2024), An Introduction to derivative securities, financial markets, and risk management, World Scientific.

Jarrow, R., Li, H. and Zhao, F. (2007), 'Interest rate caps âĿœsmileâĿž too! but can the libor market models capture the smile?', *T*he Journal of Finance **62**(1), 345–382.

Jegadeesh, N. and Titman, S. (1993), 'Returns to buying winners and selling losers: Implications for stock market efficiency', *T*he Journal of Finance **48**(1), 65–91.

Kendall, M. G. and Hill, A. B. (1953), 'The analysis of economic time-series-part i: Prices', *J*ournal of the Royal Statistical Society. Series A (General) **116**(1), 11–34.

Kim, J. H. and Ji, P. I. (2011), 'Mean-reversion in international real interest rates', *E*conomic Modelling **28**(4), 1959–1966.

Klebaner, F. C. (2012), *I*ntroduction to stochastic calculus with applications, World Scientific Publishing Company.

Klenke, A. (2020), *P*robability theory: A comprehensive course, 3rd ed., Springer.

Leland, H. E. (1985), 'Option pricing and replication with transactions costs', *T*he Journal of Finance **40**(5), 1283–1301.
Litterman, R. B., Scheinkman, J. and Weiss, L. (1991), 'Volatility and the yield curve', *T*he Journal of Fixed Income **1**(1), 49–53.
Lo, A. (2018*a*), *D*erivative pricing: A problem-based primer, Chapman and Hall/CRC.
Lo, A. W. (2018*b*), Efficient markets hypothesis, in '*T*he new palgrave dictionary of economics', Springer, pp. 3543–3560.
Longstaff, F. A. and Schwartz, E. S. (1992), 'Interest rate volatility and the term structure: A two-factor general equilibrium model', *T*he Journal of Finance **47**(4), 1259–1282.
Malkiel, B. G. (2023), *A* Random Walk Down Wall Street: The Best Investment Guide That Money Can Buy (Thirteenth), WW Norton & Company.
Mansuy, R. (2022), The origin and multiple meanings of martingale, in '*T*he splendors and miseries of martingales: their history from the casino to mathematics', Springer, pp. 3–14.
Mazzoni, T. (2018), *A* first course in quantitative finance, Cambridge University Press.
McDonald, R. L. (2013), *D*erivatives markets, Pearson.
McInerney, D. and Zastawniak, T. (2015), *S*tochastic interest rates, Cambridge University Press.
Merton, R. C. (1973), 'Theory of rational option pricing', *T*he Bell Journal of Economics and Management Science **4**(1), 141–183.
URL: *http://www.jstor.org/stable/3003143*
Miller, I. and Miller, M. (2012), *J*ohn E. Freund's mathematical statistics with applications, 8th ed., Pearson.
NY Fed (2021), *A*n Updated UserâŁ™s Guide to SOFR, The Alternative Reference Rates Committee, Federal Reserve Bank of New York.
Oksendal, B. (2013), *S*tochastic differential equations: An introduction with applications, Springer Science & Business Media.
Pirie, W. L. (2021), *D*erivatives, CFA Institute, John Wiley & Sons.
Poterba, J. M. and Summers, L. H. (1988), 'Mean reversion in stock prices: Evidence and implications', *J*ournal of Financial Economics **22**(1), 27–59.
Rebonato, R. (1996), *I*nterest-rate option models: Understanding, analysing and using models for exotic interest-rate options, Wiley.
Rebonato, R. (2004), *V*olatility and correlation: The perfect hedger and the fox, John Wiley & Sons.
Ross, S. M. (2020), *A* first course in probability, Pearson Harlow.
Russo, V. and Fabozzi, F. J. (2025), 'Bootstrapping risk-free rates in the post-ibor environment.', *J*ournal of Fixed Income **35**(1).
Shiller, R. J. (2001), *I*rrational Exuberance, Princeton University .
Shiller, R. J. (2015), Irrational exuberance: Revised and expanded third edition, in '*I*rrational exuberance', Princeton University Press.
Shreve, S. (2005), *S*tochastic calculus for finance I: The binomial asset pricing model, Springer Science & Business Media.
Shreve, S. E. (2004), *S*tochastic calculus for finance II: Continuous-time models, Vol. 11, Springer.
Simmons, G. F. (1995), Calculus with analytic geometry, 2nd ed., McGraw-Hill Education.
Spivak, M. (2006), *C*alculus, Cambridge University Press.
Strang, G. (1991), *C*alculus, Vol. 1, SIAM.
Sullivan, E. J. and Weithers, T. M. (1991), 'Louis bachelier: The father of modern option pricing theory', *T*he Journal of Economic Education **22**(2), 165–171.

Sundaram, R. K. and Das, S. R. (2011), *Derivatives: Principles and practice*, McGraw-Hill Irwin.

Svoboda, S. (2004), Brace, Gatarek and Musiela model, in '*Interest rate modelling*', Springer, pp. 213–226.

Tuckman, B. and Serrat, A. (2022), *Fixed income securities: Tools for today's markets*, John Wiley & Sons.

Veronesi, P. (2010), *Fixed income securities: Valuation, risk, and risk management*, John Wiley & Sons.

Wilmott, P. (2013), *Paul Wilmott on quantitative finance*, John Wiley & Sons.

Index

For Product Safety Concerns and Information please contact our
EU representative GPSR@taylorandfrancis.com Taylor & Francis
Verlag GmbH, Kaufingerstraße 24, 80331 München, Germany